Studies in Logic and Computation

Series editor

D. M. GABBAY

Studies in Logic and Computation

1. Intensional Logics for Programming
   Edited by Luis Fariñas del Cerro
   and Martti Penttonen

2. Substructural Logics
   Edited by Peter Schroeder-Heister
   and Kosta Došen

Erratum

The non-alphabetical naming of the editors on the title page and cover is due to an error by the publisher. It does not express any rank order.

Substructural Logics

Edited by

PETER SCHROEDER-HEISTER
Professor of Logic and Philosophy of Language, Universität Tübingen

and

KOSTA DOŠEN
Professor of Mathematical Logic Mathematical Institute, Belgrade

CLARENDON PRESS · OXFORD
1993

Oxford University Press, Walton Street, Oxford OX2 6DP

Oxford New York Toronto
Delhi Bombay Calcutta Madras Karachi
Kuala Lumpur Singapore Hong Kong Tokyo
Nairobi Dar es Salaam Cape Town
Melbourne Auckland Madrid

and associated companies in
Berlin Ibadan

Oxford is a trade mark of Oxford University Press

Published in the United States
by Oxford University Press Inc., New York

A catalogue record for this book is available from the British Library

Library of Congress Cataloging in Publication Data
(Data available on request)

ISBN 0–19–853777–8

Typeset by the authors using LaTeX
Printed in Great Britain
on acid-free paper by
Bookcraft Ltd, Midsomer Norton, Avon

Preface

The term *substructural logics*, a designation for logics with restricted structural rules, made its first public appearance at the conference dedicated to such logics that we organized in the *Seminar für natürlich-sprachliche Systeme* of the University of Tübingen on 7–8 October, 1990. The program of this conference was the following:

The Lambek Calculus

Joachim Lambek: Some remarks about residuated monoids

Johan van Benthem: Categorial structure and dynamic logic

Sergei Soloviev: Categories and logics with restricted structural rules

Relevant Logic

Michael Dunn: Gaggle theory — A common abstraction of Galois connections and residuation with applications to substructural logics

Kosta Došen: The first substructural logic

Robert Meyer: The semantics of entailment 0

Linear Logic

Jean-Yves Girard: Proof-nets — A new approach to syntax

Giovanni Sambin: The semantics of pretopologies

Andre Scedrov: Decision problems for propositional linear logic

BCK Logic

Viacheslav Grishin: Logic without contraction

Peter Schroeder-Heister: Proofs and assumptions

Hiroakira Ono: Phase structures and quantales

After the conference we invited our speakers and other experts to prepare papers for a collection of readings on substructural logics. The papers we received make this book.

Substructural logics belong to different traditions, they have various motivations, and the style of the authors writing about them varies considerably, though the results they have reached overlap to a great extent. Our ambition with the book of readings was, as with the conference, to bring together papers from these different traditions. We supposed that if they appeared within the covers of a single volume, it should become clear that they are dealing with the same subject.

Some of the papers in this book belong more or less clearly to one of the traditions delimiting the fields covered by our conference. Lambek's papers are about logics obtained by extending his syntactic calculus with other connectives. Their inspiration is in category theory and universal algebra, but they may also be taken as papers in proof theory.

The papers by Belnap, Kron, Meyer and Mares are in the tradition of relevant logic. Belnap's is a polemical piece defending the distributivity of the lattice of conjunction and disjunction in an area where Gentzen's rules suggest distribution should fail. This distribution is present in the relevant systems without contraction investigated proof-theoretically by Kron. His paper continues work on systems, not far removed from linear logic, that have been investigated in the relevant logic community since the early eighties. Meyer's and Mares' paper is the last (though it bears the least number) in a series of papers on Kripke-style models for relevant logic. It concentrates on systems with Boolean negation and modal operators.

Troelstra's paper is an introduction to linear logic in the recent tradition started by Girard.

The remaining papers are more difficult to assign to a single field.

Bunder's paper is about proof-theoretical matters related to linear implication, but it is inspired by the older tradition of BCK logic. Soloviev's paper is also related to linear logic, which it approaches from the point of view of category theory with a proof-theoretical flavour.

Sambin's paper investigates a semantics for linear and other substructural logics motivated by a constructivist approach to point-free topology. Related to that is Ono's study of algebraic and Kripke-style models for substructural logics, which covers logics in all the four fields. Dunn's paper is part of his project to generalize algebraic ideas that underlie substructural logics.

Gabbay's contribution illustrates how investigations in non-monotonic logics may profit from an association with substructural logics. Finally, van Benthem is concerned with the interplay of ideas between substructural logics and related fields like theoretical linguistics.

We wish to thank first of all the authors that accepted our invitation to speak at the Tübingen conference or send us a paper for this collection. We are especially grateful to those of them who by their comments and suggestions helped us in editing this collection. We have received such help also from Branislav Boričić, Milan Božić, Ross Brady, Wojciech Buszkowski, Djordje Čubrić, Roy Dyckhoff, André Fuhrmann, Jörg Hudelmaier and Heinrich Wansing. We would like to thank them all for their care and patience. We would also like to thank Stephan Blocher, Markus Jurinka, Knut Peter and especially Mirko Janc for their help in processing and typesetting the manuscripts. The first-mentioned among us also acknowledges support from the Alexander von Humboldt Foundation and from the Mathematical Institute in Belgrade (through Grant 0401A of the Science Fund of Serbia).

Tübingen
January, 1993

Kosta Došen
Peter Schroeder-Heister

Contents

A Historical Introduction to Substructural Logics

Kosta Došen

Mathematical Institute, Belgrade

The common denominator of several important nonclassical logics is that in their sequent formulation they reject or restrict some of Gentzen's structural rules, whereas rules for logical constants need not differ from rules that may be given for classical logic. It is as if the structural part of logic were more fundamental: to change logic, we have to change this part. Logical constants are in principle secondary: they are invariant, they play the same role in different structural contexts.

We shall call logics that can be obtained in this manner, by restricting structural rules, *substructural* logics. This term may convey that something has been subtracted in the structural part of logic, but it should also convey, according to the standard meaning of the word 'substructural', that the restricted logic is the product of interventions in the supporting structures of logic.

The most important substructural logic is intuitionistic logic. That it is such a logic was discovered by Gentzen. Other important substructural logics investigated up to now, which make the subject of this introduction and of the rest of this book, are relevant logic, BCK logic, linear logic and the Lambek calculus of syntactic categories. These logics have various inspirations. Intuitionistic logic is a product of the constructivist side in the great debate on the foundations of mathematics. Relevant logic (though its earliest roots are mixed with the roots of intuitionistic logic; see §3 below) has a more philosophical inspiration: it grew out of attempts to formalize an implication devoid of the unintuitive features of material implication. The first reason for constructing the older contractionless logic, called BCK logic, was to evade set-theoretical paradoxes while keeping unrestricted comprehension (systems with restricted contraction have also been studied in many-valued logic and in connexion with extensions of intuitionistic logic; see §4 below). The younger contractionless logic, linear logic, draws its inspiration, and much of the attention it has attracted, from computer science. Finally, the Lambek calculus belongs primarily to mathematical linguistics, but it has also deep connexions with category theory. In the production of substructural logics, there has also been, no doubt, a certain amount of experimentation, of combinatorial playing, with the possibilities of the apparatus that makes the base of Gentzen's formalism (this is especially clear in some papers on BCK logic; see §4 below). There are infinitely many possibilities here, but we concentrate only on those that have proved their worth through prolonged serious investigation.

The substructural logics of this book do not differ only in inspiration: they are also separated by time and space. They have been discovered and rediscovered during the last sixty odd years by logicians in places wide apart, who would often not know about the work of the others. In spite of a great deal of overlapping in their results, their styles are sometimes so different as to prevent understanding. This book is produced in the hope that separation and ignorance may end and a united field of substructural logics may emerge.

The present introduction is about the earlier history of substructural logics. For newer developments and current research, one may consult the references listed at the end, or papers that follow in this book. In such a brief review, I can hardly do much more than mention those earlier works I know about and find worth mentioning. I shall dwell longer on some less well known, less cited, less accessible, papers than on some better known papers that are bound to receive their due.

1 Structural Rules

A very important discovery in Gentzen's thesis [1935] is that in logic there are rules of inference that don't involve any logical constant. Gentzen called such rules *structural*. The discovery of structural rules should not be ascribed uniquely to Gentzen but also to Hertz, who influenced Gentzen (see [Hertz 1922, 1928, 1929] and Gentzen's first paper [1932]). A prefiguration of structural rules may also be found in the axiom system for logic that Hilbert presented often in the twenties (see below). Of course, structural rules are so fundamental to logic that it must be possible to find an implicit conception of them from the very beginning of logic. However, a full and clear conception appears first in [Gentzen 1935]. Let us now see how Gentzen formulated these rules.

Let L be a language for whose formulae we use the schematic letters $A, B, C, \ldots$; for sequences of formulae of L, let us use the schematic letters $X, Y, Z, \ldots$. A *sequent* is, as in [Gentzen 1935], an expression of the form $X \vdash Y$. If Y consists of a single formula or is empty, $X \vdash Y$ is a *single-conclusion* sequent; otherwise, it is a *multiple-conclusion* sequent. If we interpret a single-conclusion sequent $X \vdash C$ as asserting that there is an inference from the premises in X to the conclusion C, ordinary rules of inference may be formulated by sequent-schemata. For example, *modus ponens* may be formulated by the sequent-schema $A,\ A \to B \vdash B$. Gentzen's structural rules are different from these ordinary rules because their instances have inferences as premises and conclusions. We may say that ordinary rules are of *first level*, whereas Gentzen's structural rules are of *second level*. In [Gentzen 1935] we find the prototypes of the following structural rules, appropriate for a single-conclusion sequent system:

Thinning $$\frac{X, Y \vdash Z}{X, A, Y \vdash Z}$$

Contraction $$\frac{X, A, A, Y \vdash Z}{X, A, Y \vdash Z}$$

Permutation $$\frac{X, A, B, Y \vdash Z}{X, B, A, Y \vdash Z}$$

Cut $$\frac{X \vdash A \qquad U, A, Y \vdash Z}{U, X, Y \vdash Z}$$

Thinning may already be found in Hertz's papers mentioned above (Hertz calls it *unmittelbarer Schluß*, i.e. immediate inference). The term *Thinning* as a translation of Gentzen's *Verdünnung* may be found in Kleene's book [1952, §77], whereas the alternative translation *Weakening* may be found in Curry's books [1950, chapter II.5] and [1963, chapter 5C]. The idea is that $X, A, Y \vdash Z$ is a weaker, diluted, assertion when compared with $X, Y \vdash Z$. More recently, some authors have given the name *Monotonicity* to Thinning. This usage, probably derived from Scott's [1971] and some other of his papers from the same period, is justified by the equivalence of Thinning with the monotonicity of Tarski's consequence operation:

$$\textit{if } \mathcal{A} \subseteq \mathcal{B}, \quad \textit{then } \mathrm{Cn}(\mathcal{A}) \subseteq \mathrm{Cn}(\mathcal{B})$$

(cf. [Tarski 1930, 1930a]). It has become entrenched in the area of nonmonotonic logics (however, these logics may lack other structural rules besides unrestricted Thinning; see the end of §3 below).

The term *Contraction* seems to be universally accepted, whereas an alternative to the term *Permutation*, found in Curry's works, is Kleene's *Interchange* (some authors say *Exchange*). So, our *Thinning* and *Permutation* is a compromise between Kleene's and Curry's usage (the translation in [Gentzen 1969] follows Kleene's usage). The rule Cut, which Gentzen introduced in [1932], was prefigured by Hertz. Those who call Thinning

Monotonicity tend to call Cut *Transitivity* (and Identity below they would call *Reflexivity*).

A prefiguration of structural rules may be found in an axiom system of Hilbert for intuitionistic implication, conjunction and disjunction, which Hilbert and Bernays in [1934, chapter III, §3, p. 67] and [1939, supplement III, p. 442] call *positive logic*. By adding appropriate axioms for negation, Hilbert obtains classical logic from positive logic. The implicational axioms of positive logic in the version of [Hilbert 1922, 1923, 1928], which was often cited and has influenced Gentzen (as [Gentzen 1935, section V, §2] witnesses), are given with the following names and in the following order:

Introduction of an Assumption:	$A \to (B \to A)$
Omission of an Assumption:	$(A \to (A \to B)) \to (A \to B)$
Interchange of Assumptions:	$(A \to (B \to C)) \to (B \to (A \to C))$
Elimination of a Proposition:	$(B \to C) \to ((A \to B) \to (A \to C))$

(for the English names of these axiom-schemata, we follow the translation of [van Heijenoort 1967, p. 465]). The first axiom-schema corresponds to Thinning, the second to Contraction and the third to Permutation. The fourth corresponds to the structural rule called *Association*, which we shall consider in §6 below, but it is also related to Cut. Cut is also related to the rule of *modus ponens*, which Hilbert assumes in addition to the axiom-schemata above (we have insisted upon the order in which these axiom-schemata are presented in Hilbert's papers because this order is matched by the order in which Gentzen introduces structural rules in [Gentzen 1935, section III, §1.21]: Thinning, Contraction, Permutation, Cut). The axioms for conjunction and disjunction of Hilbert's axiomatization of positive logic are also closely related to Gentzen's rules for these connectives.

By a natural extension of Gentzen's usage, we could take the term *structural* to cover some first-level rules as well as the second-level rules above. For all structural rules it would be characteristic that they can be formulated so that no constant of the language L occurs in their formulation. A structural rule of first level is given by the sequent-schema:

$$\textit{Identity:} \qquad A \vdash A$$

which is an axiom-schema in Gentzen's sequent systems. Another structural rule of first level, not treated in [Gentzen 1935], could be given by the sequent-schema:

$$\textit{Substitution:} \qquad A \vdash A^x_a$$

where A^x_a stands for the formula obtained from A by replacing a variable x by an expression a of the same syntactic category as x. This rule may be formulated for every language L that has variables of the syntactic category of x and a. However, in the context of Gentzen's sequent systems, the first-level rule of Substitution is incorrect and we should not assume it (in other contexts, where introduction of implication on the right, i.e. the deduction theorem, is restricted, it can make sense; cf. [Kleene 1952, §23, p. 101]). Instead, we could assume the analogous second-level rule:

$$\frac{\vdash A}{\vdash A^x_a}$$

No constant of L occurs either in the sequent-schema of Identity or in the sequent-schema or second-level rule of Substitution. So, these rules are indeed structural.

Gentzen's and our structural rules should not be confused with what is often called *structural rules* in the Polish logic literature. There, a rule is called structural if its instances are closed under substitution (cf. [Wójcicki 1988, §2.1.7]).

Thinning, Contraction and Permutation as formulated above enable us to transform only the left-hand side of a sequent. In addition to these rules, Gentzen formulates in [1935] analogous structural rules for transforming the right-hand side of a sequent, and his Cut is like the following multiple-conclusion form of Cut:

$$\frac{X \vdash V, A, W \qquad U, A, Y \vdash Z}{U, X, Y \vdash V, Z, W}$$

All these structural rules, on the left and on the right, and the multiple-conclusion form of Cut are assumed for Gentzen's multiple-conclusion sequent system for classical logic. Gentzen passes from this system to the single-conclusion sequent system for intuitionistic logic by restricting the language to single-conclusion sequents.

However, the same effect can be obtained in another manner. Sequent systems for classical and intuitionistic logic can be so formulated that passing from the system for classical logic to the system for intuitionistic logic consists in restricting the rule of Thinning on the right to:

Intuitionistic Thinning on the right $\quad \dfrac{X \vdash}{X \vdash A}$

Everything else, and, in particular, assumptions about logical constants, will be the same in both systems. For example, for implication we assume the following Gentzen-style rules:

$$(\rightarrow L)\ \frac{X \vdash V, A, W \qquad U, B, Y \vdash Z}{U, X, A \rightarrow B, Y \vdash V, Z, W} \qquad (\rightarrow R)\ \frac{A, X \vdash V, B, W}{X \vdash V, A \rightarrow B, W}$$

We define $\neg A$ as $A \rightarrow \bot$ (as in [Gentzen 1935, section II, §5.2]) and for $\bot$ we assume:

$$(\bot\ L)\ \ \bot\ \vdash \qquad (\bot\ R)\ \frac{X \vdash}{X \vdash \bot}$$

If we reject even Intuitionistic Thinning on the right, we obtain a system for Kolmogorov's and Johansson's minimal intuitionistic logic (see §2 below). So, intuitionistic logic may be conceived as a logic obtained by restricting the structural rules of classical logic.

Our proposal is to call logics that can be obtained in this manner, by restricting structural rules, *substructural* logics. The logic whose structural rules we restrict in order to obtain substructural logics is in principle classical logic. A substructural logic obtained by restricting the structural rules of intuitionistic logic is ultimately substructural relative to classical logic. Canonically, we should assume the same rules for logical constants in the logic whose structural rules we restrict, i.e. classical logic, and in the resulting substructural logic (according to [D. 1989], all alternative logics should arise by changing only the structural component of logic; assumptions about logical constants should be invariant). However, in practice, the situation is not always so clear-cut, either because a weaker structural context may make classically equivalent constants split into nonequivalent constants (see §3 below) or because we may consider various extensions of our logics (for example, modal extensions), which are not standard in classical logic. Sometimes, rules for logical constants may carry hidden structural assumptions, as rules appropriate for classical negation may transform intuitionistic logic into classical without explicit interventions concerning Thinning on the right. Such rules are inappropriate for formulating substructural logics that reject the hidden structural assumptions.

We don't insist that the Gentzen formulation of classical logic whose structural rules we restrict in order to obtain a substructural logic should be the standard one. It could as well be a nonstandard formulation with sequents whose left-hand and right-hand sides are not sequences of formulae of L but some other sort of structure involving formulae of L, like sets, or multisets, or trees, or sets of sets, or sets of multisets etc. (cf. the sequents of §6 below introduced for the nonassociative Lambek calculus; Dunn's sequents of [Anderson and Belnap 1975, §28.5] are related to sequents based on sets of multisets of sets etc.).

2 The First Axiomatizations of Intuitionistic Logic

The oldest substructural logics are older than [Gentzen 1935]. Intuitionistic logic was already axiomatized when Gentzen was writing his thesis. In [1925], Kolmogorov made the first attempt to axiomatize this logic by taking from [Hilbert 1923] Hilbert's axioms for implicational positive logic mentioned above and supplementing them with a *minimal*, i.e. purely implicational, intuitionistic negation (minimal negation $\neg A$ may be defined as $A \to \bot$ without any postulate being assumed about $\bot$, as in [Johansson 1936]). Kolmogorov's paper is interesting for us here because, in addition to rejecting $A \to (\neg A \to B)$, which corresponds to Intuitionistic Thinning on the right, he also considers briefly whether $A \to (B \to A)$, which corresponds to Thinning on the left, should be kept in intuitionistic logic (see [1925, section II, §4]). He decides that it should.

Glivenko hesitated in accepting $A \to (B \to A)$ and $A \to (\neg A \to B)$ before being convinced by Heyting that they should be taken into the axiomatization of intuitionistic propositional logic he produced in [1929]. Both of these principles appear as axioms in Heyting's axiomatization of [1930] (produced in 1928). Glivenko's axiomatization is built on a version of Hilbert's axioms for positive logic (Glivenko does not refer to Hilbert's positive logic or to Kolmogorov). In addition to Hilbert's first three implicational axioms, he has $A \to A$, which corresponds to the first-level structural rule of Identity ($A \to A$ is not independent from the other postulates of positive logic), and $(A \to B) \to ((B \to C) \to (A \to C))$ instead of $(B \to C) \to ((A \to B) \to (A \to C))$ (Hilbert also made this last change in [1931] and [Hilbert and Bernays 1934, chapter III, §3, p. 65], which makes Interchange of Assumptions dependent on the other postulates). Glivenko's axioms for conjunction and disjunction are as in [Hilbert and Bernays 1934]. In [1935, section V, §2], Gentzen preferred to reproduce Glivenko's axioms for intuitionistic propositional logic rather than Heyting's, presumably because these axioms are easier to connect with his systems (Heyting's axiomatization is closer in spirit to the axiom system of *Principia Mathematica*, from which it was derived; see [Troelstra 1990]).

3 Relevant Logic

In relevant logic (also called *relevance* logic) we reject Thinning on both the left and right. So, we will lack both $A \to (B \to A)$ and $A \to (\neg A \to B)$. It seems that the canonical version of relevant logic should be included in intuitionistic logic, since Thinning on the right is rejected. However, whether in this way we will end up with something included in intuitionistic logic depends on the exact rules for logical constants we have started from.

In particular, rules for negation may be formulated in such a way that, even after rejecting Thinnning on the right, we will obtain a multiple-conclusion sequent system with a nonintuitionistic, *involutive*, negation (i.e., in addition to $A \rightarrow \neg\neg A$, we will have $\neg\neg A \rightarrow A$), which also satisfies De Morgan's laws. Historically, relevant logic with an involutive negation was dominant from the very beginning.

Relevant logic antedates [Gentzen 1935] and is as old as intuitionistic logic. Its first inception is in [Orlov 1928], which unfortunately was unknown to later authors that developed relevant logic. In his paper, Orlov axiomatized the implication-negation fragment of the best-known relevant logic, later named R. Negation in this system is involutive. Orlov's aim was to formalize intuitionistic ideas about logic by further adding to his system an S4-type necessity operator. If we know about Kolmogorov's and Glivenko's reticence concerning $A \rightarrow (B \rightarrow A)$ and $A \rightarrow (\neg A \rightarrow B)$, we may find it natural that an author from their milieu, working on the formalization of intuitionistic logic, should appear with a system that lacks both of these principles.

With Orlov's paper we may illustrate something typical for substructural logics; namely, the splitting of some old connectives. The fewer structural rules there are, the more distinctions can be made, and connectives equivalent in classical logic can now be distinguished from each other (the same phenomenon occurs already with intuitionistic logic, where, for example, $A \rightarrow B$ is not equivalent to $\neg(A \wedge \neg B)$). With the definition:

$$A \cdot B \quad =_{df} \quad \neg(A \rightarrow \neg B)$$

Orlov obtains a connective that resembles conjunction, but is now distinguished from it. This connective, which later acquired the name *fusion* in relevant logic, is important because it imitates the comma on the left-hand side of sequents. The essential principle governing fusion and implication is:

$$A \cdot B \vdash C \quad \textit{iff} \quad B \vdash A \rightarrow C$$

which does not depend on having, as Orlov, an involutive negation. Algebraically, this means that the operation corresponding to implication is the *residual* of a multiplication corresponding to fusion; i.e., we have:

$$a \cdot b \leq c \quad \textit{iff} \quad b \leq a \rightarrow c.$$

With classical and intuitionistic logic, this multiplication corresponds to conjunction and is a meet operation; with weaker substructural logics, it may lack some of the meet properties. With relevant logic, we lack $a \cdot b \leq a$ and $a \cdot b \leq b$, because we lack Thinning.

If an intuitive reading is asked for fusion, which would be distinct from *and*, we may try making a connective from *with* (or *together with*). This should mirror the conjoining of premises fusion performs. Whereas, A *and* B always implies A, in relevant logic A *with* B need not imply A, since B may be irrelevant.

With an involutive negation, we may define, as Orlov did, the De Morgan dual of fusion, later named *fission* in relevant logic:

$$A + B \quad =_{df} \quad \neg A \to B.$$

Fission resembles disjunction, but is different from it. As fusion imitates the comma on the left-hand side of sequents, so fission imitates the comma on the right-hand side of sequents. Implication, fusion, fission and negation make the so-called *intensional* fragment of relevant logic; nowadays, a more popular term than *intensional* seems to be *multiplicative*, which we borrow from [Girard 1987]. Conjunction and disjunction, which algebraically behave like respectively finite meet and finite join in a lattice (this lattice need not be distributive; see below), were later added to the multiplicative fragment. They are called *extensional*, or in the more fashionable terminology of [Girard 1987], which we follow here, *additive* connectives. A similar splitting between multiplicative and additive occurs for the propositional constants, i.e. nullary connectives, $\top$ and $\bot$. Multiplicative $\top$ and $\bot$, symbolized in [Anderson and Belnap 1975, §27.12, pp. 342–343] by respectively t and f (in [Girard 1987], we have instead 1 and $\bot$) behave algebraically like unit elements with respect to fusion and fission multiplication; i.e., we have $\mathrm{t} \cdot a = a$ and $\mathrm{f} + a = a$. Additive $\top$ and $\bot$, symbolized in [Anderson and Belnap 1975, *loc. cit.*] by T and F (in [Girard 1987], we have $\top$ and 0), behave like the greatest and least element in the lattice that corresponds to conjunction and disjunction. On the other hand, only additive versions of the quantifiers may be found in the existing systems of substructural logics. A multiplicative universal quantifier would presumably be something like an infinite fusion, but I am not aware that the idea has been investigated more thoroughly.

The reason why relevant logic appeared before the other substructural logics is probably that Thinning stands apart from other structural rules. Contraction and Permutation just serve to reduce sequences of formulae to sets of formulae, and Cut is something we try to eliminate as a primitive rule, but our systems should nevertheless be closed under it (systems not closed under Cut appear in the literature, but they are rather unusual). In Tarski's axiomatization of the consequence operation, which is an operation on sets of formulae, no axiom corresponds to Contraction and Permutation, but the monotonicity axiom, mentioned in §1 above, corresponds to Thinning. In [1936, §5.244], Gentzen finds he must justify Thinning, whereas other structural rules are accepted without further ado. He says for Thinning that:

> at first, it may seem somewhat strange; yet, for example, if a proposition is true, we are forced to admit that in that case it also holds on the basis of an *arbitrary assumption* (if we were to stipulate that this may be asserted only in cases where a 'factual dependence' ['*tatsächliche Abhängigkeit*'] exists, considerable difficulties would arise because of the possibility of proofs in which only an *apparent* use of an assumption is made) [Gentzen 1969, p. 152].

The first authors after Orlov prepared to tackle the difficulties of Gentzen's '*tatsächliche Abhängigkeit*', i.e. relevance, were Moh Shaw-Kwei in [1950] and Church in [1951]. They axiomatized the implicational fragment of R guided by a deduction theorem for a notion of deduction from hypotheses where every hypothesis is used in obtaining the conclusion. Church's interest in relevant implication should be related to his predilection for the λ-I-calculus, where $\lambda x.t$ is not well-formed if x does not occur free in the term t (cf. [Curry and Craig 1953]). In [Church 1941], the λ-I-calculus is the main lambda calculus, whereas the ordinary lambda calculus, called λ-K-calculus, is secondary. Thinning, Contraction and Permutation are related respectively to the combinators K, W and C via the Curry-Howard, formulae-as-types, interpretation, which encodes natural-deduction proofs with typed lambda terms (cf. [Lambek and Scott 1986, part I, §§1-15] and [Girard et al. 1989, chapter 3]). Natural-deduction proofs in substructural logics lacking, for example, Thinning will be coded by typed lambda terms that exclude a closed term corresponding to K, and similarly with other structural rules and combinators.

The later history of relevant logic is tied mainly to the work of Ackermann, Anderson, Belnap, Dunn, Meyer, Routley, Urquhart and logicians writing in response to their work. With these authors, relevant logic acquired its name by being welded with earlier attempts to formalize the intuitive non-truth-functional implication based on a connexion of meaning (modal logic has in its roots a similar philosophical motivation). This may seem like a hopelessly unmathematical enterprise, like fighting against zero on the ground that zero is not intuitively a number (cf. [Curry 1963, chapter 5, §2]). However, the best results of relevant logic don't bear so much on an implication in tune with intuitions (as shown by Lewis' famous argument from [Lewis and Langford 1932, p. 250; cf. Read 1988, p. 31] that A and not A implies B, these intuitions may well be contradictory beyond remedy), but on an implication arising from deductions where the assumptions are really involved in getting the conclusion. It is, after all, considered to be a mathematical improvement if a superfluous assumption is removed from a theorem (though the theorem was not wrong before the removal and though the advance is often judged in aesthetical terms). A related subject, which might seem more mathematical, is functions that really depend on their arguments — something akin to Church's λ-I-calculus.

The literature produced by the authors of the preceding paragraph is so big and has so many developments that we cannot possibly do it justice in such a brief introduction. One may consult [Dunn 1986], a shorter introductory text, and the encyclopaedic [Anderson and Belnap 1975], which has recently acquired a second volume [Anderson, Belnap and Dunn 1992]. Let us mention, however, that to the uninformed reader it may not be immediately clear that relevant logic should be characterized, as it was here, by the rejection of Thinning. Such a characterization, which is clearly given in [Kripke 1959], is sometimes hidden under the complications of rather nonstandard Gentzen formulations of the mainstream relevant logics. These complications are mainly due to the fact that in the school of Anderson and Belnap it is considered that we must assume distribution of conjunction over disjunction, i.e.:

$$(C \wedge (A \vee B)) \;\rightarrow\; ((C \wedge A) \vee (C \wedge B))$$

(by $\wedge$ and $\vee$ we symbolize additive conjunction and disjunction). If we simply reject Thinning from classical or intuitionistic logic, we will lack this principle. Instead, fusion will distribute over disjunction; i.e., we shall have:

$$(C \cdot (A \vee B)) \;\rightarrow\; ((C \cdot A) \vee (C \cdot B))$$

(the same thing happens if we reject Contraction). Distribution of conjunction over disjunction is behind many technically unwholesome properties of relevant logic, such as undecidability of propositional logic and failure of interpolation (which were quite difficult to establish). But, most important to us here, the wish to assimilate it may hide the substructural nature of relevant logic. (The substructural logics without Contraction of §§4–5 below also lack distribution of conjunction over disjunction and have only distribution of fusion over disjunction.)

The central relevant logic from the substructural perspective should be a version of R minus distribution of conjunction over disjunction, with either an involutive or an intuitionistic purely implicational negation (this logic has recently gained some status among relevant logicians in [Thistlewaite et al. 1988], though it has been known from Meyer's thesis since the middle sixties; see also [Smirnov 1972, chapter 6] and [Smirnov 1973]). The relevant logics E and T, which in the sixties and seventies have received much attention in the Anderson and Belnap school, do not engender clear-cut substructural logics, as R does. An S4-style modal extension of R, prefigured already by Orlov in [1928], is a better candidate than E (but without $(\Box A \wedge \Box B) \;\rightarrow\; \Box(A \wedge B)$, which, though favoured in [Anderson and Belnap 1975, §27.1.3, pp. 343–344], is very much like distribution of conjunction over disjunction: in the absence of Thinning, it does not follow from standard S4-style sequent rules, but it follows when conjunction is replaced by fusion).

However, it is worth mentioning several logics in the neighbourhood, characterized by various restrictions on Thinning and possibly other structural rules. Such is the so-called *mingle* extension of relevant logic, with the following restricted version of Thinning, converse to Contraction:

$$\textit{Expansion} \quad \frac{X, A, Y \vdash Z}{X, A, A, Y \vdash Z}$$

(more information on mingle systems can be found in [Dunn 1986], [Anderson and Belnap 1975] and [Avron 1987]).

Somewhat farther from relevant logic, but still comparable to it, is modal logic based on strict implication. The strict implication $\rightarrow$ of S4 can be characterized by restricting Thinning and Permutation to:

$$\textit{S4 Thinning} \quad \frac{X, B \rightarrow C, Y \vdash Z}{X, A, B \rightarrow C, Y \vdash Z}$$

$$\textit{S4 Permutation} \quad \frac{X, B \rightarrow C, A, Y \vdash Z}{X, A, B \rightarrow C, Y \vdash Z}$$

In addition to these *quasi-structural* rules, we have Identity, unrestricted Contraction on the left, Cut, $(\rightarrow L)$ and $(\rightarrow R)$ from §1 above.

It seems that nonmonotonic logics should be connected with relevant logic, with which they share the rejection of unrestricted Thinning, but it also seems that, up to now, this connexion has not been much exploited (cf. [Thistlewaite et al. 1988, §1.2, pp. 9–11]). However, nonmonotonic logics may require some specific structural rules, like the following restricted versions of Thinning and Cut:

$$\frac{X, Y \vdash A \qquad X, Y \vdash B}{X, A, Y \vdash B}$$

$$\frac{X, Y \vdash A \qquad X, A, Y \vdash B}{X, Y \vdash B}$$

whereas unrestricted Thinning and Cut are rejected. Some additional Thinning would come through assuming $X, A, Y \vdash A$ rather than $A \vdash A$ (see [Gabbay 1985]). Note that this, together with the restricted Thinning above, yields Expansion as follows:

$$\frac{X, A, Y \vdash A \qquad X, A, Y \vdash B}{X, A, A, Y \vdash B}$$

With restrictions on structural rules such as we find in nonmonotonic logics, we are abandoning the combinatory inspiration that characterizes all the main substructural logics. Namely, after rejecting a structural rule that corresponds to a well-known combinator, we are introducing something not inspired by such a combinator. Presumably, nonmonotonic logics would be better understood if methods developed for related substructural logics were applied in their investigation.

4 BCK Logic

In BCK logic we reject Contraction on both the left and right, and keep Thinning. Systems of many-valued logic (for example, the systems of Łukasiewicz) may lack Contraction, or have only restricted forms of Contraction. In this respect they resemble BCK logic. However, as a common distrust of Thinning does not suffice to conflate relevant logic with modal logic, so BCK logic is not to be conflated with many-valued logic. It is doubtful whether one should look for the roots of BCK logic in the works of the founders of many-valued logic, Vasil'ev, Łukasiewicz and Post.

We are nearer to the roots of BCK logic with a work contemporaneous with [Gentzen 1935]; namely, Fitch's doctoral thesis of 1934, from which [Fitch 1936] was adapted. Fitch's ambition at that time, as well as later, was to obtain a type-free system of logic in which paradoxes would not arise. However, at the propositional level, Fitch's system of [1936] is classical propositional logic. Contraction is rejected at the level of predicate logic, which is built in the manner of Schönfinkel and Curry, without type distinctions. Fitch's system, unlike ordinary predicate logic, does not have for every formula with two free occurrences of a variable an equivalent formula with a single free occurrence of this variable (in ordinary predicate logic, we have $A(x,x) \leftrightarrow \forall y(x = y \rightarrow A(y,y)))$. This blocks the derivation of Russell's paradox. It is well-known from Curry's work that Contraction and the corresponding combinator are essentially involved in Russell's paradox (see [Curry and Feys 1958, chapters 5G, 8A-B, and p. 338]). A system with ambitions similar to Fitch's, but without Contraction and other structural rules at the propositional level, was later introduced in a series of papers by Ackermann starting with [1950].

In the middle thirties, the implicational fragment of BCK logic was considered by Tarski. He proved in [1935] that all implicational calculuses containing the implicational BCK calculus, which is obtained from Hilbert's implicational positive logic by leaving out $(A \rightarrow (A \rightarrow B)) \rightarrow (A \rightarrow B)$, have the classical implicational calculus as their only consistent Post-complete extension closed under substitution for propositional variables. (To prove this result of Tarski, note that:

$$((A \rightarrow A) \rightarrow (A \rightarrow A)) \rightarrow (A \rightarrow A)$$

$$((A \rightarrow A) \rightarrow A) \rightarrow A$$

$$(A \to (A \to A)) \to (A \to A)$$

and the converse implications are provable in the implicational BCK calculus. If B is not a classical tautology, there must be a valuation v such that $v(B) = \bot$. For every propositional variable p in B, substitute $A \to A$ for p if $v(p) = \top$, and A if $v(p) = \bot$. It follows by replacement of equivalents that the extension of the implicational BCK calculus with B is not consistent. This proof shows that in Tarski's result we can replace the implicational BCK calculus by a weaker system.)

A system where Contraction is restricted to:

$$\frac{X, A, A, A, Y \vdash Z}{X, A, A, Y \vdash Z}$$

was investigated by Nelson in [1959], in connexion with intuitionistic logic extended with *constructible falsity*, i.e. an involutive negation that satisfies De Morgan's laws, which however does not make intuitionistic logic collapse into classical logic (see [Rasiowa 1974, chapters V and XII] and references therein; this negation resembles the negation of another type-free system of Fitch from [1952]). A similar system with restricted Contraction and involutive negation was investigated starting from the sixties by Zaslavskiĭ (these investigations are reported in his monograph [1978]). Zaslavskiĭ finds that his system is an intuitionistic analogue of Łukasiewicz's three-valued logic. The implication $\Rightarrow$ of this system may be defined by:

$$A \Rightarrow B \quad =_{df} \quad (A \to B) \wedge (\sim B \to \sim A)$$

where $\to$ is intuitionistic implication and $\sim$ is involutive negation (cf. the definition of $\Rightarrow$ in [Rasiowa 1974, pp. 68, 280]).

Next, we mention two papers from the sixties, dealing with implicational BCK logic. In [1963], Jaśkowski showed that a formula is provable in this logic iff it is obtained by identifying variables of a classical implicational tautology in which every variable occurs once or twice. In the second paper, [Meredith and Prior 1963], the name 'BCK' makes one of its first appearances (see also [Prior 1962, appendix I, p. 316]). This name is derived from the combinators corresponding to the axioms of implicational BCK logic according to the Curry-Howard interpretation (cf. §3 above).

In the middle sixties, Iséki started the investigation of algebras related to implicational BCK logic (see [Iséki and Tanaka 1978]). Apart from Japanese mathematicians, Wroński, his collaborators in Cracow and Bunder made the main contributions in this area (references are too numerous to be given here: I have counted more than twenty papers only in *Mathematica Japonica* for the first half of the eighties).

Grishin seems to be the first logician that investigated BCK logic extensively without restricting himself to the implicational fragment (a decidable sequent system of predicate logic without Contraction is investigated in [Smirnov 1972, chapter 5]; however, this system, which is not closed under Cut, does not amount to BCK logic). Grishin formulates BCK logic by rejecting Contraction from Gentzen's sequent system for classical predicate logic and obtains a system with an involutive negation that satisfies De Morgan's laws. Since we don't restrict Thinning, this, rather than an intuitionistic version, should be the canonical version of BCK logic (an intuitionistic version of BCK logic would restrict Thinning on the right). So, Grishin's approach is properly in the spirit of substructural logics. His initial motivation for considering BCK logic was similar to Fitch's, but he does not refer to Fitch, Curry or Ackermann. He sees himself as continuing the work of Bochvar, Skolem, Chang and Fenstad on set theory with unrestricted comprehension based on many-valued logic (similar investigations may be found in the area of paraconsistent logics; however, so far as I know, these investigations, though they resemble things in the relevant logic and BCK logic literature, don't seem to be in the spirit of substructural logics). In [1974], Grishin shows via cut elimination that unrestricted comprehension can be consistently added to his BCK predicate logic. He also concludes from cut elimination that BCK predicate logic, a first-order, possibly many-sorted, system, is decidable (a remark concerning this fact may be found in [Wang 1962, chapter IX, §1, p. 228]).

Moreover, Grishin introduces for every natural number n a translation t_n from BCK predicate logic into classical predicate logic, whose essential clause is:

$$t_n(\forall xA) \quad = \quad \forall x t_n(A) \cdot \ldots \cdot \forall x t_n(A)$$

where $\forall xA$ is a negative, i.e. antecedent, subformula and $\forall x t_n(A)$ is repeated n times in the fusion on the right-hand side (cf. [Ono 1990, §3]). He shows that A is provable in classical predicate logic iff, for some n, $t_n(A)$ is provable in BCK predicate logic. Since the latter system is decidable, the existential quantifier 'for some n' on the right-hand side of this equivalence is the culprit for the undecidability of classical predicate logic. Via the translation t_n, we see that this quantifier is tied to Contraction involving formulae of the form $\forall xA$ on the left-hand side of sequents. When, after having established cut elimination for classical predicate logic, we attempt to use it to show decidability, we shall run into unsurmountable difficulties with Contraction involving formulae of the form $\forall xA$ on the left-hand side (and, dually, with Contraction involving formulae of the form $\exists xA$ on the right-hand side).

In [1974], Grishin remarks that BCK logic is properly included in Łukasiewicz's infinite-valued logic, the latter not being recursively axiomatizable. In the same paper, a formulation of Łukasiewicz's three-valued logic is obtained by extending BCK logic with the following restricted version of Contraction on the left:

$$\frac{X, A+A, A+A, Y \vdash Z}{X, A+A, Y \vdash Z}$$

and the analogous quasi-structural rule on the right (the fission $A + A$ stands for $\neg A \rightarrow A$, as in §3 above). A similar, but more involved, restriction on Contraction, introduced at the end of [Grishin 1976], gives Łukasiewicz's four-valued logic. In subsequent papers [1976, 1979, 1981, 1983, 1985], Grishin studied the algebraic models of BCK logic, an analogue of Herbrand's theorem, and the effect of adding both unrestricted comprehension and extensionality to BCK logic. This effect is that we regain classical logic and the paradoxes (cf. [White 1979] for a similar result concerning Łukasiewicz's infinite-valued logic).

In [Dardžaniá 1977], results concerning decidability and translation similar to Grishin's are presented for an intuitionistic version of BCK predicate logic. A decision procedure for a fragment of Grishin's BCK predicate logic, whose propositional part is purely multiplicative, is described in [Ketonen and Weyhrauch 1984]. We shall not review in detail subsequent papers on BCK logic by Ono and Komori, of which the main is their joint paper [1985] (further references are in Ono's survey [1990]). These papers are devoted mainly to Kripke-style models for BCK and related substructural logics (see §7 below).

5 Linear Logic

In linear logic, we reject both Thinning and Contraction, on both sides. In this logic we have no structural rules except Identity and Cut if sequents are based on *multisets* of formulae instead of sequences of formulae as in [Gentzen 1935].

In a discussion of Church's idea of a minimal logic in [Curry 1954], one finds implicational systems obtained from the implicational fragment of the relevant logic R by restricting Contraction to:

$$\frac{X, A^{k+1}, Y \vdash Z}{X, A^{k}, Y \vdash Z}$$

where A^k stands for a sequence of k occurrences of A, for some $k \geq 2$. The suggestion to reject Contraction completely, as well as Thinning, to arrive at a minimal logic, may be found in [Kripke 1959]. An implication-negation calculus based on the same rejection is in [Smiley 1959, §5]; its implicational postulates are:

$$(B \to C) \to ((A \to B) \to (A \to C))$$

$$(A \to (B \to C)) \to (B \to (A \to C))$$

$$A \to A$$

and *modus ponens*. In [Jaśkowski 1963], the theorems of this implicational calculus receive a characterization analogous to the characterization given in the same paper for implicational BCK logic (cf. [Smiley 1959, footnote 21]). This implicational calculus is called BCI in [Meredith and Prior 1963] and [Prior 1962, appendix I, p. 316], according to the combinators corresponding to the three implicational axiom-schemata above (cf. §4 above and [Meredith 1978]). Systems without Thinning and Contraction have also been considered in relevant and BCK quarters (see [Meyer and McRobbie 1982] and [Komori 1986]) and in the survey [D. 1988] (for additional references see [D. 1992a]).

Linear logic acquired its name in [Girard 1987] (according to Girard, there is an analogy between this logic and linear algebra). Girard is motivated by applications he foresees in computer science, but his project has also a more philosophical side expounded in [Girard 1989]. A novelty in [Girard 1987] is that besides the usual logical connectives, implication, negation, conjunction, disjunction (appearing in a rather colourful notation), and the less usual ones, fusion, fission and multiplicative and additive propositional constants, which were already considered in relevant and BCK logic, he has S4-like modal operators that serve to mimic the missing structural rules. Namely, we have restricted forms of Thinning and Contraction for modalized formulae. This makes it possible to embed intuitionistic and classical logic into linear logic by a modal translation (of course, analogous embeddings are available for other substructural logics, provided we add the requisite modal operators; see [D. 1992a]). However, linear propositional logic with the modal operators is undecidable (this is proved in [Lincoln et al. 1992]), though linear predicate logic without the modal operators is decidable, as BCK predicate logic is. Another novelty of Girard's paper is a new representation of proofs by what he calls *proof nets*. In [1987], Girard considers linear logic with an involutive negation, but, subsequently, intuitionistic versions of linear logic were also considered. In [Troelstra 1992], one may find a detailed survey of developments that have arisen from Girard's paper in a surprisingly short time.

6 The Lambek Calculus

In the calculus introduced by Lambek in [1958], besides Thinning and Contraction, we reject Permutation too. This gives the *associative* variant of the Lambek calculus. In this calculus, sequences of formulae in sequents are really taken seriously: there are no structural rules to reduce them to multisets or sets of formulae. The only structural rules are Identity and Cut.

In the *nonassociative* variant of the calculus, introduced by Lambek in [1961], sequents are not based on sequences of formulae, but on terms of formulae built with a binary comma, which we shall call *premise terms* (Belnap in [1982, 1990] exploits the same idea). More precisely, all formulae are premise terms, and if X and Y are premise terms, so is (X, Y). By $X[Y]$ we denote a premise term in which Y is a subterm, and $X[Z]$ is obtained from $X[Y]$ by replacing a *single* occurrence of Y by Z. In a sequent $X \vdash A$, we take X to be a premise term and not a sequence of formulae. However, these sequents will not differ essentially from single-conclusion sequents based on sequences of formulae if we assume the following structural rules:

$$\textit{Association} \quad \frac{X[((Y_1, Y_2), Y_3)] \vdash C}{X[(Y_1, (Y_2, Y_3))] \vdash C} \qquad \frac{X[(Y_1, (Y_2, Y_3))] \vdash C}{X[((Y_1, Y_2), Y_3)] \vdash C}$$

(we might also need a premise term to stand for the empty sequence and structural rules to insert and delete this premise term). In the nonassociative Lambek calculus we reject Association. What remains of structural rules is Identity and Cut in the following form:

$$\frac{Y \vdash A \qquad X[A] \vdash C}{X[Y] \vdash C}$$

In both versions of the Lambek calculus, we have only single formulae on the right-hand side of sequents, and as connectives we have fusion $\cdot$ and two implications, $\rightarrow$ and $\leftarrow$ (Lambek's notation for the two implications is $\backslash$ and $/$). The essential principles governing these connectives are:

$$A \cdot B \vdash C \quad \textit{iff} \quad B \vdash A \rightarrow C,$$
$$A \cdot B \vdash C \quad \textit{iff} \quad \vdash C \leftarrow B.$$

Algebraically, this means that we have a left and right residual. With Permutation, the fusion connective, which stands for the comma of premise terms, becomes commutative and the two implications become equivalent.

As Thinning, Contraction and Permutation correspond to the combinators K, W and C respectively, so the Association rules correspond to the combinator B. With the first Association rule above we can prove:

(1) $A \cdot (B \cdot C) \vdash (A \cdot B) \cdot C$ and $B \rightarrow C \vdash (A \rightarrow B) \rightarrow (A \rightarrow C)$

and with the second:

(2) $(A \cdot B) \cdot C \vdash A \cdot (B \cdot C)$ and $C \leftarrow B \vdash (B \leftarrow A) \leftarrow (C \leftarrow A)$.

The implicational principles in (1) and (2) are related to the fourth implicational axiom of Hilbert's positive logic (see §1 above).

In the beginning, Lambek proposed his calculus not as a new logic but as an instrument underlying *categorial grammars* (he was also inspired by multilinear algebra and noncommutative ring theory). In this linguistic application, $A, B, C, \ldots$ stand for grammatical categories, or syntactic types, or, simply, types; $A \cdot B$ is the type of an expression obtained by concatenating an expression of type A with an expression of type B, whereas $A \rightarrow B$ and $B \leftarrow A$ are types of an expression that concatenated with an expression of type A on respectively the left-hand side and right-hand side gives an expression of type B. The sequent $A \vdash B$ means that every expression of type A is of type B.

As a linguistic tool, the Lambek calculus started attracting greater attention in the last decade. Two logicians, Buszkowski and van Benthem, have special merit for promoting investigations into the theoretical aspects of the Lambek calculus from the point of view of linguistics (see [Buszkowski et al. 1988], [Buszkowski 1989] and [van Benthem 1991] for surveys of the literature; for something closely related to the Lambek calculus in the work of the relevant logicians, see [Meyer and Routley 1972]).

Lambek's own interest in his calculus was later motivated not so much by linguistics as by category theory (in his recent papers [1988, 1989, 1989a] and in this book, he considers both aspects of the matter). The investigations of the Lambek calculus and related Gentzen-like calculuses in category theory are closely related to logic if following Lambek we understand categories as a special kind of deductive system. In a deductive system, we introduce terms that encode proofs and we are interested in relations between these terms. Categorical axioms given as identities between these terms correspond to reduction steps in a normalization procedure. These terms are also related to terms in a typed lambda calculus and the identities between terms are related to lambda conversion. This approach, which unites Gentzen-style proof theory, category theory and combinatory logic, and incorporates the Curry-Howard interpretation, is expounded for intuitionistic logic in [Lambek and Scott 1986]. For the Lambek calculus and linear logic, one may consult [Mints 1977], [Szabo 1978] and [Girard and Lafont 1987]. Presumably, this approach can be extended to other substructural logics, including relevant and BCK logic (cf. [Helman 1977], [Meredith 1978] and [Szabo 1983]). Of all these works, [Szabo 1978], which in addition to intuitionistic logic considers both the Lambek calculus and fragments of linear logic, is most open to experimenting with various structural rules and may be taken as a general work on substructural logics.

Although the Lambek calculus was not proposed as a new logic, we suppose that one may view the nonassociative Lambek calculus as a minimal multiplicative logic (multiplicative, in the sense of [Girard 1987]). From it, we pass to the stronger logics not by changing anything in the rules for fusion and implications, but only by adding structural rules. These structural rules simplify and make more manageable work with hypotheses. They enable us to consider hypotheses as given in sequences, or multisets, or sets. But if we want more complete information about hypotheses arranged at the top of proof-trees, then we should work with sequents based on premise terms that correspond to the proof-trees; and with binary proof-trees the appropriate logic should be the nonassociative Lambek calculus.

To make it more of a logic, we should extend the Lambek calculus with other connectives, as in Lambek's papers in this book. The additive connectives and the multiplicative constant t (i.e. Girard's 1) don't seem to be problematic (though they need not yield conservative extensions). However, negation may be problematic. As implication has split into two implications, so negation may split into two negations, a left one and a right one. If $\neg_L A$ and $\neg_R A$ are equivalent with respectively $A \rightarrow \bot$ and $\bot \leftarrow A$, where no postulate is assumed about $\bot$, then in the nonassociative Lambek calculus we shall have $A \vdash \neg_L \neg_R A$ and $A \vdash \neg_R \neg_L A$, which correspond to $A \vdash \neg\neg A$. To get something like an involutive negation, we should presumably consider assuming the converse sequents.

7 The Semantics of Substructural Logics

Substructural logics are proof-theoretically motivated, and their syntax seems to be better understood than their semantics. This is true even for intuitionistic logic, though model theory is incomparably more developed for intuitionistic logic than for the weaker substructural logics. In this section, I will try to say something about models for these weaker logics.

All these logics have algebraic models, but, as usual, these models are not far removed from the syntax. The core of these models serves for the multiplicative fragment and is some sort of residuated groupoid. We also have a unary operation for negation, which is involutive if negation is involutive. The lattice ordering of a residuated groupoid serves to model the additive connectives. An involutive negation operation satisfies De Morgan's laws with respect to lattice meet and join.

Somewhat farther removed from syntax, but closer to it than seems desirable, are Kripke-style models. We take as Kripke-style models something where formulae are mapped by valuations to sets of *worlds*, and relations and operations on worlds are used to formulate semantic clauses for the connectives; in other words, we are defining an algebraic model in the power set of the set of worlds. Kripke models for intuitionistic logic are so suggestive because everything is reduced to a quasi-ordering of the set of worlds and the semantic clauses for the connectives are quite natural.

In relevant logic, the standard is Kripke-style models of Routley and Meyer with a ternary relation R on worlds such that $Rxyz$ means something like $x \cdot y \leq z$, where $\cdot$ is a binary operation on worlds corresponding to fusion and $\leq$ is a partial ordering of worlds. The ternary relation R serves to model the multiplicative fragment. The semantic clause for implication is:

$$y \models A \to B \quad \textit{iff}, \quad \textit{for every } x \textit{ and } z, \textit{ if } Rxyz \textit{ and } x \models A, \textit{ then } z \models B,$$

which is related to the clause:

$$y \models A \to B \quad \textit{iff}, \quad \textit{for every } x, \textit{ if } x \models A, \textit{ then } x \cdot y \models B.$$

The involutive negation brings in an additional operation on worlds. Kripke's clauses for conjunction and disjunction produce a distributive lattice structure (this is presumably one of the reasons why relevant logicians want distribution of conjunction over disjunction). If we don't want distribution, we can introduce an additional binary operation on worlds to serve for disjunction. The whole thing starts looking *ad hoc*, and comes too close to algebra and syntax. The Kripke-style models for BCK logic of [Ono and Komori 1985] are quite close in spirit to models for relevant logic, and the phase models of [Girard 1987] don't seem to be far removed (Girard's development of Scott domains, called *coherence spaces*, in [Girard 1987] and [Girard et al. 1989] is a different matter). Standard models for the Lambek calculus are inspired by linguistics, but they are also related to the Kripke-style models of relevant, BCK and linear logic (see Ono's paper in this book and [D. 1988, 1992]).

Perhaps Kripke-style set-theoretic semantics is not what is needed for substructural logics (one may have doubts about this semantics even for intuitionism). A different sort of semantics, inspired by proof theory and category theory, as indicated in §6 above, may be more congenial to the subject matter. However, to talk about such things, we would have to abandon the historical perspective of this introduction and look into current investigations of substructural logics.

Acknowledgements. I would like to thank Jim Lambek and Peter Schroeder-Heister, who read an earlier version of this introduction and made helpful comments.

References

Ackermann, Wilhelm

1950 Widerspruchsfreier Aufbau der Logik I: Typenfreies System ohne tertium non datur, *The Journal of Symbolic Logic 15*, 33–57.

Anderson, Alan Ross, and Belnap, Nuel D. Jr

1975 *Entailment: The Logic of Relevance and Necessity, Vol. I*, Princeton University Press, Princeton.

Anderson, Alan Ross; Belnap, Nuel D. Jr, and Dunn, Jon Michael

1992 *Entailment: The Logic of Relevance and Necessity, Vol. II*, Princeton University Press, Princeton.

Avron, Arnon

1987 A constructive analysis of RM, *The Journal of Symbolic Logic 52*, 939–951.

Belnap, Nuel D. Jr

1982 Display logic, *Journal of Philosophical Logic 11*, 375–417.

1990 Linear logic displayed, *Notre Dame Journal of Formal Logic 31*, 14–25.

Buszkowski, Wojciech

1989 Principles of categorial grammar in the light of current formalisms, in: K. Szaniawski (ed.), *The Vienna Circle and the Lvov-Warsaw School*, Kluwer, Dordrecht, 113-137.

Buszkowski, Wojciech; Marciszewski, Witold, and van Benthem, Johan (eds)

1988 *Categorial Grammar*, Benjamins, Amsterdam.

Church, Alonzo

1941 *The Calculi of Lambda-Conversion*, Princeton University Press, Princeton.

1951 The weak theory of implication, in: A. Menne et al. (eds), *Kontrolliertes Denken: Untersuchungen zum Logikkalkül und zur Logik der Einzelwissenschaften*, Alber, Munich, 23–37 (abstract: The weak positive implicational calculus, *The Journal of Symbolic Logic 16* (1951), 238).

Curry, Haskell Brooks

1950 *A Theory of Formal Deducibility*, University of Notre Dame Press, Notre Dame.

1954 Generalizations of the deduction theorem, *Proceedings of the International Congress of Mathematicians, Vol. 2*, Noordhoff, Groningen, 399–400.

1963 *Foundations of Mathematical Logic*, McGraw-Hill, New York.

Curry, Haskell Brooks, and Craig, William

1953 Review of [Church 1951], *The Journal of Symbolic Logic 18*, 177–178.

Curry, Haskell Brooks, and Feys, Robert

1958 *Combinatory Logic, Vol. I*, North-Holland, Amsterdam.

Dardžaniá, G. K.

1977 Intuitionistic system without contraction, *Polish Academy of Sciences, Bulletin of the Section of Logic 6*, 2–8 (I know this paper only from references and the *Mathematical Reviews*).

Došen, Kosta

1988 Sequent systems and groupoid models, I and II, *Studia Logica 47*, 353–385, *48* (1989), 41–65, *49* (1990), 614.

1989 Logical constants as punctuation marks, *Notre Dame Journal of Formal Logic 30*, 362–381.

1992 A brief survey of frames for the Lambek calculus, *Zeitschrift für mathematische Logik und Grundlagen der Mathematik 38*, 179–187.

1992a Modal translations in substructural logics, *Journal of Philosophical Logic 21*, 283–336.

Dunn, Jon Michael

1986 Relevance logic and entailment, in: D. Gabbay and F. Guenthner (eds), *Handbook of Philosophical Logic, Vol. III: Alternatives to Classical Logic*, Reidel, Dordrecht, 117–224.

Fitch, Frederic Brenton

1936 A system of formal logic without an analogue to the Curry *W* operator, *The Journal of Symbolic Logic 1*, 92–100.

1952 *Symbolic Logic*, Ronald, New York.

Gabbay, Dov M.

1985 Theoretical foundations for non-monotonic reasoning in expert systems, in: K. R. Apt (ed.), *Logics and Models of Concurrent Systems*, Springer, Berlin, 439–457.

Gentzen, Gerhard

1932 Über die Existenz unabhängiger Axiomensysteme zu unendlichen Satzsystemen, *Mathematische Annalen 107*, 329–350 (English translation in [Gentzen 1969], 29–52).

1935 Untersuchungen über das logische Schließen, *Mathematische Zeitschrift 39*, 176–210, 405–431 (English translation in [Gentzen 1969], 68–131).

1936 Die Widerspruchsfreiheit der reinen Zahlentheorie, *Mathematische Annalen 112*, 493–565 (English translation in [Gentzen 1969], 132–213).

1969 *The Collected Papers of Gerhard Gentzen*, M. E. Szabo (ed.), North-Holland, Amsterdam.

Girard, Jean-Yves

1987 Linear logic, *Theoretical Computer Science 50*, 1–102.

1989 Towards a geometry of interaction, in: J. W. Gray and A. Scedrov (eds), *Categories in Computer Science and Logic*, American Mathematical Society, Providence, 69–108.

Girard, Jean-Yves, and Lafont, Yves

1987 Linear logic and lazy computation, in: H. Ehrig et al. (eds), *TAPSOFT '87, Vol. 2*, Springer, Berlin, 52–66.

Girard, Jean-Yves; Taylor, Paul, and Lafont, Yves

1989 *Proofs and Types*, Cambridge University Press, Cambridge.

Glivenko, Valeriĭ Ivanovich

1929 Sur quelques points de la Logique de M. Brouwer, *Académie Royale de Belgique, Bulletin de la Classe des Sciences (5e série) 15*, 183–188.

Grishin, Vyacheslav Nikolaevich

1974 A nonstandard logic and its application to set theory (in Russian), in: *Studies in Formalized Languages and Nonclassical Logics* (in Russian), Nauka, Moscow, 135–171 (*Mathematical Reviews* 56 # 8331, 1978).

1976 The algebraic semantics of logic without contractions (in Russian), in: *Studies in Set Theory and Nonclassical Logics* (in Russian), Nauka, Moscow, 247–264 (*Mathematical Reviews* 57 # 9533, 1979).

1979 Herbrand's theorem for logic without contractions (in Russian), in: *Studies in Nonclassical Logics and Set Theory* (in Russian), Nauka, Moscow, 316–329 (*Mathematical Reviews* 82d:03021).

1981 Predicate and set-theoretic calculuses based on logic without contractions (in Russian), *Izvestiya Akademii Nauk SSSR, Seriya Matematicheskaya 45*, 47–68 (English translation in *Math. USSR-Izv. 18* (1982), 41–59).

1983 A generalization of the Ajdukiewicz-Lambek system (in Russian), in: *Studies in Nonclassical Logics and Formal Systems* (in Russian), Nauka, Moscow, 315–334 (*Mathematical Reviews* 85f:03068).

1985 The impossibility of defining the class of L_0-algebras by means of identities (in Russian), *Matematicheskie Zametki 38*, 641–651 (English translation in *Mathematical Notes 38* (1985), 861–867).

Helman, Glen Harry

1977 *Restricted Lambda Abstraction and the Interpretation of some Non-Classical Logics*, doctoral dissertation, University of Pittsburgh.

Hertz, Paul

1922 Über Axiomensysteme für beliebige Satzsysteme, I and II, *Mathematische Annalen 87*, 246–269, *89* (1923), 76–102.

1928 Reichen die üblichen syllogistischen Regeln für das Schließen in der positiven Logik elementarer Sätze aus?, *Annalen der Philosophie und Philosophische Kritik 7*, 272–277.

1929 Über Axiomensysteme für beliebige Satzsysteme, *Mathematische Annalen 101*, 457–514.

Heyting, Arend

1930 Die formalen Regeln der intuitionistischen Logik, *Sitzungsberichte der Preußischen Akademie der Wissenschaften, Physikalisch-mathematische Klasse*, 42–56.

Hilbert, David

1922 Neubegründung der Mathematik (Erste Mitteilung), *Abhandlungen aus dem mathematischen Seminar der Hamburgischen Universität 1*, 157–177 (reprinted in [Hilbert 1935], 157–177).

1923 Die logischen Grundlagen der Mathematik, *Mathematische Annalen 88*, 151–165 (reprinted in [Hilbert 1935], 178–191).

1928 Die Grundlagen der Mathematik, *Abhandlungen aus dem mathematischen Seminar der Hamburgischen Universität 6*, 65–85 (English translation in [van Heijenoort 1967], 464–479).

1931 Die Grundlegung der elementaren Zahlenlehre, *Mathematische Annalen 104*, 485–494 (reprinted in [Hilbert 1935], 192–195).

1935 *Gesammelte Abhandlungen, Vol. 3*, Springer, Berlin.

Hilbert, David, and Bernays, Paul

1934 *Grundlagen der Mathematik, Vol. 1*, Springer, Berlin.

1939 *Grundlagen der Mathematik, Vol. 2*, Springer, Berlin.

Iséki, Kiyoshi, and Tanaka, Shōtarō

1978 Introduction to the theory of BCK-algebras, *Mathematica Japonica 23*, 1–26.

Jaśkowski, Stanislaw

1963 Über Tautologien, in welchen keine Variable mehr als zweimal vorkommt, *Zeitschrift für mathematische Logik und Grundlagen der Mathematik 9*, 219–228.

Johansson, Ingebrigt

1936 Der Minimalkalkül, ein reduzierter intuitionistischer Formalismus, *Compositio Mathematica 4*, 119–136.

Ketonen, Jussi, and Weyhrauch, Richard

1984 A decidable fragment of predicate calculus, *Theoretical Computer Science 32*, 297–307.

Kleene, Stephen Cole

1952 *Introduction to Metamathematics*, North-Holland, Amsterdam.

Kolmogorov, Andrei Nikolaevich

1925 On the principle of excluded middle (in Russian), *Matematicheskiĭ sbornik 32*, 646–667 (English translation in [van Heijenoort 1967], 414–437).

Komori, Yuichi

1986 Predicate logics without the structure rules, *Studia Logica 45*, 393–404.

Kripke, Saul A.

1959 The problem of entailment (abstract), *The Journal of Symbolic Logic 24*, 324.

Lambek, Joachim

1958 The mathematics of sentence structure, *The American Mathematical Monthly 65*, 154–170 (reprinted in [Buszkowski et al. 1988], 153–172).

1961 On the calculus of syntactic types, in: R. Jacobson (ed.), *Structure of Language and its Mathematical Aspects*, American Mathematical Society, Providence, 166–178, 264–265.

1988 Categorial and categorical grammars, in: R. T. Oehrle et al. (eds), *Categorial Grammars and Natural Language Structures*, Reidel, Dordrecht, 297–317.

1989 Multicategories revisited, in: J. W. Gray and A. Scedrov (eds), *Categories in Computer Science and Logic*, American Mathematical Society, Providence, 217–239.

1989a On a connection between algebra, logic and linguistics, *Diagrammes 22*, 59–75.

Lambek, Joachim, and Scott, Philip J.

1986 *Introduction to Higher Order Categorical Logic*, Cambridge University Press, Cambridge.

Lewis, Clarence Irving, and Langford, Cooper Harold

1932 *Symbolic Logic*, Century, New York.

Lincoln, Patrick; Mitchell, John; Scedrov, Andre, and Shankar, Natarajan

1992 Decision problems for propositional linear logic, *Annals of Pure and Applied Logic 56*, 239–311.

Meredith, Carew Arthur, and Prior, Arthur Norman

1963 Notes on the axiomatics of the propositional calculus, *Notre Dame Journal of Formal Logic 4*, 171–187.

Meredith, David

1978 Positive logic and λ-constants, *Studia Logica 37*, 269–285.

Meyer, Robert K., and McRobbie, Michael A.

1982 Multisets and relevant implication, I and II, *Australasian Journal of Philosophy 60*, 107–139, 265–281.

Meyer, Robert K., and Routley, Richard

1972 Algebraic analysis of entailment, I, *Logique et Analyse (N.S.) 15*, 407–428.

Mints, Grigori E.

1977 Closed categories and proof theory (in Russian), *Zapiski Nauchnykh Seminarov Leningradskogo Otdeleniya Matematicheskogo Instituta 68*, 83–114 (English translation in *Journal of Soviet Mathematics 15* (1981), 45–62).

Moh Shaw-Kwei

1950 The deduction theorems and two new logical systems, *Methodos 2*, 56–75.

Nelson, David

1959 Negation and separation of concepts in constructive systems, in: A. Heyting (ed.), *Constructivity in Mathematics*, North-Holland, Amsterdam, 208–225.

Ono, Hiroakira

1990 Structural rules and a logical hierarchy, in: P. P. Petkov (ed.), *Mathematical Logic*, Plenum, New York, 95–104.

Ono, Hiroakira, and Komori, Yuichi

1985 Logics without the contraction rule, *The Journal of Symbolic Logic 50*, 169–201.

Orlov, I. E.

1928 The calculus of compatibility of propositions (in Russian), *Matematicheskiĭ sbornik 35*, 263–286 (review in: K. Došen, The first axiomatization of relevant logic, *Journal of Philosophical Logic 21* (1992), 339–356).

Prior, Arthur Norman

1962 *Formal Logic*, second edition, Clarendon Press, Oxford.

Rasiowa, Helena

1974 *An Algebraic Approach to Non-Classical Logics*, North-Holland, Amsterdam.

Read, Stephen

1988 *Relevant Logic*, Blackwell, Oxford.

Scott, Dana S.

1971 On engendering an illusion of understanding, *The Journal of Philosophy 68*, 787–807.

Smiley, Timothy John

1959 Entailment and deducibility, *Proceedings of the Aristotelian Society 59*, 233–254.

Smirnov, Vladimir A.

1972 *Formal Deduction and Logical Calculuses* (in Russian), Nauka, Moscow.

1973 An absolute first-order predicate calculus, *Polish Academy of Sciences, Bulletin of the Section of Logic 2*, 38–45 (I know this paper only from the *Mathematical Reviews*).

Szabo, Manfred Egon

1978 *Algebra of Proofs*, North-Holland, Amsterdam.

1983 The continuous realizability of entailment, *Zeitschrift für mathematische Logik und Grundlagen der Mathematik 29*, 219–233.

Tarski, Alfred

1930 Über einige fundamentale Begriffe der Metamathematik, *Comptes Rendus des Séances de la Société des Sciences et des Lettres de Varsovie 23*, 22–29 (English translation in [Tarski 1956], 30–37).

1930a Fundamentale Begriffe der Methodologie der deduktiven Wissenschaften, I, *Monatshefte für Mathematik und Physik 37*, 361–404 (English translation in [Tarski 1956], 60–109).

1935 Über die Erweiterungen der unvollständigen Systeme des Aussagenkalküls, *Ergebnisse eines mathematischen Kolloquiums 7* (1934/35), 51–57 (English translation in [Tarski 1956], 393–400).

1956 *Logic, Semantics, Metamathematics: Papers from 1923 to 1938*, J. H. Woodger (ed.), Clarendon Press, Oxford (second edition, J. Corcoran (ed.), Hackett, Indianapolis, 1983).

Thistlewaite, Paul B.; McRobbie, Michael A., and Meyer, Robert K.

1988 *Automated Theorem-Proving in Non-Classical Logics*, Pitman, London.

Troelstra, Anne S.

1990 On the early history of intuitionistic logic, in: P. P. Petkov (ed.), *Mathematical Logic*, Plenum, New York, 3–17.

1992 *Lectures on Linear Logic*, Center for the Study of Language and Information, Stanford.

van Benthem, Johan

1991 *Language in Action: Categories, Lambdas and Dynamic Logic*, North-Holland, Amsterdam.

van Heijenoort, Jean (ed.)

1967 *From Frege to Gödel: A Source Book in Mathematical Logic, 1879–1931*, Harvard University Press, Cambridge, Massachusetts.

Wang, Hao

1962 *A Survey of Mathematical Logic*, Science Press, Peking (North-Holland, Amsterdam, 1964 ; reprinted under the title *Logic, Computers and Sets*, Chelsea, New York, 1970).

White, Richard B.

1979 The consistency of the axiom of comprehension in the infinite-valued predicate logic of Łukasiewicz, *Journal of Philosophical Logic 8*, 509–534.

Wójcicki, Ryszard

1988 *Theory of Logical Calculi: Basic Theory of Consequence Operations*, Kluwer, Dordrecht.

Zaslavskiĭ, I. D.

1978 *Symmetric Constructive Logic* (in Russian), Armenian Academy of Sciences, Erevan.

Life in the Undistributed Middle

Nuel Belnap

Department of Philosophy, University of Pittsburgh

Some logics have no connectives with the lattice properties. Some logics have lattice connectives satisfying distribution as well. What this paper wonders about is those logics in the middle: They have lattice connectives that do *not* satisfy the distributive laws.

By the 'lattice properties' I mean the familiar properties of conjunction and disjunction that can be stated in Gentzen fashion with no reference to structural context.

Structure-free rules

Left rules

$$\frac{A \vdash Y}{A\&B \vdash Y} \qquad \frac{B \vdash Y}{A\&B \vdash Y}$$

$$\frac{A \vdash Y \qquad B \vdash Y}{A \vee B \vdash Y}$$

Right rules

$$\frac{X \vdash A \qquad X \vdash B}{X \vdash A\&B}$$

$$\frac{X \vdash A}{X \vdash A \vee B} \qquad \frac{X \vdash B}{X \vdash A \vee B}$$

In the statement of these rules, A and B are formulas, and X and Y are perhaps complex structures. In display logic[1] these lattice rules are called 'structure-free.'

They are in contrast to rules such as the following:

Structure-dependent rules

$$\frac{A, B \vdash Y}{A\&B \vdash Y} \qquad \frac{X \vdash A \qquad Y \vdash B}{X, Y \vdash A\&B}$$

$$\frac{A \vdash X \qquad B \vdash Y}{A \vee B \vdash X, Y} \qquad \frac{X \vdash A, B}{X \vdash A \vee B}$$

1 Belnap 1982. A slightly amended version appears in Anderson and Belnap and Dunn 1992 (henceforth cited as ABD). See also Belnap 1990 for some improvements relevant to the present discussion. Another study with an entirely different flavor but with overlapping applicability is Dunn's gaggle theory.

In these 'structure-dependent' rules the meaning of & is evidently (but perhaps not obviously) dependent on the meaning of the structuring comma. This contrast is in boldest relief in display logic, whose defining characteristic — the Display property — is that each constituent of each consecution can be 'displayed' as either a stand-alone antecedent or a stand-alone consequent of an equivalent consecution. In the context of the Display property, *all* rules can be stated in simple context-free versions such as the above. In such circumstances one decides with ease whether a rule is structure-dependent (not structure-free): Look to see whether its statement exhibits any structural elements such as commas.

Connectives governed by structure-dependent rules can also have the lattice properties — if analogs of these properties are given to the structure itself (e.g., weakening). In either case I will say that & and $\vee$ are the 'lattice connectives'.

By 'distribution' I intend the distribution of conjunction over disjunction, say

Dist 1. $A\&(B \vee C) \vdash (A\&B) \vee (A\&C)$,

sometimes put more briefly in the equivalent form

Dist 2. $A\&(B \vee C) \vdash (A\&B) \vee C$,

as well as in various equivalent ways, often in terms of the provability of an implicational statement, e.g.

Dist 3. $\vdash A\&(B \vee C) \longrightarrow (A\&B) \vee C$.

Though perhaps it wasn't so clear to Peirce, everyone now knows that distribution does not follow from the lattice properties. This paper is about logics with lattice connectives but without distribution.[2] The principle examples of which I happen to be aware of distributionless logics are the following:

Quantum logic.

'Relevance logics without distribution', which one obtains from an axiomatization of a relevance logic by dropping the distributive axiom. These have been considered briefly and sporadically since Meyer 1966 proved **R** without distribution decidable by extending the procedure of Kripke 1959 (described in §13 of Anderson and Belnap 1975).

2 Logics altogether without lattice connectives are also of interest as objects of study; but it seems to me doubtful that such a logic could actually be applied without supplementation with lattice connectives, and in any event these logics are not the topic of this paper.

The 'linear logic' proposed and defended in Girard 1987 as applicable to parallel processing.

The logic **RMI** of Avron 1990 advertised as 'sufficient for all our needs'.[3]

The **BCK** logics that arise by withholding contraction are another example, but I am not familiar with that literature.

Distributionless logics are worth thinking about. Motivation for such an enterprise can be either *material* or *formal.* A 'material' example is the suggestion that the logic of quantum mechanics should be nondistributive because, as we are said to learn from consideration of Hilbert spaces, that is the nature of the lattice of propositions of micro-physical theory: that lattice is 'orthomodular' all right, but not fully modular, much less distributive. This interesting question I put aside, except for a single brief reference.

Sometimes, however, more 'formal' complaints are lodged against distribution, perhaps especially against distribution in the context of relevance logics. By '*standard* relevance logics' I refer historically and parochially but clearly to the bulk of systems treated in Anderson and Belnap 1975. Standard relevance logics tend to contain both intensional connectives and extensional connectives. Of first importance among the intensional connectives has been an implicational connective $A \to B$, with fusion $A \circ B$ and intensional disjunction $A + B$ being others. In standard relevance logic negation, $\sim$, also falls on the intensional side. Some key connections are these:

$$A \circ B \vdash C \;\; \textit{iff} \;\; A \vdash B \to C \;\; \textit{iff} \;\; A \vdash \sim B + C$$

$$A \circ \sim B \vdash C \;\; \textit{iff} \;\; A \vdash B + C$$

The first equivalence on the first line is *residuation,* which comes up again. In these logics $B + C$ is always commutative, whereas $A \circ B$ may or may not be commutative.

The standard extensional connectives have been conjunction $A \& B$ and disjunction $A \vee B$. Standard relevance logics tend to postulate that the extensional connectives satisfy Dist 3, with the arrow taken as the implicational connective; and as a matter of narrow convenience, for the scope of this paper I hereby license the dropping of 'tend to', making the postulation of Dist 3 a defining character of '*standard* relevance logic'.

A number of points arise about distribution or its absence. Each of them should be familiar in kind; it may nevertheless help to have a selection gathered in one place.

---

3 The mentioned paper cites a number of other papers by Avron on the same or related topics.

1. *Historically* distributionless logics were first introduced into the mainstream by Kripke as part of his investigation of problems raised by standard relevance logics.

1.1. Every relevance logician knows about distributionless systems, including the proof-theoretical naturalness with which they arise. The abstract Kripke 1959 alludes to a number of ideas and results that people are still exploring. Kripke laid out the Gentzen details in the late fifties partly on blackboards and partly in a personal communication now irretrievably lost. Anderson and I passed on Kripke's ideas to all of our friends and students. Meyer 1966 is, I think, the earliest extant writing on the topic. It would be interesting to determine the earliest published accounts.

1.2. As well as I can recall, what Kripke indicated in that letter of so long ago was this. Start with the Gentzen system for $\mathbf{R}_{\rightarrow\sim}$, and think about rules for & and $\vee$. For each there are two sane pairs of rules, where it is cut-elimination that guards sanity. One is a structure-free rule such as is displayed in the first display of this paper, and one is a structure-dependent rule such as is described in the second display. There are therefore four distinct systems combining & and $\vee$. Let an inference be counted as a 'valid entailment' (I do not recall if this was precisely his definiendum) if it holds in any one. Kripke observed that distribution fails the test.

2. The distributionless systems are themselves relevance logics even if not 'standard'.

2.1. They are generally treatable by the same kinds of techniques that have been used in the investigation of the standard relevance logics. Of equal interest is the fact that their study has given rise to fresh approaches.

2.2. They share a large measure of generalized motivation with standard relevance logics. For this reason the differences in rationale are likely to be especially interesting. The story I hear and tell most is that a relevance logic may be useful when contradictions threaten. An application for linear logic is said to be found in the context of parallel processing. **RMI** is said to be good for everything. Perhaps these motivations can be connected by more than offhand allusions; the attempt to do so would be difficult, and correspondingly worthwhile.

3. Relevance logicians have not emphasized these logics, partly for reasons and partly not needing a reason.

3.1. It is equally true that logics *with* distribution arise with great naturalness. The context of first degree entailments (tautological entailments) makes distribution seem almost indispensable: This logic has distributive normal forms in its very conception.

3.2. Relevance logicians have seldom taken the formal problems that distribution causes to be reasons for excluding it; instead, they have sometimes seen these problems as posing a legitimate challenge to research. Certainly Dunn's reaction to the fact that distribution failed the 'Kripke test' was to continue to look for alternate Gentzen systems for distributive relevance logics.

3.2.1. Dunn was successful, as independently was Mints; see ABD. There was, however, a limitation: their methods are not known to apply in the presence of negation.

3.2.2. Display logic later offered a Gentzen formulation of relevance logic including negation, but only with the additional help of Boolean negation (known to be conservatively addible). It is not known if display logic can be adapted to provide a formulation without Boolean negation.

3.2.3. More generally, it is not known if there is a way to Gentzenize the standard relevance logics in a standard vocabulary (e.g. $\rightarrow, \sim, \circ, \&, \vee$). What a deeply interesting technical result it would be if someone could show it impossible! Whether the result would have any conceptual significance, however, would surely depend on exactly what it was and how it was proved. Whatever the outcome, it is hard to believe that anyone would propose the matter relevant to the evaluation of the usefulness of the standard relevance logics.

3.3. Let us mention the generally good idea that we should let quite a few flowers bloom. But it should not be part of that idea that every gardener is obligated to admire all of the neighbors' dandelions, much less to cultivate every species.

3.4. Even beyond letting flowers bloom, those interested in relevance logic have generally exhibited a positive interest in the tilling of new fields by others. Unlike your classicalists and your intuitionists, who tend to monology, relevance logicians typically hold that to search for variations is a good thing; no one knows what may turn up. This point cannot be made without the ring of trite; but it is significant.

4. Applications. Distributionless systems are sometimes thought to have applications in areas in which the standard relevance logics do not. This seems likely; why shouldn't it be so? On the other hand, although sometimes the abstract possibility of or outline of such an application is described, one is not given a particular case. A good negative analogy comes from the quantum logic literature. There one is indeed told abstractly that a certain lattice derived from Hilbert spaces is nondistributive, but in *addition* the famous 'two slit experiment' is laid out as a definite, concrete counterexample to distribution. One can then decide what to think about the matter without having to buy a pig in a poke.

All that follows is that there may be work to be done. What is not needed is to describe some abstract lattice, note that it is not distributive, and conclude that we have an 'application' for a distributionless logic. Instead one wants to see the detailed case of an application in which it is made obvious that there is a concrete counterexample to distribution.

5. Distribution is forced on any semantics satisfying certain conditions.

(1) *The semantics evaluates sentences as true or false at indices.* A double contrast is intended: there are just the two truth-values, and the semantics is 'indexical' rather than algebraic or purely propositional.[4]

(2) *The semantics treats & and* $\vee$ *extensionally.* That is, for each index the truth value of the compound at that index depends only on the truth values of the components at that same index.

(3) *The semantics takes logical consequence to be preservation of truth at every index.*

And these principles, though certainly not inscribed in granite, are themselves deeply natural. The Kripke semantics for intuitionism is a perfect example.[5] It is well to keep in mind that any semantics not verifying distribution must give up at least one of them.

6. Distribution is sometimes vilified as leading to *undecidability.*

6.1. Once a useful language is defined, the logician should of course ask if there is a mechanical way of telling the good arguments from the bad — *if* he or she can. Such results are often interesting even when the decision procedures provided are mathematical abstractions rather than usable procedures.

6.2. There is, however, no argument that a language is more useful if it (or perhaps its propositional part) is decidable, even if some whose profession involves the provision of decision procedures are likely to disagree.

6.3. And certainly no one thinks that we should abstain from using undecidable languages. Examples spring to mind: everyone agrees that we would be foolish to give up the use of quantifiers because their logic is undecidable, or even to give up elementary arithmetic because its concept of good argument is not only undecidable but even unaxiomatizable. That

4 Neither this nor the other conditions are intended as necessary. It is for instance transparently all right from a purely formal point of view to admit numerous truth values if they form a distributive lattice, or to be purely algebraic or propositional if the target is distributive. But the purely formal point of view is not to be recommended if one wishes to secure other than purely formal results.

5 In the Beth semantics for intuitionism, $\vee$ is not extensional — and one sees exactly how and why this should be so.

the arguer is a programmed computer is irrelevant. Suppose elementary arithmetic becomes decidable if we leave out distribution of multiplication over addition $(a \times (b + c)) = (a \times b) + (a \times c)$; that is an exceedingly bad reason for touting the value of a 'distributionless arithmetic'.

6.4. It seems to follow that (mere) decidability is irrelevant to the design of useful languages.

7. Residuation. Let $\circ$ be a binary commutative positive connective. Such an operation is said to be *residuated* in a logic if there is another connective, negative on the left and positive on the right, such that

$$B \circ A \vdash C \quad \textit{iff} \quad A \vdash B \to C,$$

so that something like a 'deduction theorem' unites the two connectives. From the point of view of display logic, the purpose of residuation is to permit the 'display' of A as a self-standing antecedent.[6] Where $\circ$ corresponds to a structural element in some weak logic, it is typical that $\circ$ be residuated, and suspect if it is not. In the latter case one always thinks of expressive incompleteness.

A common fact is that any logic that gives & and $\vee$ the usual lattice properties becomes distributive if a residual is supplied for &. The nerve is that residuation allows one to 'display' $B \vee C$ in (say) Dist 2, so that lattice properties can take over. That is, Dist 2 is by residuation equivalent to $B \vee C \vdash A \to ((A\&B) \vee C)$, which by simple lattice properties follows from $B \vdash A \to ((A\&B) \vee C)$ and $C \vdash A \to ((A\&B) \vee C)$. And these are each derivable from a combination of simple lattice and residuation properties.

7.1. It follows that if a logic is not distributive, then & has no residual. One should properly think of both algebraic awkwardness and expressive incompleteness. Since each of awkwardness and incompleteness is a cost, a justification of twice the price is in order.

7.2. Residuation and intuitionism. For background, observe as first fact that relevance logic in its usual forms has no residual for its &, but it does have a *quasi-residual*, namely, its own intuitionistic implication, $A \supset B$, sometimes defined as $\exists p\,(p\&((p\&A) \to B))$ and sometimes as $A\&t \supset B$, by means of which it can be seen to contain intuitionism.[7] Here t is a left identity for the arrow: $t \to A \leftrightarrow A$. This quasi-residual is a true

6 In this and other contexts in both display logic and Dunn's gaggle theory, the restriction to binary commutative positive connectives is misleading. I lay on the restriction here just for expository convenience.

7 This is laid out most accessibly in ABD.

residual relative to a new relation that one could define by conjoining t to the antecedent:

$$A \vdash_t B =_{df} t \& A \vdash B$$

Then one has an equivalence between $A \& B \vdash_t C$ and $A \vdash_t B \supset C$, and intuitionism is found. My own view is that the technical result alone is not so interesting. It is just too easy for one logic to contain another. On the other hand, I do think that the conceptual discussion of Anderson and Belnap 1961, reprinted in ABD, shows how the containment result clarifies the motivation for both relevance logic and intuitionism; each illuminates the other.

A second fact is that a conservative extension of relevance logic, sometimes called 'classical relevance logic', contains a true residual for &. This residual is defined by $\neg A \vee B$, where $\neg A$ is Boolean negation.[8] Since relevance logic is anyhow distributive, the availability of a residual for & does not strike one dumb.

In parallel to the first fact, linear logic, which is distributionless, also contains intuitionism, with $A \supset B$ defined using the 'exponential' $!A$ of linear logic by $!A \to B$. How can that be? The trick is the same: linear logic's horseshoe is *not* a true residual for &, so that the nondistributivity of linear logic is not an issue. Its horseshoe is instead only a quasi-residual in the same sense as that of relevance logic, i.e., it is a true residual relative to $\vdash_t$. And just as for relevance logic, intuitionism is found.[9] It is then a practical matter if in a particular application the intuitionist fragment is useful, and it is a conceptual matter if the containment is interesting. I do not know how to prejudge either of these questions. However, whoever finds interesting the containment of intuitionism in linear logic by means of exponentials ought to find even more interesting the containment of intuitionism in punctual logic by means of exponentials.[10]

In contrast to the second fact is that there is no conservative extension of linear logic containing a residual for its &. For if there were, one could prove distributivity. Thus the absence of a residual for & is not just a matter of expressive incompleteness that could be fixed up in some conservative extension.

The import of this fact might be put as follows:

No logic with distributionless lattice connectives (i.e., no logic that lives

8 The result is due to Meyer; see ABD for references and a discussion that includes an expression of a common distaste for plastic, mouthwash, and Boolean negation.

9 I learned from Dunn to see the matter this way.

10 See Belnap 1990 for a discussion of punctual logic and its containment of intuitionism. I am sure punctual logic also has received other names in the substructural literature, but they aren't as pointed.

in the undistributed middle) can be conservatively extended to contain a set of classical connectives.

For example, it is a feature of display logic that there can be many, many logics all living harmoniously, with suitable interactions and suitable independences. Intuitionism, relevance logics of various kinds, different modal logics, and even classical logic can all take their places in the same framework. And display logic can also contain distributionless logics. It cannot, however, contain both a distributionless logic and a classical logic; adding classical logic automatically adds distribution for any logic with lattice connectives.

8. Enough for all our needs. Distributionless systems are sometimes said to give us all that we need. It is hard to know what to make of such claims when they are thrown off casually and defended abstractly. Here are relevant points.

8.1. Use of distribution in simple arguments about the distributionless systems, e.g. cut-elimination. Distributionless systems are frequently motivated by the techniques of a proof theory of which Gentzen's cut-elimination is the centerpiece. Here is a typical case. The rules adopted for & have at least as special cases the Left and Right rules displayed at the very beginning of this paper. There is then the problem of carrying out the case of the cut-elimination theorem in the case when one is given the provability of (1) and (2) below as conclusions known to follow by means of the rules just mentioned. One needs in this case to show the provability of (3):

$$\frac{(1)\ X \vdash A\&B \qquad\qquad (2)\ A\&B \vdash Y}{(3)\ X \vdash Y}$$

The following seems the only available argument. 'Since it is given that (1) comes by the Right rule for &, one must have (P1a) $X \vdash A$ and (P1b) $X \vdash B$ both provable, and since it is given that (2) comes by a Left rule for &, one must have either (P2a) $A \vdash Y$ or (P2b) $B \vdash Y$ provable. But then, *by distribution*, there are two cases: all of (P1a), (P1b), and (P2a) are provable, or else all of (P1a), (P1b), and (P2b) are provable. In either case an appeal to an appropriate inductive hypothesis yields the desired provability of (3).' It therefore cannot be a *casual* truth that we can obtain all we need without distribution.

There is a kind of general purpose counterargument that needs exposing in order that a *reductio* can license its neglect. The argument is that such and such a symbolic system can *casually* be seen to include great riches 'in principle'. That such casual arguments merit no consideration can be easily seen by reflecting on Gödel's once striking but now familiar insight that the proof-theoretical aspects of any symbolic system can be embedded in the system of numerals. We could if we wished give up all other symbolic systems, and just go around calling out — or even publishing — Gödel numbers. In principle.

8.2. Sufficient and necessary conditions. Suppose your retrieval system contains each of (1) A and (2) $B \vee C$, and also each of (3) $A\&B \rightarrow Q$ and (4) $C \rightarrow Q$. Suppose you ask your system whether Q; will you let it answer 'That's what I've been told'? I would, as follows. First combine (1) and (2) into (5) $A\&(B \vee C)$. Then combine (3) and (4) into (6) $((A\&B) \vee C) \rightarrow Q$. Distribution now warrants the answer I propose. Since without distribution, it seems impossible to answer the question of whether Q, one wants an account of why the system refuses our question in such a simple case.

I am thinking of an example such as the following. $A \leftrightarrow$ (You can get from 1 to 2 by plane). $B \leftrightarrow$ (You can get from 2 to 3 by bus). $C \leftrightarrow$ (You can get from 1 to 3 by automobile). $Q \leftrightarrow$ (You can get from 1 to 3). My intent is to describe a retrieval system in which each of A and $B \vee C$ are kept in entirely separate ('irrelevant') subsystems, to be combined only when driven to do so by a question such as that concerning Q.

9. Comparison with claims of relevance logic. Belnap and Dunn 1981 (reprinted in ABD) distinguish relevance logicians, those who like Dunn and myself are students of relevance logic, from relevantists, who practice it. It is quite generally the duty of the logician to spell out not only the benefits but also the costs of using a certain logic as organon. As relevance logicians we therefore make the case that the way of a relevantist is heroic but *hard.* The linear logician or the **RMI** logician should not pretend that life in the undistributed middle is *easy.*

References

Anderson, A. R., and Belnap, N. D., 'Enthymemes.' *The Journal of Philosophy 58* (1961), 713–723.

———, *Entailment: The Logic of Relevance and Necessity. Vol. I*, Princeton (Princeton University Press), 1975.

Anderson, A. R., Belnap, N. D., and Dunn, J. M., *Entailment: The Logic of Relevance and Necessity. Vol. II*, Princeton (Princeton University Press), 1992.

Avron, A., 'Relevance and paraconsistency — a new approach. Part II: The formal systems.' *Notre Dame Journal of Formal Logic 31* (1990), 169–202.

Belnap, N. D., 'Display logic.' *Journal of Philosophical Logic 11* (1982), 375–417.

———, 'Linear logic displayed.' *Notre Dame Journal of Formal logic 31* (1990), 14–25.

Belnap, N. D., and Dunn, J. M., 'Entailment and the disjunctive syllogism.' In: G. Fløistad and G. H. von Wright (eds.), *Philosophy of Language/ Philosophical Logic*, The Hague (Martinus Nijhoff), 1981, 337–366.

Girard, J.-Y., 'Linear logic.' *Theoretical Computer Science 50* (1987), 1–102.

Kripke, S., 'The problem of entailment' (abstract). *The Journal of Symbolic Logic 24* (1959), 324.

Meyer, R. K., *Topics in Modal and Many-Valued Logic.* Ph. D. thesis, University of Pittsburgh, 1966.

Theorems in Classical Logic are Instances of Theorems in Condensed BCI Logic

Martin W. Bunder

Department of Mathematics, University of Wollongong

Any proof in classical logic can be rewritten as a proof in intuitionistic implicational logic with the constant formula f and two additional axioms, *ex falso quodlibet* ($\mathbf{f_1}$) and *double negation* ($\mathbf{f_2}$). We show here that any classical proof in this form can be transformed into a proof of a theorem of condensed **BCI** logic, from which the original theorem can be recovered by means of a series of simple replacements.

The fact that, for a given classical theorem, only a finite number of such replacements are possible from its condensed **BCI** counterpart allows us to formulate decision procedures for the intermediate systems $\mathbf{BCIf_1}$, $\mathbf{BCIf_2}$, $\mathbf{BCKf_1}$, $\mathbf{BCKf_1f_2}$, $\mathbf{BCIWf_1}$, $\mathbf{BCIWf_2}$ and $\mathbf{BCKWf_1}$ in terms of known decision procedures for **BCI**, **BCK**, **BCIW**, and **BCKW** implicational logics.

1 BCK and BCI Logics

The logic **BCK** has modus ponens as its only rule and has as axioms all substitution instances of the following:

$$\begin{array}{ll} \mathbf{B} & \vdash (a \to b) \to (c \to a) \to c \to b \\ \mathbf{C} & \vdash (a \to b \to c) \to b \to a \to c \\ \mathbf{K} & \vdash a \to b \to a. \end{array}$$

Note that we use association to the right and that throughout the paper $a, b, c, \ldots$ are propositional variables; $A, B, C, \ldots$ will be used for formulas. The logic **BCI** has the same postulates except that **K** is replaced by:

$$\mathbf{I} \qquad \vdash a \to a.$$

The axioms are named after the combinators **B**, **C**, **K** and **I**, and are in fact their 'functional characters' or 'principal type schemes' (see Curry and Feys [8] or Hindley and Seldin [12]). The first to be interested in these logics was C. A. Meredith, who, in Lemmon, Meredith and Prior [14], was also the first to formulate explicitly the condensed detachment rule, which we will state below. There has been much interest in **BCK** and **BCI** algebras, which (roughly) have $\to$ as a binary operation and replace '$\vdash$' by '$0 =$ ' and modus ponens by substitution of equality. For details on the relations between the logics and the algebras see [5] and Blok and Pigozzi [2].

Extensions of **BCK** and **BCI** logic that include axioms for additional connectives are related to linear logic (see Girard [9] and Ono and Komori [16]). They are all characterized by the absence of the axiom

$$\mathbf{W} \qquad \vdash (a \to a \to b) \to a \to b.$$

Such logics have important applications to computer science, as programs based on them can be executed in linear time. Logics with **W** include **BCKW** logic, which is intuitionistic implicational logic, and **BCIW** logic, which is the relevance logic $R_{\to}$.

Classical logic can be represented as **BCKW** logic with the constant formula f and the extra axiom schemes:

$$\begin{array}{ll} \mathbf{f_1} & \vdash \mathrm{f} \to A \\ \mathbf{f_2} & \vdash ((A \to \mathrm{f}) \to \mathrm{f}) \to A. \end{array}$$

Negation, disjunction and conjunction can be defined as follows:

$$\begin{array}{rcl} {\sim}A & = & A \to \mathrm{f} \\ A \vee B & = & {\sim}A \to B \\ A \& B & = & {\sim}(A \to {\sim}B). \end{array}$$

When a formula of classical logic is transformed by eliminating $\sim$, $\vee$ and & by means of these definitions, we will say that it is in $(\to, \mathrm{f})$ *form*.

We will be interested in various subsystems of $\mathbf{BCKWf_1f_2}$; those without $\mathbf{f_2}$ are intuitionistic, those without $\mathbf{f_1}$ paraconsistent, those without **K** relevant and those without **W** linear. In [4] we showed that $\mathbf{BCKf_1f_2}$ with an unrestricted comprehension axiom provides a basis for multiset theory.

Note that in **BCK** logic **I** is provable, in $\mathbf{BCIf_1f_2}$, **K** is provable, and in $\mathbf{BKf_2}$, $\mathbf{f_1}$ is provable (this was shown in [6]). (This also applies to the 'condensed' versions of these logics.) In each subsystem only a restricted set of properties of $\sim$, $\vee$ and & can be derived. Alternative definitions, such as $A \wedge B = {\sim}((A \to B) \to {\sim}A)$, can give properties of conjunction not possessed by the connectives as defined above.

2 Condensed Logics

Each implicational logic has a *condensed* counterpart with the same *axioms* (rather than axiom schemes) and modus ponens replaced by the *condensed detachment rule* (D). Before stating this rule we require the notion of most general unification.

A *unifier* of a pair of formulas A and B with no variables in common is a substitution σ such that $\sigma(A) = \sigma(B)$. The substitution σ is a *most general unifier* (m.g.u.) of A and B and $\sigma(A)$ is the *most general unification* (m.g.u.) if every other unification is a substitution instance of $\sigma(A)$.

We now state the condensed detachment rule:

(D) If $\vdash A$ and $\vdash B \to C$, where A and $B \to C$ have no variables in common and σ is an m.g.u. of A and B chosen so that no variables in C that are not in B appear in $\sigma(B)$, then $\vdash \sigma(C)$.

Much of the recent material on these condensed logics can be found in Hindley and Meredith [11], Hindley [10] and Meyer and Bunder [15].

3 Standardization of Classical Proofs

An outcome of the work in [15] was that in [6], which showed that a classical proof based on the postulates given above can be rewritten so that it consists of the proof of a condensed **BCI** theorem followed by uses of axioms $\mathbf{f_1}$, $\mathbf{f_2}$, **W** and **K** as minor premises in modus ponens, and perhaps a variable-for-variable substitution. In this paper we consider a different standardization procedure, with some different applications. A natural-deduction version of our logics will be most convenient.

4 Natural-Deduction Formulations of Condensed **BCI** and **BCK** Implicational Logics

We will give two natural-deduction systems for condensed **BCI** and **BCK** logics, the first being based on the standard structural rules without weakening or contraction, but with a generalized cut rule. The following are the postulates:

$*\mathbf{I}*$	$a \vdash a$
$\to_{\mathbf{e}}$	$a \to b, a \vdash b$
$*\mathbf{C}*$	If $\Delta, A, B, \Gamma \vdash C$, then $\Delta, B, A, \Gamma \vdash C$.
$\to_{\mathbf{i}}$	If $\Delta, A \vdash B$, then $\Delta \vdash A \to B$.

Cut If $\Delta, A \vdash B$ and $\Gamma \vdash C$, where $\Delta \cup \{A, B\}$ and $\Gamma \cup \{C\}$ have no variables in common, and σ is an m.g.u. of A and C chosen so that no variables of B that are not in A appear in $\sigma(A)$, then $\sigma(\Delta), \sigma(\Gamma) \vdash \sigma(B)$.

Note that as this system does not have the contraction rule:

$*$**W**$*$ If $\Delta, A, A, \Gamma \vdash B$, then $\Delta, A, \Gamma \vdash B$,

the formulas on the left of any $\vdash$ form a multiset.

The natural-deduction version of condensed **BCI** logic has the 2-*property*; i.e.

If $\Delta \vdash A$, then every variable appears in $\Delta \uplus \{A\}$ (where $\uplus$ is multiset union) exactly twice and no variable occurs more than once in any element of Δ.

This property was first proved for the axiomatic system by Belnap in [1]. Another version in [3] is to form part of the revision of [15]. It is easy to see that this property holds for $*$**I**$*$ and $\rightarrow_{\mathbf{e}}$ and that it is preserved by $*$**C**$*$ and $\rightarrow_{\mathbf{i}}$. The proof that the property is preserved by **Cut** is similar to the proof for (D) in [3].

The natural-deduction version of condensed **BCK** logic has the postulates of condensed **BCI** logic with the weakening rule:

$*$**K**$*$ If $\Delta \vdash A$, then $\Delta, b \vdash A$.

Condensed **BCIW** and **BCKW** logics have $*$**W**$*$ as an additional rule. Meyer-Bunder [15] shows that these are equivalent to **BCIW** and **BCKW** logics respectively. Such 'uncondensed' logics have arbitrary formulas for the variables in $*$**I**$*$, $\rightarrow_{\mathbf{e}}$ and $*$**K**$*$.

We define *the number of uses* of a formula A in a deduction ending in $\Delta \vdash B$ to be the number of times A (or a formula of which A becomes a substitution instance through **Cut**) is introduced to the left of $\vdash$ through a rule other than $*$**K**$*$ minus the number of such formulas that are moved to the right of $\vdash$ by $\rightarrow_{\mathbf{i}}$. In a logic without $*$**W**$*$ or $*$**K**$*$ this number of uses is the number of times A appears in Δ.

An alternative form of condensed **BCI** logic has $*$**I**$*$ and $*$**C**$*$ but has $\rightarrow_{\mathbf{e}}$ and **Cut** replaced by the following counterpart to (D):

(D′) (i) If $\Delta \vdash C \rightarrow D$ and $\Gamma \vdash E$, where $\Gamma \cup \{E\}$ and $\Delta \cup \{C \rightarrow D\}$ have no variables in common and σ is an m.g.u. of C and E chosen so that no variables in C that are not in B appear in $\sigma(D)$, then $\sigma(\Delta), \sigma(\Gamma) \vdash \sigma(\Delta)$.

(ii) If $\Delta \vdash a$ and $\Gamma \vdash E$, and if b is a variable not in Δ, Γ or E and $\Delta \cup \{a\}$ and $\Gamma \cup \{E\}$ have no variables in common, then, with σ given by $\sigma(a) = E \to b$, we have $\sigma(\Delta), \Gamma \vdash b$.

The following introduction rule replaces $\to_{\mathrm{i}}$:

($\to_{\mathrm{iBCI}}$) If $\Delta, A \vdash B$, and if A appears in $\Delta \uplus \{A\}$ exactly as often as it is used in the deduction and no variable appears more than once in A, then $\Delta \vdash A \to B$.

We now prove:

Theorem 1. *The two natural-deduction versions of (condensed)* **BCI** *logic produce the same theorems.*

Proof. It is easy to prove by induction on the number of non $\to_{\mathrm{i}}$ steps in the proof that every theorem of the system with **Cut** is a theorem of the system with (D′). The reverse is proved by induction on the number of non $\to_{\mathrm{iBCI}}$ steps.

Before showing that the axiomatic and natural-deduction systems produce the same theorems, we need to prove a lemma.

Lemma 1. *If $n \geq 1$ and $k \geq 1$, then in* **BC** *logic:*

$$\vdash (a_1 \to \ldots \to a_n \to c) \to (b_1 \to \ldots \to b_k \to a_n)$$
$$\to (a_1 \to \ldots \to a_{n-1} \to b_1 \to \ldots \to b_k \to c).$$

Proof. By a double induction on n and k.

Theorem 2. *The axiomatic version of condensed* **BCI** *logic is equivalent to the natural-deduction version.*

Proof. It is easy to show that all the postulates of the axiomatic system are derivable in the natural-deduction version. In the axiomatic system we prove, by induction on the number of applications of **Cut** and using Lemma 1, that $\vdash A_1 \to \ldots \to A_n \to B$ whenever $A_1, \ldots, A_n \vdash B$ holds in the natural-deduction system.

Theorem 3. *The axiomatic version of* **BCK** *logic is equivalent to the natural-deduction version.*

Proof. As for Theorem 2.

5 Natural-Deduction Formulation of $\mathbf{f_1}$ and $\mathbf{f_2}$

In systems with the constant formula f, the axiom scheme $\mathbf{f_1}$ is clearly equivalent to the rule:

$$\text{fe}\ \frac{\mathrm{f}}{A}$$

The axiom scheme $\mathbf{f_2}$ is equivalent to the rule:

$$\text{ffe}\ \frac{A \to \mathrm{f} \to \mathrm{f}}{A}$$

Hereafter we use the following abbreviations:

$$\begin{aligned} A \to_0 B &= A \to^0 B = A \\ A \to_{n+1} B &= (A \to_n B) \to B \\ A \to^{n+1} B &= A \to (A \to^n B), \end{aligned}$$

whereas $[A/B]C$ is the result of replacing a particular occurrence of B in C by A (which occurrence will be clear from the context).

6 The Second Standardization Procedure

We will assume that all classical formulas are written in the $(\to, \mathrm{f})$ form (i.e. in terms of $\to$ and f only) and that all deductions use the $\to$ and f postulates. To standardize deductions it will be convenient to represent these in the second natural-deduction form.

Normalization. A deduction is said to be *normalized* if all possible simplifications of the following types are made:

$$\text{(I)}\qquad \frac{\text{fe}\ \dfrac{\begin{matrix}\mathrm{D}_1\\ \mathrm{f}\end{matrix}}{A \to B} \qquad \begin{matrix}\mathrm{D}_2\\ A\end{matrix}}{B} \qquad \text{becomes} \qquad \text{fe}\ \frac{\begin{matrix}\mathrm{D}_1\\ \mathrm{f}\end{matrix}}{B}$$

$$\text{(II)}\qquad \frac{\text{fe}\ \dfrac{\begin{matrix}\mathrm{D}_1\\ \mathrm{f}\end{matrix}}{A} \qquad \begin{matrix}\mathrm{D}_2\\ A \to B\end{matrix}}{B} \qquad \text{becomes} \qquad \text{fe}\ \frac{\begin{matrix}\mathrm{D}_1\\ \mathrm{f}\end{matrix}}{B}$$

Note. If in (I) and (II) the left-hand deduction appears in a proof, and, below B, hypotheses that appear in D_2, but not elsewhere, are cancelled, this can be duplicated in the new proof containing the right-hand deductions by application of $\rightarrow_{\mathrm{iBCK}}$ (i.e. $\rightarrow_{\mathrm{i}}$ cancelling hypotheses introduced by $*\mathbf{K}*$). In $\mathbf{BCIf_1}$ and $\mathbf{BCIWf_1}$ logics however this procedure is not available: this normalization procedure would produce a proof of a different theorem and so should not be carried out.

(III)
$$\mathrm{fe}\ \dfrac{\begin{array}{c}\mathrm{D}_1\\ \mathrm{f}\end{array}}{\mathrm{f}} \qquad \text{becomes} \qquad \begin{array}{c}\mathrm{D}_1\\ \mathrm{f}\end{array}$$

(IV)
$$\begin{array}{l}\mathrm{fe}\\ \mathrm{ffe}\end{array}\ \dfrac{\dfrac{\begin{array}{c}\mathrm{D}_1\\ \mathrm{f}\end{array}}{A \rightarrow_2 \mathrm{f}}}{A} \qquad \text{becomes} \qquad \mathrm{fe}\ \dfrac{\begin{array}{c}\mathrm{D}_1\\ \mathrm{f}\end{array}}{A}$$

(V)
$$\mathrm{ffe}\ \dfrac{\begin{array}{c}\mathrm{D}_1\\ \mathrm{f} \rightarrow_2 \mathrm{f}\end{array}}{\mathrm{f}} \qquad \text{becomes} \qquad \dfrac{\rightarrow_{\mathrm{i}}\ \dfrac{\begin{array}{c}(\not{1})\\ \mathrm{f}\end{array}}{\mathrm{f} \rightarrow \mathrm{f}}\ \text{-(1)} \qquad \begin{array}{c}\mathrm{D}_1\\ \mathrm{f} \rightarrow_2 \mathrm{f}\end{array}}{\mathrm{f}}$$

(VI)
$$\mathrm{ffe}\ \dfrac{\begin{array}{c}\mathrm{D}_1\\ A \rightarrow_3 \mathrm{f}\end{array}}{A \rightarrow \mathrm{f}} \qquad \text{becomes} \qquad \rightarrow_{\mathrm{i}}\ \dfrac{\dfrac{\begin{array}{c}\mathrm{D}_1\\ A \rightarrow_3 \mathrm{f}\end{array} \qquad \rightarrow_{\mathrm{i}}\ \dfrac{\dfrac{\begin{array}{c}(\not{1})\\ A\end{array} \quad \begin{array}{c}(\not{2})\\ A \rightarrow \mathrm{f}\end{array}}{\mathrm{f}}}{A \rightarrow_2 \mathrm{f}}\ \text{-(2)}}{\mathrm{f}}}{A \rightarrow \mathrm{f}}\ \text{-(1)}$$

(VII)
$$\dfrac{\rightarrow_{\mathrm{i}}\ \dfrac{\begin{array}{c}(\not{1})\\ A\\ \mathrm{D}_1\\ B\end{array}}{A \rightarrow B}\ \text{-(1)} \qquad \begin{array}{c}\mathrm{D}_2\\ A\end{array}}{B} \qquad \text{becomes} \qquad \begin{array}{c}\mathrm{D}_2\\ A\\ \mathrm{D}_1\\ B\end{array}$$

$$\text{(VIII)} \qquad \rightarrow_i \dfrac{\dfrac{\overset{(\cancel{1})}{A} \qquad \overset{D_2}{A \to B}}{B}}{A \to B}\ \text{-}(1) \qquad \text{becomes} \qquad \begin{array}{c} D_2 \\ A \to B \end{array}$$

Transforming a classical deduction into a standardized one. Standardization consists of the following processes applied to a normalized proof, each followed, if necessary, by further normalization:

(A) Replace $\rightarrow_i$ cancelling multiple hypotheses (or $*\mathbf{W}*$ and $\rightarrow_i$) by putting the lower subdeduction for the upper one:

$$\dfrac{\rightarrow_i \dfrac{\begin{array}{c} (\cancel{1}) \ldots (\cancel{n_1}) \ldots ((\cancel{\Sigma_{i=1}^{k-1} n_i)} + 1) \ \ldots \ (\cancel{\Sigma_{i=1}^{k} n_i}) \\ A_1 \ \ldots \ A_1 \ \ldots \qquad A_k \qquad \ldots \quad A_k \\ D_1 \\ B \end{array}}{A_k \to \ldots \to A_1 \to B} \qquad \begin{array}{c} D_2 \\ (A_k \to \ldots \to A_1 \to B) \to C \end{array}}{C}$$

$$\dfrac{\rightarrow_i \dfrac{\begin{array}{c} (\cancel{1}) \ldots (\cancel{n_1}) \ldots ((\cancel{\Sigma_{i=1}^{k-1} n_i)} + 1) \ \ldots \ (\cancel{\Sigma_{i=1}^{k} n_i}) \\ A_1 \ \ldots \ A_1 \ \ldots \qquad A_k \qquad \ldots \quad A_k \\ D_1 \\ B \end{array}}{A_k \to^{n_k} \ldots \to A_1 \to^{n_1} B} \qquad \begin{array}{c} [(A_k \to^{n_k} \ldots \to A_1 \to^{n_1} B)/ \\ (A_k \to \ldots \to A_1 \to B)]\, D_2 \\ (A_k \to^{n_k} \ldots \to A_1 \to^{n_1} B) \to C \end{array}}{C}$$

Here we assume at least one n_i is greater than 1. (By Theorem 4 below, the replacement above can always be made.)

We will call a consecutive sequence of applications of $\rightarrow_i$ including at least one cancellation of multiple occurrences of a hypothesis an m $\rightarrow_i$ rule.

(B) Replace ffe steps in the deduction by putting the right subdeductions for the left ones:

$$\text{(i)} \qquad \dfrac{\text{ffe}\dfrac{\begin{array}{c} D_1 \\ A \to_2 \mathrm{f} \end{array}}{A} \qquad \begin{array}{c} D_2 \\ A \to B \end{array}}{B} \qquad\qquad \dfrac{\dfrac{\begin{array}{c} D_1 \\ A \to_2 \mathrm{f} \end{array} \qquad \dfrac{\dfrac{\dfrac{\overset{(\cancel{1})}{A} \qquad \overset{D_2}{A \to B}}{B} \qquad \overset{(\cancel{2})}{B \to \mathrm{f}}}{\mathrm{f}}}{A \to \mathrm{f}}\ \text{-}(1)}{\mathrm{f}}}{B \to_2 \mathrm{f}}\ \text{-}(2)$$

(ii)

$$\dfrac{\text{ffe}\ \dfrac{\overset{\mathrm{D}_1}{(A\to B)\to_2 \mathrm{f}}}{A\to B}\qquad \overset{\mathrm{D}_2}{A}}{B}
\qquad\qquad
\dfrac{\dfrac{\overset{\mathrm{D}_1}{(A\to B)\to_2 \mathrm{f}}\qquad \dfrac{\dfrac{\overset{(\not 2)}{B\to \mathrm{f}}\qquad \dfrac{\overset{(\not 1)}{A\to B}\qquad \overset{\mathrm{D}_2}{A}}{B}}{\mathrm{f}}}{(A\to B)\to \mathrm{f}}\ \text{-(1)}}{\mathrm{f}}}{B\to_2 \mathrm{f}}\ \text{-(2)}$$

(iii)

$$\dfrac{\text{ffe}\ \dfrac{\overset{\overset{(\not 1)}{B}}{\overset{\mathrm{D}_1}{A\to_2 \mathrm{f}}}}{A}}{B\to A}\ \text{-(1)}
\qquad\qquad
\dfrac{\overset{\overset{(\not 1)}{B}}{\overset{\mathrm{D}_1}{A\to_2 \mathrm{f}}}}{B\to A\to_2 \mathrm{f}}\ \text{-(1)}$$

In (iii), except possibly in $\mathbf{BCIf_1}$ and $\mathbf{BCIWf_1}$ logics, A cannot be f as the deduction has been normalized. (By Theorem 5 below, the replacements above can always be made after (A) has been applied.)

(C) Replace each application of fe in the deduction by putting the lower subdeduction for the upper one:

$$\dfrac{\to_{\mathrm{i}}\ \dfrac{\text{fe}\ \dfrac{\overset{\overset{(\not 1)\ \dots\ (\not k)}{A_1\ \dots\ A_k}}{\overset{\mathrm{D}_1}{\mathrm{f}}}}{B}}{A_1\to\dots\to A_k\to B}\ \text{-(1)},\dots,(k)\qquad \overset{\mathrm{D}_2}{(A_1\to\dots\to A_k\to B)\to C}}{C}$$

$$\dfrac{\to_{\mathrm{i}}\ \dfrac{\overset{\overset{(\not 1)\ \dots\ (\not k)}{A_1\ \dots\ A_k}}{\overset{\mathrm{D}_1}{\mathrm{f}}}}{A_1\to\dots\to A_k\to \mathrm{f}}\ \text{-(1)},\dots,(k)\qquad \overset{[\mathrm{f}/B]\,\mathrm{D}_2}{(A_1\to\dots\to A_k\to \mathrm{f})\to C}}{C}$$

N.B. Some of the A_i's may be identical and we may have $\mathrm{D}_1=\emptyset$, $k=1$ and $A_1=\mathrm{f}$.

If normalizations (I) and (II) have not been carried out, the deduction in (I) becomes:

$$\frac{\begin{array}{cc} [(A \to B)/\mathrm{f}]\,\mathrm{D}_1 & \mathrm{D}_2 \\ A \to B & A \end{array}}{B}$$

and that in (II) becomes:

$$\frac{\begin{array}{cc} \mathrm{D}_1 & [\mathrm{f}/A]\,\mathrm{D}_2 \\ \mathrm{f} & \mathrm{f} \to B \end{array}}{B}$$

(By Theorem 6 below the replacements above can always be made after (A) has been applied.)

(D) Replace each application of $\to_{\mathrm{i}}$ cancelling a hypothesis not used in the deduction by putting the lower subdeduction for the upper one:

$$\frac{\to_{\mathrm{i}}\dfrac{\begin{array}{c} \begin{array}{ccc} (\not 1) & \dots & (\not k) \\ C_1 & \dots & C_k \end{array} \\ \mathrm{D}_1 \\ B \end{array}}{C_1 \to \dots \to C_k \to A \to B}\text{-}(1),\dots,(k) \qquad \begin{array}{c} \mathrm{D}_2 \\ (C_1 \to \dots \to C_k \to A \to B) \to D \end{array}}{D}$$

$$\frac{\to_{\mathrm{i}}\dfrac{\begin{array}{c} \begin{array}{ccc} (\not 1) & \dots & (\not k) \\ C_1 & \dots & C_k \end{array} \\ \mathrm{D}_1 \\ B \end{array}}{C_1 \to \dots \to C_k \to B}\text{-}(1),\dots,(k) \qquad \begin{array}{c} [B/A \to B]\,\mathrm{D}_2 \\ (C_1 \to \dots \to C_k \to B) \to D \end{array}}{D}$$

Some of the C_i's may be identical. (By Theorem 7 the replacement above can always be made after (A) has been applied.)

(E) (i) Consider steps of the following type

$$\frac{\begin{array}{cc} \mathrm{D}_1 & \mathrm{D}_2 \\ A & A \to B \end{array}}{B}$$

in a deduction where there are no $\to_{\mathrm{e}}$ steps in D_1 and D_2. Change all of these to

$$\begin{array}{cc} (1) & (2) \\ a & a \to b \\ \hline \multicolumn{2}{c}{b} \end{array}$$

where a and b are variables not used elsewhere so far in the old or new deduction; a and $a \to b$ are now treated as new hypotheses.

(ii) If in a sequence of steps

$$\begin{array}{c} A \\ \mathrm{D}_1 \\ B \\ \hline A \to B \end{array}$$

A is changed to A', B to B', and D_1 to D_1', change the sequence to

$$\begin{array}{c} A' \\ \mathrm{D}_1' \\ B' \\ \hline A' \to B' \end{array}$$

If the hypothesis A has been eliminated by the changes, the sequence of steps changes to:

$$\begin{array}{c} \mathrm{D}_1' \\ B' \end{array}$$

(iii) If in a sequence of steps

$$\begin{array}{cc} \mathrm{D}_1 & \mathrm{D}_2 \\ A & A \to B \\ \hline \multicolumn{2}{c}{B} \end{array}$$

A is changed to A_1 and $A \to B$ to $A_2 \to B_2$, apply the function σ, the m.g.u. of A_1 and A_2, to all parts of it. If $A \to B$ is changed to a variable a, then

$$\begin{aligned} \sigma(a) &= A_1 \to b \qquad \text{where } b \text{ is a new variable} \\ \sigma(A_1) &= A_1. \end{aligned}$$

(Theorem 8 will show that the changes in (E) can always be made if (A) has been performed.)

An example. We take the classical proof

$$
\to_{\mathrm{i}}\dfrac{\to_{\mathrm{i}}\dfrac{\mathrm{ffe}\,\dfrac{\to_{\mathrm{i}}\dfrac{\to_{\mathrm{e}}\dfrac{\to_{\mathrm{e}}\dfrac{\to_{\mathrm{e}}\dfrac{\to_{\mathrm{i}}\dfrac{\mathrm{fe}\,\dfrac{\to_{\mathrm{e}}\dfrac{\overset{(\not 1)}{A} \qquad \overset{(\not 2)}{A \to \mathrm{f}}}{\mathrm{f}}}{B}}{A \to B}\,\text{-(1)} \qquad \overset{(\not 3)}{(A \to B) \to B}}{B} \qquad \overset{(\not 4)}{B \to A}}{A} \qquad \overset{(\not 5)}{A \to \mathrm{f}}}{\mathrm{f}}}{(A \to \mathrm{f}) \to \mathrm{f}}\,\text{-(2), (5)}}{A}}{(B \to A) \to A}\,\text{-(4)}}{((A \to B) \to B) \to (B \to A) \to A}\,\text{-(3)}
$$

and eliminate the last ffe step by (B), which gives:

$$
\to_{\mathrm{i}}\dfrac{\to_{\mathrm{i}}\dfrac{\to_{\mathrm{i}}\dfrac{}{(A \to \mathrm{f}) \to \mathrm{f}}\,\text{-(2), (5)}}{(B \to A) \to (A \to \mathrm{f}) \to \mathrm{f}}\,\text{-(4)}}{((A \to B) \to B) \to (B \to A) \to A}\,\text{-(3)}
$$

As this was the only ffe step, we now have a $\mathbf{BCKWf_1}$ proof. Now we eliminate the m $\to_{\mathrm{i}}$ step by (A), which gives:

$$
\to_{\mathrm{i}}\dfrac{\to_{\mathrm{i}}\dfrac{\to_{\mathrm{i}}\dfrac{\to_{\mathrm{i}}\dfrac{}{(A \to \mathrm{f}) \to \mathrm{f}}\,\text{-(2)}}{(A \to \mathrm{f}) \to (A \to \mathrm{f}) \to \mathrm{f}}\,\text{-(5)}}{(B \to A) \to (A \to \mathrm{f}) \to^2 \mathrm{f}}\,\text{-(4)}}{((A \to B) \to B) \to (B \to A) \to (A \to \mathrm{f}) \to^2 \mathrm{f}}\,\text{-(3)}
$$

We now have a $\mathbf{BCKf_1}$ proof, and in fact a $\mathbf{BCIf_1}$ proof, as **K** is not used. Next we transform this, using (D), into a **BCI** proof:

$$
\to_{\mathbf{iBCI}}\dfrac{\to_{\mathbf{iBCI}}\dfrac{\to_{\mathbf{iBCI}}\dfrac{\dfrac{\dfrac{\dfrac{\overset{(\not 2)}{A \to \mathrm{f}} \qquad \overset{(\not 3)}{(A \to \mathrm{f}) \to B}}{B} \qquad \overset{(\not 4)}{B \to A}}{A} \qquad \overset{(\not 5)}{A \to \mathrm{f}}}{\mathrm{f}}}{(A \to \mathrm{f}) \to^2 \mathrm{f}}\,\text{-(5), (2)}}{(B \to A) \to (A \to \mathrm{f}) \to^2 \mathrm{f}}\,\text{-(4)}}{((A \to \mathrm{f}) \to B) \to (B \to A) \to (A \to \mathrm{f}) \to^2 \mathrm{f}}\,\text{-(3)}
$$

and then, using (E), into a condensed **BCI** proof:

$$\dfrac{\dfrac{\dfrac{\overset{(\not 2)}{a} \qquad \overset{(\not 3)}{a \to b}}{b} \qquad \overset{(\not 4)}{b \to c}}{c} \qquad \overset{(\not 5)}{c \to \mathrm{f}_1}}{\mathrm{f}_1}$$

$$\to_{\mathbf{iBCI}}\ \dfrac{\mathrm{f}_1}{(c \to \mathrm{f}_1) \to \mathrm{f}_1}\ \text{-}(5)$$

$$\to_{\mathbf{iBCI}}\ \dfrac{}{a \to (c \to \mathrm{f}_1) \to \mathrm{f}_1}\ \text{-}(2)$$

$$\to_{\mathbf{iBCI}}\ \dfrac{}{(b \to c) \to a \to (c \to \mathrm{f}_1) \to \mathrm{f}_1}\ \text{-}(4)$$

$$\to_{\mathbf{iBCI}}\ \dfrac{}{(a \to b) \to (b \to c) \to a \to (c \to \mathrm{f}_1) \to \mathrm{f}_1}\ \text{-}(3)$$

Before showing that this second form of standardization can always be carried out, we need to look at the structure of a deduction in a little more detail.

The *spine* of a deduction is defined as follows. The bottom formula in a deduction is in the spine. If A is derived by fe or ffe, then f or $A \to_2 \mathrm{f}$ is in the spine. If A is derived by $\to_{\mathbf{e}}$ from $B \to A$ and B, then $B \to A$ is in the spine. If A ($= C \to D$) is derived by $\to_{\mathrm{i}}$, then D is in the spine. The spine of a subdeduction of a deduction will be called a *subspine*.

Any subdeduction that has as its bottom formula a minor premise for $\to_{\mathbf{e}}$ whose major premise is in the spine has its *spine distance* 1 from the spine of the deduction. Another subdeduction that has spine distance 1 from the spine of this subdeduction has *spine distance* 2 from the spine of the original, etc.

If a formula A appears as a positive part of a formula, it remains in a positive position in formulas below unless it is eliminated or unless it originally appeared in a cancelled hypothesis. This is easy to show by induction on the length of a deduction, noting that formulas that appeared in positive positions in hypotheses appear in negative positions in the formula resulting by $\to_{\mathrm{i}}$ from the cancellation of the hypothesis, and negative positions in hypotheses became positive positions after $\to_{\mathrm{i}}$. We will say that a formula A is in a *positive position in a deduction* if it is in a positive position in the bottom formula or in a negative position in an uncancelled hypothesis. (If a deduction consists of a single hypothesis A, then A is in a positive as well as a negative position.)

Theorem 4. *If the last* m $\to_{\mathrm{i}}$ *or* ffe *rule applied in a normalized* (*sub*)*classical deduction is an* m $\to_{\mathrm{i}}$ *rule, the deduction can be transformed using* (A) *so that this rule is eliminated. Some positive subformulas* $A \to B$ *may have been replaced by* $A \to^n B$.

Proof. If the last m $\to_{\mathbf{i}}$ or ffe rule in a normalized deduction is m $\to_{\mathbf{i}}$, the only rule that can follow it is $\to_{\mathbf{e}}$, as shown in (A).

The step prior to $(A_k \to \ldots \to A_1 \to B) \to C$ in D_2 cannot be $\to_{\mathbf{i}}$ or fe as the proof is normalized. If $(A_k \to \ldots \to A_1 \to B) \to C$ comes by $\to_{\mathbf{e}}$ and ffe steps, these start at some first major premise P that is a hypothesis and has $(A_k \to \ldots \to A_1 \to B) \to C$ as a positive part, and so $(A_i \to \ldots \to A_1 \to B)$ as a negative part (for $i = 1$ to k). The change from $(A_k \to \ldots \to A_1 \to B)$ to $(A_k \to^{n_k} \ldots A_1 \to^{n_1} B)$ can therefore be carried out affecting only the major premises in $\to_{\mathbf{e}}$ (and any in ffe) up to at most one hypothesis.

Below C there need to be no change to the deduction until the hypothesis P is cancelled by $\to_{\mathbf{i}}$. As the sequence of $\to_{\mathbf{i}}$ steps we are considering is the last where multiple copies of a hypothesis are cancelled, P is in fact cancelled by $\to_{\mathbf{iBCI}}$ and the altered version P' of P can be cancelled in the same way. Since P and its altered version are now in a negative position in the formula, $(A_k \to \ldots \to A_1 \to B)$ is in a positive position in the formula. It and P stay in this position to the end of the proof (thus verifying the theorem) unless it is eliminated by being part of a minor premise in an $\to_{\mathbf{e}}$ step. In this case the major premise needs to be changed in the same way as $(A_k \to \ldots \to A_1 \to B) \to C$ was changed again leading at most to a change to a hypothesis cancelled singly later on. Clearly, this process terminates, leading to a deduction altered in the required way, so that the last set of consecutive applications of $\to_{\mathbf{i}}$ that include cancellation of multiple uses of a hypothesis is eliminated.

Note that if we applied (B) before (A), the changes to some hypothesis could prevent some $\to_{\mathbf{i}}$ steps cancelling the same hypothesis in several places from being carried out. The process (B) could be adapted to do part of (A)'s job in these cases.

Theorem 5. *If the last* m $\to_{\mathbf{i}}$ *or* ffe *rule applied in a normalized (sub)classical deduction is an* ffe *rule, the deduction can be transformed using* (B) *so that this rule is eliminated. Some positive subformulas A may have been replaced by $A \to_2$* f.

Proof. If the last m $\to_{\mathbf{i}}$ or ffe rule in a normalized deduction is ffe, it can only be followed by $\to_{\mathbf{e}}$ or $\to_{\mathbf{i}}$ as in (B).

If the ffe step in the spine is as in (B)(i) or (B)(ii), it can be pushed downward, and then perhaps be eliminated by normalization. Thus, if the last ffe step in the spine cannot be eliminated by this procedure, it is at the end of the deduction or it is one of type B(iii). Note that the $B \to A$ in B(iii) is not followed by an $\to_{\mathbf{e}}$ step as, being in the spine, it cannot be a minor premise and being in a normalized proof it cannot be a major premise. The $B \to A$ is therefore followed only by $\to_{\mathbf{i}}$ steps and these same steps can be applied when the deduction is transformed as in B(iii). The result of the transformation is a positive occurrence of A replaced by $A \rightarrow_2$ f. This transformed deduction has eliminated the final application of ffe.

If the deduction had no ffe steps in the spine, consider the last such step minimally distant from the spine. This can be pushed down as in B(i) or B(ii) until it reaches the spine and can then be treated as above. If this ffe step is of type B(iii), it can be followed by several $\rightarrow_{\mathrm{i}}$ steps to give $B_1 \rightarrow \ldots \rightarrow B_k \rightarrow A$, and then by an $\rightarrow_{\mathrm{e}}$ step with $(B_1 \rightarrow \ldots \rightarrow B_k \rightarrow A) \rightarrow C$ as major premise. This change to the major premise is handled as in the proof of Theorem 4. Note that, as there is no m $\rightarrow_{\mathrm{i}}$ step below the ffe step, a change in a hypothesis does not effect the operation of the $\rightarrow_{\mathrm{i}}$ step that cancels it later.

Theorem 6. *Any normalized (sub)classical deduction can be transformed by* (A) *and* (B), *resulting in a deduction in a system without* **W** *or* ffe *in which some positive subformulas* $A \rightarrow B$ *and* C *may have been replaced by* $A \rightarrow^n B$ *and* $C \rightarrow_2$ f *respectively.*

Proof. This follows from Theorems 4 and 5.

Theorem 7. *Any normalized (sub)classical deduction, already transformed by* (A) (*and perhaps* (B)) *so that all* m $\rightarrow_{\mathrm{i}}$ *steps have been eliminated, can be transformed, as in* (C), *resulting in a paraconsistent deduction in the corresponding system without* $\mathbf{f_1}$ *in which some positive subformulas are replaced by* f.

Proof. The possible fe transformations of the proofs all require a change to a major premise in $\rightarrow_{\mathrm{e}}$ just as in the last case considered in the proof of Theorem 4. The cases are handled in the same way.

Theorem 8. *Any normalized (sub)classical deduction, already transformed by* (A) (*and perhaps* (B)) *so that all* m $\rightarrow_{\mathrm{i}}$ *steps have been eliminated, can be transformed, as in* (D), *resulting in a deduction in the corresponding system without* **K** (*or* ∗**K**∗) *in which some positive parts* $A \rightarrow B$ *are replaced by* B.

Proof. As for Theorem 5.

Theorem 9. *Any* **BCI**-*deduction can be transformed by* (E) *into a condensed* **BCI**-*deduction that is not a substitution instance of any other deduction.*

Proof. We prove by induction on the length of the deduction that (E) can be carried out, that there is a substitution σ replacing the variables of the deduction by formulas giving the old deduction and that the new deduction is not a nontrivial substitution instance of any other deduction. If we by applying (E) to the subdeductions

$$\begin{array}{ccc} \mathrm{D}_1 & & \mathrm{D}_2 \\ & \text{and} & \\ A & & A \to B \end{array}$$

we have obtained

$$\begin{array}{ccc} \mathrm{D}_1' & & \mathrm{D}_2' \\ & \text{and} & \\ A_1 & & A_2 \to B_2 \end{array}$$

where an $\to_{\mathbf{e}}$ step is to follow, we show that we can replace $\to_{\mathbf{e}}$ by a (D′) step. By the induction hypothesis, there are substitutions σ_1 and σ_2 such that $\sigma_1(A_1) = \sigma_2(A_2) = A$; so A_1 and A_2 have a common unifier. By the work of Robinson [17] it follows that there is an m.g.u. given by $\sigma(A_1) = \sigma(A_2)$ (note that as A_1 and A_2 have no variables in common, the functions mapping them into the common term can be a single function). The substitution σ (replacing variables by formulas) can clearly be applied to D_1' and D_2', and to B_2, to give a substitution instance of the hypotheses, and $\sigma(B_2)$ as the result of applying (D′). As σ provides an m.g.u. of A_1 and A_2, there can be no simpler substitution that will allow (D′) to be performed; so the deduction obtained is not a nontrivial substitution instance of any other deduction.

As σ is the m.g.u. of A_1 and A_2 and σ_1 and σ_2 unify these, the latter can be expressed as $\sigma_3\sigma$ and $\sigma_4\sigma$. D_1' and D_2' are disjoint; so $\sigma(\mathrm{D}_1')$ and $\sigma(\mathrm{D}_2')$ will have only the variables of $\sigma(A_1) = \sigma(A_2)$ in common. Thus the only variables for which σ_3 and σ_4 are both defined are those in $\sigma(A_1)$ (and $\sigma_3(\sigma(A_1)) = \sigma_4(\sigma(A_2)) = A$). Therefore, $\sigma_3 \cup \sigma_4$ maps the new deduction down to $\sigma(B)$ into the old down to B.

If by applying (E) to a subdeduction we have transformed

$$\begin{array}{ccc} (1) & & (1) \\ A & & A' \\ & \text{into} & \\ \mathrm{D}_1 & & \mathrm{D}_1' \\ B & & B' \end{array}$$

the same transformation will map

$$\begin{array}{ccc} (\not{1}) & & (\not{1}) \\ A & & A' \\ \mathrm{D}_1 & & \mathrm{D}_1' \\ \to_{\mathrm{i}} \dfrac{B}{A \to B}\ \text{-(1)} & \text{into} & \to_{\mathrm{i}} \dfrac{B'}{A' \to B'}\ \text{-(1)} \end{array}$$

Thus the theorem holds.

It is possible to extend (E) so that it can apply to any subclassical deduction. The result will be the corresponding condensed subclassical deduction.

7 Decision Procedures

Before giving, as a preliminary, some known results (mainly from Hindley [10]) for (condensed) **BCK** and **BCI** logics we need some definitions.

A formula has the 2-*property* if every propositional variable appears exactly twice in it. A formula has the 1-2-*property* if no propositional variable appears more than twice in it.

A formula is a C' tautology if it takes on the values T and $\frac{1}{2}$ when evaluated by the following table:

$\rightarrow$	T	$\frac{1}{2}$	F
T	T	F	F
$\frac{1}{2}$	T	$\frac{1}{2}$	F
F	T	T	T

Theorem 10. *A purely implicational formula is a theorem of condensed* **BCI** *logic iff it is a C' tautology with the* 2*-property.*

Proof. Follows from Hindley [10], Theorems 3.2.1, 3.2.3 and 3.5.2 (the proofs are largely due to Jaśkowski [13]).

Theorem 11. *A purely implicational formula is a theorem of* **BCI** *logic iff it is a variable-for-variable substitution instance of a theorem of condensed* **BCI** *logic.*

Proof. Meyer and Bunder [15].

Theorem 12. *A purely implicational formula is a theorem of condensed* **BCK** *logic iff it is a classical tautology with the* 1-2*-property that is not a variable-for-variable substitution instance of another classical tautology.*

Proof. Hindley [10], Theorem 2.5.2 and 2.2.1 (proofs from Jaśkowski [13], Meyer and Bunder [15] or [3]).

Theorem 13. *A purely implicational formula is a theorem of* **BCK** *logic iff it is a variable-for-variable substitution instance of a classical tautology with the* 1-2-*property.*

Proof. See Jaśkowski [13].

Each of the above characterizations provides a simple finite algorithm for determining whether a given formula is a theorem of the given logic. For algorithms that also provide a proof see Hindley [10]. The following characterizations for the intermediate logics between **BCI** and classical logic similarly provide decision procedures. The formulas we consider will be written in the $(\to, \mathrm{f})$ form, i.e. in terms of $\to$ and f only.

Theorem 14. *An* $(\to, \mathrm{f})$ *formula is a theorem of* $\mathbf{BCIf_1}$ *logic iff it can be transformed into a theorem of* **BCI** *logic by replacing one or more positive subformulas by* f.

Proof. If A is a theorem of $\mathbf{BCIf_1}$ logic, in its proof it has no applications of $\to_{\mathrm{i}}$ cancelling multiple hypotheses; so by Theorem 7 the result follows.

Theorem 15. *An* $(\to, \mathrm{f})$ *formula is a theorem of* $\mathbf{BCKf_1}$ *logic iff it can be transformed into a theorem of* **BCK** *logic by replacing one or more positive subformulas by* f.

Proof. As for Theorem 14.

Theorem 16. *An* $(\to, \mathrm{f})$ *formula is a theorem of* $\mathbf{BCIf_2}$ *logic iff it can be transformed into a theorem of* **BCI** *logic by replacing one or more positive subformulas* A *by* $A \to_2 \mathrm{f}$.

Proof. As for Theorem 14, but using Theorem 5.

Theorem 17. *An* $(\to, \mathrm{f})$ *formula is a theorem of* $\mathbf{BCKf_2}$ $(= \mathbf{BCKf_1f_2} = \mathbf{BCIf_1f_2})$ *logic iff it can be transformed into a theorem of* **BCK** *logic by replacing one or more positive subformulas* A *by* $A \to_2 \mathrm{f}$.

Proof. As for Theorem 16.

It is clear that **BCKW** and **BCIW** theorems could be characterized as those that lead to **BCK** and **BCI** theorems after a (bounded) number of replacements of the form $A \to^n B$ for $A \to B$. It is however, at this stage, not clear how the values of n can be bounded.

References

[1] Belnap, N. D., The two property, *Relevance Logic Newsletter 1* (1976), 173–180.

[2] Blok, W. J., and Pigozzi, D., Algebraizable logics, *Memoirs of the American Mathematical Society 77* (1989), No 396.

[3] Bunder, M. W., The 2 and 1-2 properties in condensed **BCI** and **BCK** logics, Typescript, University of Wollongong, 1989.

[4] ———, **BCK** predicate logic as a foundation for multiset theory, Preprint, University of Wollongong.

[5] ———, **BCK** and related algebras and their corresponding logics, *The Journal of Non-Classical Logic 2* (1983), 15–24.

[6] ———, Proofs in classical logic can be reduced to **BCI**-implicational proofs, to appear in the Proceedings of the Prior Memorial Conference (Christchurch 1989).

[7] ———, Combinatory logic and lambda calculus with classical types, Preprint, University of Wollongong, 1991.

[8] Curry, H. B., and Feys, R., *Combinatory Logic, Vol. I*, North-Holland, Amsterdam, 1958.

[9] Girard, J.-Y., Linear logic, *Theoretical Computer Science 50* (1987), 1–101.

[10] Hindley, J. R., **BCK** and **BCI** logics, condensed detachment and the 2-property: A summary, to appear in the *Notre Dame Journal of Formal Logic.*

[11] Hindley, J. R., and Meredith, D., Principal type-schemes and condensed detachment, *The Journal of Symbolic Logic 55* (1990), 90–105.

[12] Hindley, J. R., and Seldin, J. P., *Introduction to Combinators and λ-Calculus*, Cambridge University Press, Cambridge, 1986.

[13] Jaśkowski, S., Über Tautologien, in welchen keine Variable mehr als zweimal vorkommt, *Zeitschrift für Mathematische Logik und Grundlagen der Mathematik 9* (1963), 219–228.

[14] Lemmon, E. J., Meredith, C. A., Meredith, D., Prior, A. N., and Thomas, I., Calculi of pure strict implication, Canterbury University College, Christchurch, 1957; reprinted in *Philosophical Logic*, J. W. Davis, D. J. Hockney and W. K. Wilson (eds), Reidel, Dordrecht, 1969, 215–250.

[15] Meyer, R. K., and Bunder, M. W., Condensed detachment and combinators, Automated Reasoning Project, Australian National University, Technical Report, 1988.

[16] Ono, H., and Komori, Y., Logics without the contraction rule, *The Journal of Symbolic Logic 50* (1985), 169–201.

[17] Robinson, J. A., A machine-oriented logic based on the resolution principle, *Journal of the Association of Computing Machinery 12* (1965), 23–41.

Partial Gaggles Applied to Logics with Restricted Structural Rules

J. Michael Dunn

Departments of Philosophy and Computer Science
Indiana University[1]

1 Introduction

In Dunn (1991) a class of algebraic structures called 'gaggles' were investigated with an idea to abstracting out an algebraic structure common to a wide class of logics. But one feature, although common to many of the logics and abstracted into the notion of a gaggle, still is inessential to the main ideas. This feature is that there be a distributive lattice underlying the family of gaggle operations. This is an appropriate assumption to make for many logics, and leads to a representation theory where 'conjunction' (meet) is interpreted as intersection and 'disjunction' (join) as union.

But there are other logics where the distributive law relating conjunction and disjunction fails (*e.g.*, linear logic), and it is the aim of this paper to develop the main ideas of gaggle theory without this assumption. There are even formalisms which have neither conjunction nor disjunction, and so we shall here presuppose only an underlying partial ordering (a 'partial gaggle'). The case where the underlying order structure is an arbitrary lattice can be handled by way of the basic representation theorem for lattices of Urquhart (1978), and Allwein and Dunn (199+) explore the essence of this idea restricted to binary operations in view of the applications to linear logic and other related logics.

1 I acknowledge once more the help of people I thanked in Dunn (1991), Nuel Belnap, Gary Hardegree and Gerard Allwein, listed in the temporal order in which I have benefitted from discussing the 'Gaggle Project' with them. The work of Bob Meyer has most heavily influenced the Project. Not only was his work with Routley (1972) the key to generalizing Jónsson and Tarski (1951–52), but also I owe to him the underlying view of this paper that premisses can be regarded as 'bunched together' by a groupoid operation.

The reader is advised to at some point read Dunn (1991) for a sometimes parallel and sometimes divergent presentation of the same ideas, but in the setting of distributive lattices. Dunn (1991) also spends more time in placing these ideas in the literature. But the present paper is more or less self-sufficient in terms of logical development, and the arguments and techniques are in fact much more elementary.[2] Indeed, the present paper can be viewed as a 'primer' for Dunn (1991).

Thus rather than beginning with the abstraction of a partial gaggle, we first investigate the 'concrete' case of a residuated p.o. groupoid, showing how various properties of implication are captured by various properties of the groupoid operation (which can be thought of as an operation combining premisses). Much of this material is to be found scattered through other sources (see below), but it is important motivationally to have it set out a certain way. We begin by motivating the connection between substructural logics and residuated p.o. groupoids by examining Gentzen systems and the relationship of structural rules to properties of the groupoid operation.

The relationship between implication and residuation has been known for years (cf. *e.g.*, Birkhoff (1940)). And one can find many correspondences between laws for fusion and laws for residuation by browsing through say Birkhoff (esp. the 1967 edition), Certaine (1943), Fuchs (1963), Curry (1963) and Dunn (1966). Perhaps the single best source is Meyer and Routley (1972), where they provide a long list of such correspondences. But since they assume fusion is commutative (as it is in most usual logics), they have no distinction between the left and right residual. There are applications where one would not want this assumption, *e.g.*, Lambek calculus and quantales. Good recent work emphasizing implication as residuation is to be found in Došen (1988–1989, 1992). Another good source is Routley with Plumwood, Meyer and Brady (1982).

The representation results for residuated p.o. groupoids are modeled after those of Meyer and Routley (1972), which in turn are based on their semantics for relevance logic. Meyer and Routley (1972) should be a seminal paper for substructural logics, but it seems to have been largely neglected.

2 Logics with Restricted Structural Rules

The aim of 'gaggle theory' is to provide a uniform semantical approach to a variety of logics, including classical logic, intuitionistic logic, modal logic, relevance logic, linear logic, dynamic logic, temporal logic, BCK-logics, Lambek calculus and orthologic. It turns out that one way to distinguish among many of these logics is to see them as deriving from different choices

---

2 In particular prime filters are replaced with cones, and so maximalization arguments using Zorn's Lemma are not needed.

of Gentzen's 'structural rules'. Other papers in this book undoubtedly discuss this in more specific detail, but I want to lay out some basic options.

Corresponding to each structural rule below there is a corresponding algebraic law. The fundamental formal object in a Gentzen system is a *sequent*: $\varphi_1, \varphi_2, \varphi_3 \vdash \psi$. For simplicity, as well as to make connection with the gaggle ideas, we will assume that sequents are 'singular', *i.e.*, they can have one or more sentences on their left, but exactly one sentence on their right.[3]

In order to relate the collection of premises on the left-hand side of a sequent to algebra we assume that premisses are collected together in a binary manner, rather than as sequences as did Gentzen. If we associate parentheses to the right, we can then write $\varphi_1, \varphi_2, \varphi_3 \vdash \psi$ for $(\varphi_1, \varphi_2), \varphi_3 \vdash \psi$, keeping at least the appearance of the anadic grouping of Gentzen. To get the spirit of the anadic grouping we should assume extra structural rules (associativity) which allow regrouping.

We then get the following correspondences between logic and algebra, which we label not only by the usual names from logic and algebra, but also link with the corresponding combinator:

LOGIC	ALGEBRA
Regrouping (B)	*Left and Right Associativity*
$\dfrac{(\varphi_1, \varphi_2), \varphi_3 \vdash \psi}{\varphi_1, (\varphi_2, \varphi_3) \vdash \psi}$	$a \circ (b \circ c) \leq (a \circ b) \circ c$
$\dfrac{\varphi_1, (\varphi_2, \varphi_3) \vdash \psi}{(\varphi_1, \varphi_2), \varphi_3 \vdash \psi}$	$(a \circ b) \circ c \leq a \circ (b \circ c)$
Permutation (C)	*Commutation*
$\dfrac{\Gamma, \varphi, \psi, \Delta \vdash \chi}{\Gamma, \psi, \varphi, \Delta \vdash \chi}$	$a \circ b = b \circ a$
Contraction (W)	*Square-Increasingness*
$\dfrac{\Gamma, \varphi, \varphi, \Delta \vdash \chi}{\Gamma, \varphi, \Delta \vdash \chi}$	$a \leq a \circ a$
Thinning (K)	*Lower Bound*
$\dfrac{\Gamma \vdash \varphi}{\psi, \Gamma \vdash \varphi}$	$a \circ b \leq b$

3 Our attention to singular Gentzen systems is less restrictive than it might seem (compare Belnap's 'Display Logic'), but it might be well to develop a theory of 'symmetric gaggles' to handle such cases.

There are further rules we must discuss besides these. One is the rule of Cut, which Gentzen showed was redundant by way of his famous '*Hauptsatz*'. As usually stated it looks something like this:

$$\frac{\Gamma \vdash \varphi \qquad \Delta, \varphi \vdash \chi}{\Gamma, \Delta \vdash \chi} \qquad \textit{Cut}$$

But it is clear that this presupposes a certain amount of commutation (as well as associativity). The following statement of Cut does not rely on commutation, and amounts to right or left isotonicity (when Δ or Δ' are empty, and the two of them together when neither are empty).

Cut	*Isotonicity*
$\dfrac{\Gamma \vdash \varphi \qquad \Delta, \varphi, \Delta' \vdash \chi}{\Delta, \Gamma, \Delta' \vdash \chi}$	$a \leq b \Rightarrow c \circ a \leq c \circ b$ $a \leq b \Rightarrow a \circ c \leq b \circ c$

There are also 'operational' rules for the various logical operations, one rule for introducing the operation on the left of the turnstile and another rule for introducing it on the right. These correspond in a well-known way to the introduction and elimination rules in a natural deduction system, and this is the way we shall present them here for ease of exposition. Thus, focusing attention on implication, we have the 'two-way' rule:[4]

Implication int-elim	*Residuation*
$\dfrac{\varphi, \Gamma \vdash \psi}{\Gamma \vdash \varphi \to \psi}$ (double line)	$a \circ b \leq c$ *iff* $b \leq a \to c$

There is a final technical complication caused by rules like the Implication-int rule that can empty out a left-hand side. (There would be also problems with Cut if we allowed Γ to be empty.) From the algebraic point of view we want all our statements to be inequalities, but then there would be no term on the left-hand side of the $\leq$. We thus introduce a constant proposition $\mathbf{t}$ thought of as the conjunction of all logical truths, and plug

4 The rule from top-to-bottom is the same as the Genzen rule for 'arrow on the right'. The rule from bottom-to-top corresponds (in the presence of Cut) to the Gentzen rule for 'arrow on the left' (stated so as to avoid any assumption of permutation):

$$\frac{\Gamma \vdash \varphi \qquad \Delta, \psi, \Delta' \vdash \chi}{\Delta, \varphi \to \psi, \Gamma, \Delta' \vdash \chi}$$

It is easy to see that this last depends on $\varphi \to \psi, \varphi \vdash \psi$, or, in algebraic notation on $(a \to b) \circ a \leq b$. It is routine to verify that this is equivalent to the right-to-left part of the Residuation Law.

it into the breech, thinking of it as having been there in spirit all along in virtue of $\varphi,\ \mathbf{t} \vdash \varphi$.[5]

Emptying the Left-Hand Side	*Lower Right Identity*
$\dfrac{\varphi \vdash \psi}{\mathbf{t} \vdash \varphi \to \psi}$	$a \circ e \leq a$

The general approach here, as in Dunn (1991), is algebraic, and so we will think of a logic in terms of its 'Lindenbaum algebra', formed by dividing the sentences into classes of provable equivalents, defining operations on these equivalence classes by means of the connectives applied to representatives. Thus $|\varphi|$ is the class of formulas ψ such that both $\varphi \vdash \psi$ and $\psi \vdash \varphi$ are derivable. We shall assume the rule of Cut (so in particular if $\varphi \vdash \psi$ and $\psi \vdash \chi$ are both derivable, then so is $\varphi \vdash \chi$), and so we can define a partial order by: $|\varphi| \leq |\psi|$ *iff* $\varphi \vdash \psi$ is derivable.

We shall next define an appropriate class of algebras, and represent them so that elements are mapped into sets (thought of as 'propositions', or sets of states where the sentences are true). This gives completeness results for the various logics. (Cf. Dunn (1966) for a discussion of the general relation between representation results and completeness theorems and compare Dunn (1991) for further helpful references.)

The strategy adopted in Dunn (1991) was based on the framework of Kripke-style semantics (cf. e.g. Kripke (1963)), using accessibility relations to give truth conditions for the connectives, but generalized in line with Jónsson and Tarski (1951–52) so that an n-placed operation is represented using an $n+1$ placed accessibility relation. Although Jónsson and Tarski assumed as an underlying structure a Boolean algebra, and required that the 'operators' must distribute over Boolean disjunction in each of their places, Dunn (1991) worked with 'distributoids', which considerably relax the strictures of Jónsson-Tarski.

But still gaggles were required to have underlying distributive lattices so that meet ('and') is represented as intersection and join ('or') is union. Canonically, then states are prime filters. However, this condition can be weakened to where the underlying structure is just a partial order (as with the Lambek calculus). Then states can just be principal cones and the complements of principal dual cones, and one does not need Zorn's Lemma.

5 We could in fact make it a requirement that every sequent be of the form $\Gamma,\ \mathbf{t} \vdash \varphi$, so that the 'empty' part to the immediate left of the turnstile has an explicit name. Cf. my Gentzen system for $\mathbf{R}^{+}$ in §28.5.1 of Anderson and Belnap (1975).

3 Implication and Residuation

A *partially ordered groupoid* is a structure $(S, \leq, \circ)$, where $\leq$ is a partial order on S, and $\circ$ is a binary operation on S that is isotone in each of its positions. When the partial order is a lattice ordering, we speak of a *lattice-ordered groupoid* when $\circ$ distributes over $\vee$ from both directions (in which case isotonicity becomes redundant).

The binary operation $\rightarrow$ is a *right residual* when it satisfies:

$$(\textit{Law of Right Residual}) \qquad x \circ y \leq z \;\; \textit{iff} \;\; y \leq x \rightarrow z.$$

A *left residual* satisfies:

$$(\textit{Law of Left Residual}) \qquad x \circ y \leq z \;\; \textit{iff} \;\; x \leq z \leftarrow y.$$

It is easy to see that left and right residuals are uniquely defined by the above properties.

Note a similarity between residuals and implication. Thus thinking of $\circ$ as a premiss grouping operation (we call it 'fusion' following the relevance logic literature — the term originates with R. K. Meyer) and thinking of $\leq$ as deducibility, the law of the right residual is just the deduction theorem and its converse. One could say the same about left residuals, but notice that right and left residuals differ in whether it is the formula to the left of $\circ$ or to the right of $\circ$ that is 'exported' from the premiss side to the conclusion side.

We have the following easily derivable facts:

1. $a \circ (a \rightarrow b) \leq b$, $(b \leftarrow a) \circ a \leq b$ (*'modus ponens'*),
2. $a \leq b \rightarrow (b \circ a)$, $a \leq (a \circ b) \leftarrow b$ (*fusing*),
3. *Let* $\gamma(b)$ *be any product* $a_1 \circ \ldots \circ b \circ \ldots \circ a_n$ (parentheses *ad lib*). *Then* $a \leq b \;\&\; \gamma(b) \leq c$ *only if* $\gamma(a) \leq c$ (*cut*).

It is easy to see that one may replace the law of the right residual equivalently with the first halves of (1) and (2), and similarly with the left residual and the second halves. One may also replace transitivity and isotonicity with (3).[6]

Isotonicity of $\circ$ and transitivity yield that residuals are antitone in their first arguments and isotone in their second arguments, as is stated in the following:

---

6 It is implicit in Certaine (1943) that the postulate of the isotonicity of $\circ$ is redundant when both residuals are present. Thus with the right residual, if $a \leq b$, then, since $b \leq x \rightarrow (x \circ b)$ (fusing), by transitivity we get $a \leq x \rightarrow (x \circ b)$. So by residuation, $x \circ a \leq x \circ b$. Similar moves with the left residual give isotonicity in the first position.

Fact. *If* $a \leq b$, *then* $b \to c \leq a \to c$ *(Rule Suffixing) and* $c \to a \leq c \to b$ *(Rule Prefixing), and the same for the left-residual* $\leftarrow$.

Proof. We do only the proofs for the right residual, the others being analogous:

1. $a \leq b$ (*hypothesis*)
2. $a \circ (b \to c) \leq b \circ (b \to c)$ (1, *isotonicity*)
3. $b \circ (b \to c) \leq c$ (*'modus ponens'*)
4. $a \circ (b \to c) \leq c$ (2, 3, *transitivity*)
5. $b \to c \leq a \to c$ (4, *right residuation*).

1. $a \leq b$ (*hypothesis*)
2. $c \circ (c \to a) \leq a$ (*'modus ponens'*)
3. $c \circ (c \to a) \leq b$ (1, 2 *transitivity*)
4. $c \to a \leq c \to b$ (4, *right residuation*).

It is customary to assume various familiar requirements on $\circ$, *i.e.*, associativity, commutation, idempotence, which are even likely to be taken for granted when premisses are collected together into sets. In fact these (or in the absence of associativity, slightly generalized versions) all correspond to various implicational axioms.

Thus, consider half of associativity (let us call it 'right associativity'[7]):

$$a \circ (b \circ c) \leq (a \circ b) \circ c.$$

It yields the following property of the right residual, familiar from implicational logics:

$$(\textit{Prefixing}) \qquad a \to b \leq (c \to a) \to (c \to b);$$

1. $c \circ (c \to a) \leq a$ (*'modus ponens'*)
2. $a \circ (a \to b) \leq b$ (*'modus ponens'*)
3. $[c \circ (c \to a)] \circ (a \to b) \leq b$ (1, 2, *cut*)
4. $c \circ [(c \to a) \circ (a \to b)] \leq b$ (3, *right associativity*)
5. $(c \to a) \circ (a \to b) \leq c \to b$ (4, *right residual*)
6. $(a \to b) \leq (c \to a) \to (c \to b)$ (5, *right residual*).

We can also go the other way and derive right associativity from from prefixing. Before doing so let us note the following obvious consequence of prefixing (apply the law of the right residual twice):

$$(\textit{Imported Prefixing}) \qquad c \circ [(c \to a) \circ (a \to b)] \leq b.$$

Then:

---

7 The reason for the word 'right' is to keep mnemonic linkage to the *right* residual. I would not quarrel with anyone who said that it was more natural to call this '*left* associativity', but below we will use this name for the dual of the above inequality, to keep linkage with the *left* residual.

1. $b \leq a \to (a \circ b)$ (*fusing*)
2. $c \leq (a \circ b) \to [(a \circ b) \circ c]$ (*fusing*)
3. $a \circ (b \circ c) \leq a \circ \{[a \to (a \circ b)] \circ [(a \circ b) \to ((a \circ b) \circ c)]\}$ (1,2, *isotonicity*)
4. $a \circ \{[a \to (a \circ b)] \circ [(a \circ b) \to ((a \circ b) \circ c)]\} \leq (a \circ b) \circ c$ (*imported prefixing*)
5. $a \circ (b \circ c) \leq (a \circ b) \circ c$ (3, 4, *transitivity*).

It is easy to argue symmetrically that 'left associativity'

$$(a \circ b) \circ c \leq a \circ (b \circ c)$$

is equivalent to prefixing for the left residual.

Let us next consider the relation between the 'sequent form' of

$$(\textit{Permutation}) \qquad a \to (b \to c) \leq b \to (a \to c)$$

and (contextual)

$$(\textit{Right Commutation}) \qquad a \circ (b \circ x) \leq b \circ (a \circ x).$$

First we show permutation follows from right commutation:

1. $a \circ [a \to (b \to c)] \leq b \to c$ ('*modus ponens*')
2. $b \circ \{a \circ [a \to (b \to c)]\} \leq c$ (1, *right residual*)
3. $a \circ \{b \circ [a \to (b \to c)]\} \leq c$ (2, *right commutation*)
4. $b \circ [a \to (b \to c)] \leq a \to c$ (3, *right residual*)
5. $a \to (b \to c) \leq b \to (a \to c)$ (4, *right residual*).

Now going the other way we derive right commutation from permutation:

1. $a \circ x \leq b \to [b \circ (a \circ x)]$ (*fusing*)
2. $x \leq a \to \{b \to [b \circ (a \circ x)]\}$ (1, *right residual*)
3. $x \leq b \to \{a \to [b \circ (a \circ x)]\}$ (2, *permutation*)
4. $b \circ x \leq a \to [b \circ (a \circ x)]$ (3, *right residual*)
5. $a \circ (b \circ x) \leq b \circ (a \circ x)$ (4, *right residual*).

Let us remark that it is clear that in the presence of associativity, right commutation can be replaced simply with

$$(\textit{Commutation}) \qquad a \circ b \leq b \circ a,$$

and even in the absence of associativity, the 'rule form' of permutation

$$a \leq b \to c \quad \textit{implies} \quad b \leq a \to c$$

can be shown equivalent to simple commutation. Also, of course, given a right identity element ($a \circ e = a$), the 'sequent form' of permutation can be shown equivalent to the 'theorem form':

$$e \leq [a \to (b \to c)] \to [b \to (a \to c)].$$

We shall devote a section to identity elements below, and it makes ideas simpler to assume the presence of such an identity element (and a left one too), as well as associativity. But we shall not so assume unless we explicitly indicate.

Another familiar implicational law is

$$(\textit{Contraction}) \qquad a \to (a \to b) \leq a \to b.$$

If we had associativity and a right (left) identity, contraction for the right (left) residual would just amount to

$$(\textit{Square-Increasingness}) \qquad a \leq a \circ a.$$[8]

But working in their absence, we must consider more general forms:

$$(\textit{Right Square-Increasingness}) \qquad a \circ x \leq a \circ (a \circ x),$$

$$(\textit{Left Square-Increasingness}) \qquad x \circ a \leq (x \circ a) \circ a.$$

We show that contraction for the right residual is equivalent to right square-increasingness. The corresponding property for the left residual and left square-increasingness follows by a symmetric argument.

Let us first show that contraction follows from right square-increasingness:

1. $a \circ [a \to (a \to b)] \leq a \to b$ ('*modus ponens*')
2. $a \circ \{a \circ [a \to (a \to b)]\} \leq b$ (1, *right residual*)
3. $a \circ [a \to (a \to b)] \leq b$ (2, *right square-increasingness, transitivity*)
4. $a \to (a \to b) \leq a \to b$ (3, *right residual*).

The converse goes as follows:

1. $a \circ (a \circ x) \leq a \circ (a \circ x)$ (*reflexivity*)
2. $a \circ x \leq a \to (a \circ (a \circ x))$ (1, *right residual*)
3. $x \leq a \to [a \to (a \circ (a \circ x)]$ (2, *right residual*)
4. $x \leq a \to (a \circ (a \circ x))$ (3, *contraction, transitivity*)
5. $a \circ x \leq a \circ (a \circ x)$ (4, *right residual*).

8 This may be easily seen by modifying the proofs using the more general forms below. In particular plug in e in place of x, and associate to get the context right before applying square-increasingness.

Besides the rules that correspond to thinking of premisses as collected together into sets, there is one more rule that is often taken for granted, namely dilution or thinning (sometimes called 'monotonicity'; it is the absence of this rule that delineates so-called 'non-monotonic logics'). Dilution is the rule that says it never hurts to add more premisses; algebraically it amounts to saying that $x \circ y \leq y$ ('right lower bound') and corresponds to the following principle of the right residual

$$(\textit{Positive Paradox for the Right Residual}) \qquad a \leq b \rightarrow a.$$

The proof that these are equivalent is an immediate application of the law of the right residual:

$$b \circ a \leq a \quad \textit{iff} \quad a \leq b \rightarrow a.$$

We are really examining the case of 'thinning on the right'. There is also 'thinning on the left', which algebraically amounts to saying that $x \circ y \leq x$ ('left lower bound'), and which corresponds to

$$(\textit{Positive Paradox for the Left Residual}) \qquad a \leq a \leftarrow b.$$

The various relationships we have discovered between principles of the right residual and principles for fusion all have their obvious duals for left residual.

Besides residuation giving familiar properties of either left or right arrow, it also gives 'almost familiar' properties relating the two, at least if the reader 'squints' so as not to be able to distinguish them. Thus both of the following are easy to derive:

Pseudo-Assertion:

$$a \leq (b \leftarrow a) \rightarrow b, \qquad a \leq b \leftarrow (a \rightarrow b),$$

Rule Pseudo-Permutation:

$$a \leq c \leftarrow b \ \textit{implies} \ b \leq a \rightarrow c, \qquad a \leq b \rightarrow c \ \textit{implies} \ b \leq c \leftarrow a.$$

It is easy to see that pseudo-assertion and pseudo-permutation are equivalent to each other. Thus, *e.g.*, the first variety of pseudo-assertion follows easily from the first variety of pseudo-permutation as follows:

1. $b \leftarrow a \leq b \leftarrow a$ (*reflexivity*)
2. $a \leq (b \leftarrow a) \rightarrow b$ (2, *rule pseudo-permutation*).

Of course, when $\circ$ is commutative, the two arrows are identical and hence we get ordinary assertion and rule permutation, familiar from relevance logic (cf. Anderson and Belnap (1975)). Indeed, in the presence of the commutativity of $\circ$, the principles $\rightarrow = \leftarrow$, assertion and rule permutation are all equivalent to each other.

4 Examples

By putting various conditions on fusion, we get algebras corresponding to various systems of implication in the logical literature (in the following diagram one is supposed to keep every condition below and add the new condition). Thus with associativity alone one gets the Lambek calculus (*Lambek calculus* stands here for the *associative* calculus of Lambek (1958)). If one adds commutativity, one gets linear logic (Girard). From here one has two choices naturally motivated by Gentzen's structural rules, either add square-increasingness to get relevant implication (Anderson and Belnap), or the postulate that fusion produces lower bounds ($a \circ b \leq a$) to get BCK implication (Ono and Komori). Finally one gets the properties of intuitionistic implication by collecting all these properties together. These relationships are summarized in the following diagram, where conditions below are always preserved in adding new properties above.

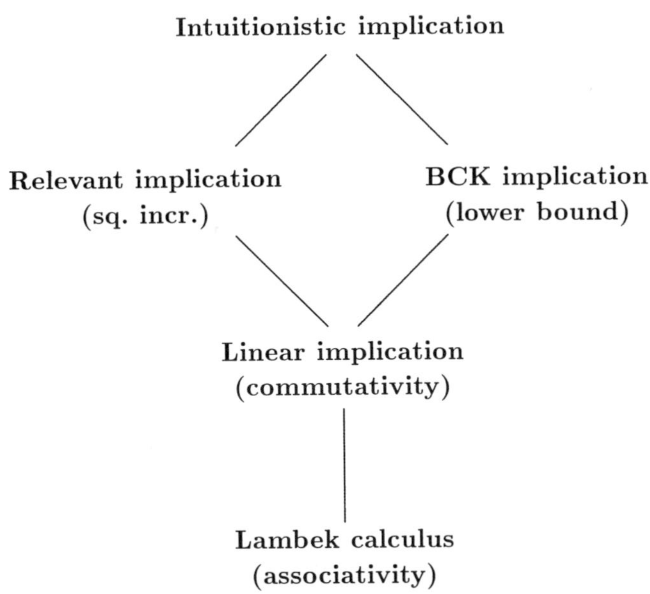

Remark 1. The Lambek calculus was first formulated as a Gentzen system, and there are two versions depending on whether one allows an empty left-hand side or not. Algebraically this corresponds to whether one assumes the existence of an identity or not. The same point arises with the other systems, and amounts to whether we are interested purely in the implication relations ($a \leq b$), or want theorems as well ($e \leq c$). If all theorems were of the form $a \to b$, this might well be thought to be a distinction

without a difference, but even when the only operations are $\circ$ and $\to$, one still gets things like

$$e \leq (a \to a) \circ (b \to b).$$

Remark 2. There is a subtlety implicit in the relationship of the algebraic systems to their parent logics that goes beyond the usual routine of seeing the logics as their 'Lindenbaum algebras' (identifying provably equivalent sentences). The problem is that almost all of the logics mentioned above had connectives other than fusion and implication in their original formulations, and some of them did not have fusion as an explicit connective. The Lambek calculus is a notable exception, since it had no other connectives than fusion and the two arrows, and although it was formulated as a Gentzen system, there is not too much involved in reading comma as fusion and adding associativity to 'defuse' the binary nature of the fusion operation.

The story with the other systems is more complicated, and it would take us too far afield here to try to work out in detail that when τ is a term containing only $\to$, we can derive $e \leq \tau$ in the algebraic system just in case the formula τ is a theorem in the logical system (for simplicity using the same symbols for sentential variables in the logic as for variables in the algebra). But by browsing through the literature one can find pure implicational fragments of all of the above logics (try Anderson and Belnap (1975) as a start — careful reading will give the clues needed for the other systems). The only question is then how to add fusion conservatively, and this we address below.

5 Residuals without Residuation

While we take the point of view that useful properties of implication can be sorted out by relating them to properties of fusing premisses, it must be acknowledged that fusion is (regretably) still a relatively arcane notion. Accordingly it is interesting to wonder just what are the minimal properties of implication that are needed to allow it to be the residual of some fusion operator. Let us arbitrarily take implication to be the right residual — it will turn out, as we shall see, that there is no way of distinguishing the left residual from the right residual in the absence of fusion. So we assume that we have a poset with a binary operation $(S, \leq, \to)$. The only properties of $\to$ we assume are that it is antitone in its first position (antecedent) and isotone in its second position (consequent), *i.e.*, rule suffixing and rule prefixing. We shall call such a structure an *implicational poset*; it is an example of a more general structure, discussed below, called a *tonoid*.

We shall now prove the following, which illustrates a more general result for tonoids:

Embedding Theorem for Implication Posets. *Every implicational poset* $(S, \leq, \to)$ *can be embedded in a right residuated p.o. groupoid* $(S', \leq, \circ, \to)$.

Proof. We prove this by way of a representation result. Let U be a non-empty set and let R be a ternary relation on U. We call the structure (U, R) a *ternary frame.* We define the following operations on subsets of U:

$$\begin{aligned} X \circ Y &= \{\chi : \exists\alpha \in X,\ \exists\beta \in Y,\ R\alpha\beta\chi\} \\ X \to Y &= \{\chi : \forall\alpha, \beta,\ \textit{if}\ R\alpha\chi\beta\ \&\ \alpha \in X,\ \textit{then}\ \beta \in Y\} \\ X \leftarrow Y &= \{\chi : \forall\alpha, \beta,\ \textit{if}\ R\chi\alpha\beta\ \&\ \alpha \in Y,\ \textit{then}\ \beta \in X\}. \end{aligned}$$

It is easy to verify that $(\mathcal{P}(U), \circ, \to, \leftarrow)$ is a residuated p.o. groupoid. Indeed, any subcollection S' closed under the operations $\circ, \to, \leftarrow$ is a residuated p.o. groupoid. We call these *concrete residuated p.o. groupoids.*

We can now state and prove the

Representation Lemma for Implicational Posets. *Every implicational poset* $(S, \leq, \to)$ *can be embedded in a concrete residuated p.o. groupoid.*

Proof. Let U be the set of all cones C on S. Next define the map $h(a) = \{C : C \textit{ is a cone and } a \in C\}$. It is easy to see that $a \leq b$ *iff* $h(a) \subseteq h(b)$ (and so h is 1–1). Left-to-right is obvious since cones are closed under $\leq$. For right-to-left, consider the principal cone $[a) = \{x : a \leq x\} \in h(a)$. Clearly $[a) \in h(a)$ and so, on our assumption that $h(a) \subseteq h(b)$, $[a) \in h(b)$, *i.e.*, $b \in [a)$, *i.e.*, $a \leq b$.

Define the *canonical accessibility* relation by RC_1CC_2 *iff* $\forall x, y$, *if* $x \in C_1$ *and* $x \to y \in C$, *then* $y \in C_2$. Using R we can now define the operation $\to$ on subsets of U as in the concrete residuated poset above.

We show that h preserves $\to$, *i.e.*, $C \in h(a \to b)$ *iff* $C \in h(a) \to h(b)$. To facilitate this we first translate the left and right sides via their definitions:

$$\begin{aligned} C \in h(a \to b) &\quad \textit{iff} \quad a \to b \in C, \\ C \in h(a) \to h(b) &\quad \textit{iff} \quad \forall C_1, C_2,\ \textit{if}\ RC_1CC_2\ \&\ a \in C_1,\ \textit{then}\ b \in C_2. \end{aligned}$$

Then it only remains to show that

$$a \to b \in C \quad \textit{iff} \quad \forall C_1, C_2,\ \textit{if}\ RC_1CC_2\ \&\ a \in C_1,\ \textit{then}\ b \in C_2.$$

But left-to-right is immediate, given the canonical definition of R, which simply says that if an implication is in C and its antecedent is in C_1, then its consequent is in C_2.

For right-to-left, we proceed contrapositively, assuming that $a \to b \notin C$. We then show that $\exists C_1, C_2, RC_1CC_2$ & $a \in C_1$ *yet* $b \notin C_2$. We simply let $C_1 = [a)$. To get C_2, first consider the principal dual cone determined by b, $(b] = \{x : x \leq b\}$, and then set $C_2 = U - (b]$ (it is easy to check that this is a cone). It is clear that $a \in C_1$ and $b \notin C_2$. What needs argument is that RC_1CC_2. Recalling the canonical definition of R, this amounts to assuming that $x \in C_1$, $x \to y \in C$ and showing $y \in C_2$. For reductio, let us suppose then that $y \notin C_2$. Our hypotheses that $x \in C_1$ and $y \notin C_2$ become $a \leq x$ and $y \leq b$. Using the rule forms of prefixing and suffixing, we can easily derive $x \to y \leq a \to b$ ($y \leq b$ *implies* $x \to y \leq x \to b$, $a \leq x$ *implies* $x \to b \leq a \to b$, and apply transitivity). Then our hypothesis that $x \to y \in C$ gives that $a \to b \in C$, contrary to our initial supposition.

Let us next consider the case where we have both 'residuals' but no fusion operation to connect them. To this end we will define a *double implicational poset* to be a structure $(S, \leq, \to, \leftarrow)$, where both $(S, \leq, \to,)$ and $(S, \leq, \leftarrow)$ are implicational posets, and the arrows interact by 'pseudo-assertion', *i.e.*,

$$\begin{aligned} a &\leq (b \leftarrow a) \to b, \\ a &\leq b \leftarrow (a \to b). \end{aligned}$$

The following shows that we have again correctly axiomatized (in this case both) residuals without residuation.

Embedding Theorem for Double Implicational Posets. *Every double implicational poset can be embedded in a concrete residuated p.o. groupoid.*

Proof. From the theorem above we know that every implicational poset is so embeddable, using the canonical relation R defined above, but let us now subscript it:

$$R_{\to}C_1C_2C_3 \quad \textit{iff} \quad \forall x, y, \textit{ if } x \in C_1 \ \& \ x \to y \in C_2, \textit{ then } y \in C_3.$$

By a symmetric argument every implicational poset is also so embeddable using the canonical relation:

$$R_{\leftarrow}C_1C_2C_3 \quad \textit{iff} \quad \forall x, y, \textit{ if } y \leftarrow x \in C_1 \ \& \ x \in C_2, \textit{ then } y \in C_3.$$

The only problem is that we seem to have two canonical relations instead of one. But it is easy to see that in fact the two relations coincide. Thus suppose $R_{\to}C_1C_2C_3$. In order to show $R_{\leftarrow}C_1C_2C_3$, we suppose $b \leftarrow a \in C_1$

and $a \in C_2$ and show $b \in C_3$. But, by virtue of $a \in C_2$ and pseudo-assertion, we obtain $(b \leftarrow a) \rightarrow b \in C_2$. But our hypothesis that $R_{\rightarrow}C_1C_2C_3$ says $\forall x, y$, *if* $x \in C_1$ & $x \rightarrow y \in C_2$, *then* $y \in C_3$. So let $x = b \leftarrow a$ and $y = b$; it follows that $b \in C_3$ as desired. The dual form of pseudo-assertion takes care of the converse.

Remark 3. The reader can easily verify that pseudo-permutation can replace pseudo-assertion in the definition of a *double implicational poset.*

6 Representation of Residuated P.O. Groupoids

We take the opportunity to show that the representations above using cones extend naturally to residuated groupoids (with fusion now present). The trick is to define yet one more relation

$$R_{\circ}C_1C_2C_3 \quad \textit{iff} \quad \forall x \in C_1, \forall y \in C_2, x \circ y \in C_3.$$

We now consider the ternary frame $(U, R_{\circ})$ and the operation $\circ$ on subsets U defined above:

$$X \circ Y = \{\chi : \exists \alpha \in X, \exists \beta \in Y, R_{\circ}\alpha\beta\chi\}.$$

Using the same function h as used in the representation above ($h(a) = \{C : a \in C\}$) we show

$$h(a \circ b) = h(a) \circ h(b), \quad \textit{i.e.},$$
$$a \circ b \in C \quad \textit{iff} \quad \exists C_1, C_2, a \in C_1 \ \ \& \ b \in C_2 \ \ \& \ R_{\circ}C_1C_2C.$$

As with the canonical definitions of the relations $R_{\rightarrow}$ and $R_{\leftarrow}$ above, one direction of this is immediate. Here however it is the direction from right-to-left. We then only consider the other direction. To this end we assume that $a \circ b \in C$, define $C_1 = [a)$, $C_2 = [b)$, and show $R_{\circ}C_1C_2C$. Thus suppose $x \in C_1$ and $y \in C_2$. Then $a \leq x$ and $b \leq y$, and so by isotonicity we get $a \circ b \leq x \circ y$. Since the first of these is in C, which as a cone is closed upward, then $x \circ y \in C$ as needed.

We are not yet through, because if we were to just add this representation of $\circ$ to the representations of the operations $\rightarrow$ and $\leftarrow$ above, we would end up using three different relations $R_{\circ}, R_{\rightarrow}, R_{\leftarrow}$, when we only want one. But just as above we can prove that the three relations in fact coincide, this time in virtue of residuation. We leave details to the reader and simply record the

Representation Theorem for Residuated P.O. Groupoids. *Every residuated p.o. groupoid is isomorphic to a concrete p.o. groupoid on some frame.*

Remark 4. Let us consider how we might in fact have been more subtle in our representations of residuated partially ordered groupoids and related structures above. Let (U, R) be a ternary frame and let $\sqsubseteq$ be a p.o. on U. We shall call the structure $(U, \sqsubseteq, R)$ an *articulated frame* if it satisfies the following 'tonicity' conditions:

$$\textit{if}\quad R\alpha\beta\gamma\ \&\ \gamma \sqsubseteq \gamma',\ \textit{then}\quad R\alpha\beta\gamma',$$
$$\textit{if}\quad R\alpha\beta\gamma\ \&\ \alpha' \sqsubseteq \alpha,\ \textit{then}\quad R\alpha'\beta\gamma,$$
$$\textit{if}\quad R\alpha\beta\gamma\ \&\ \beta' \sqsubseteq \beta,\ \textit{then}\quad R\alpha\beta'\gamma.$$

We shall say that a subset of U is *hereditary* when it satisfies the following: *if* $\alpha \in X$ & $\alpha \sqsubseteq \beta$, *then* $\beta \in X$. Let $\mathcal{P}(U)\!\uparrow$ be the class of all hereditary subsets. It is easy to see that $\mathcal{P}(U)\!\uparrow$ is closed under the concrete operations defined above (that the resultant sets are hereditary follows from the 'tonicity' conditions), and so we get another concrete example of a residuated p.o. groupoid, which we shall call *an articulated concrete residuated p.o. groupoid.*

It is easy to see that but a slight modification of the canonical frame of all cones produces an articulated frame. Again U is the set of all cones of S and $R_\circ$, $R_\rightarrow$ and $R_\leftarrow$ are defined as before and are all equivalent (we shall thus denote them ambiguously as R). The difference is that we add as well the natural partial order $\sqsubseteq$ on cones, *i.e.*, set inclusion restricted to U. Thus we get an articulated frame $(U, \subseteq, R)$. We verify that the 'tonicity' conditions above hold when $R = R_\circ$.

For the first, let us suppose that $R_\circ P_1 P_2 P_3$ & $P_3 \subseteq P_3'$, and show $R_\circ P_1 P_2 P_3'$. But this is obvious, since our first supposition says that the hypotheses $x \in P_1$ and $y \in P_2$ *imply* $x \circ y \in P_3$, and so *a fortiori* these same hypotheses imply $x \circ y \in P_3'$ (since it includes P_3).

For the second 'tonicity' condition, suppose $R_\circ P_1 P_2 P_3$ & $P_1' \subseteq P_1$. We need to show that $R_\circ P_1' P_2 P_3$. So assume $x \in P_1'$ and $y \in P_2$. Then *a fortiori* $x \in P_1$ and $y \in P_2$, and, since $R_\circ P_1 P_2 P_3$, we conclude $x \circ y \in P_3$ as needed. A similar argument takes care of the third 'tonicity' condition.

We still have to show that h carries each element $a \in S$ into a hereditary subset of the cones of S. But $h(a) = \{C : a \in C\}$ is clearly hereditary, for if $a \in C$ and $C \subseteq C'$, then $a \in C'$.

7 Assertional Residuated P.O. Groupoids

We now consider what happens when the residuated p.o. groupoid has a 'right identity element' e satisfying

$$a \circ e = a.$$

It is the presence of this element that allows us to consider the algebraic analog of 'theorems'. Thus by residuation we get

$$e \leq a \rightarrow a,$$

which can be interpreted as saying that 'a implies a' is a theorem. The element e can be interpreted as 'empty' or 'void', but it should not be interpreted as not there at all. It is better to think of it as the conjunction of all theorems.

In general, the presence of e allows us to 'push the metalinguistic deducibility relation into the object language', giving us for each 'deducibility relation' $a \leq b$ a corresponding 'object language theorem' $a \rightarrow b \geq e$. Thus $a \leq b$ *iff* $a \circ e \leq b$ *iff* $e \leq a \rightarrow b$. We shall talk of 'pushing' and 'popping' for the left and right directions of this equivalence, respectively.[9]

Note that if we assume a left identity e' with the usual law

$$(\textit{Left Identity}) \qquad e' \circ a = a,$$

as well as a right identity it can be shown that $e = e'$:

$$e = e' \circ e = e'.$$

This is pleasant because we do not have two classes of 'theorems' (the principal cone of e and also that of e'). But it is not absolutely unavoidable since for many purposes (including the most general representation theorem as well as dealing with some subtleties in application to certain weak modal logics) we might have just 'lower right and left identities',

$$a \circ e \leq a, \qquad e' \circ a \leq a,$$

and the uniqueness proof would no longer go through. Note incidentally, that all we need for 'pushing' is that e is a lower right identity. But the assumption of full equality is far from unusual and certainly simplifies things.

9 To the best of my knowledge the first explicit algebraic characterization of theorems as the cone determined by the identity element occurs in Dunn (1966) (relevant portions of which are in Anderson, Belnap *et al.* (1975), §28.2 and §28.5). The reason I say 'explicit' is this of course generalizes the well-known characterization of theorems as identical to the top element **1** of say a Boolean algebra or a Heyting lattice. But there the cone is trivial, just being $\{1\}$.

Remark 5. Amplifying the above, recall that our proof of 'push and pop' for the right residual depends on the axiom for the right identity being stated with an equality and not just the inequality, and of course similarly for the left residual and the left identity. Indeed, from push and pop from the right residual we can infer that e is a full right identity, and similarly for the left residual and e'. Showing just the first we start with $e \leq a \rightarrow (a \circ e)$ (fusing), and by 'popping' we obtain $a \leq a \circ e$. But we get $a \circ e \leq a$ by residuation from $e \leq a \rightarrow a$, which comes from $a \leq a$ by 'pushing'.

8 Assertional Implicational Posets

Again we return to the topic of 'residuals without residuation'. By an *assertional implicational poset* we shall mean a structure $(S, \leq, \rightarrow, e)$, where $(S, \leq, \rightarrow)$ is an implicational poset and e (thought of as the 'conjunction of all logical implications') satisfies

$$(\textit{Push and Pop}) \qquad a \leq b \quad \textit{iff} \quad e \leq a \rightarrow b.$$

The element e functions in the role of the right identity element in a right residuated p.o. groupoid, except for the embarrassment of their being no fusion operation present for it to be the right identity of. But it still allows us to characterize 'theorems' ('assertions') by the relationship $e \leq a$.

We next examine what happens when we have both residuals without residuation. By an *assertional double implicational poset* we shall mean a structure $(S, \leq, \rightarrow, \leftarrow, e)$, where each of $(S, \leq, \rightarrow, e)$ and $(S, \leq, \leftarrow, e)$ is an implicational poset.

Note that it is easy to prove that if we had not required the same '**t**', but instead allowed that we have e' in place of e in the second implicational poset, then we could prove $e = e'$ (and thus we have lost no generality by building uniqueness into our definition). Thus we choose to show that $e \leq e'$. To start, we derive from $e' \leq e'$, by 'push' for e, that $e \leq e' \rightarrow e'$. But next we shall derive $e' \rightarrow e' \leq e'$ (and so $e \leq e'$ results from transitivity). By pseudo-assertion we obtain $e' \leq e' \leftarrow (e' \rightarrow e')$, and by 'pop' for e' we obtain $e' \rightarrow e' \leq e'$. The thing that glues e and e' together is that they 'push and pop' the same p.o. $\leq$.

Remark 6. 'Push' can be easily seen to be equivalent to the following inequality:

$$(\textit{Self-Implication Thesis}) \qquad e \leq a \rightarrow a.$$

This inequality follows immediately from 'push' using $a \leq a$. Conversely, we obtain push by the following sequence of moves.

1. $a \leq b$ (*hypothesis*)
2. $a \to a \leq a \to b$ (1, *rule prefixing*)
3. $e \leq a \to a$ (*self-implication thesis*)
4. $e \leq a \to b$ (2, 3, *transitivity*).

It would be nice to find some inequality equivalent to 'pop', but the best I have been able to come up with assumes either rule permutation or the presence of the other (left) residual (with the equivalent properties of rule pseudo-permutation or of pseudo-assertion).

Let us work through the case where we have the addition of the left residual (the case where we have just the right residual but have rule permutation is just a special case, since rule permutation gives that the right residual is also the left residual). In this case one can postulate

$$(\textit{Specialized Pseudo-Assertion}) \qquad a \leftarrow e \leq a.$$

From this we derive 'pop' as follows:

1. $e \leq a \to b$ (*hypothesis*)
2. $a \leq b \leftarrow e$ (*rule pseudo-permutation*)
3. $a \leq b$ (2, *specialized pseudo-assertion, transitivity*).

Conversely, from 'pop' we can derive specialized pseudo-assertion:

1. $e \leq (a \leftarrow e) \to a$ (*pseudo-assertion*)
2. $a \leftarrow e \leq a$ (1, *'pop'*).

The dual form of specialized pseudo-assertion $e' \to a \leq a$ would similarly be equivalent to 'pop' for the left-residual.

9 Representations of Assertional Implicational Posets

The element e causes more problems than one might have thought. We shall first work through the representation when we have only one implication $\to$, and then work through the case of assertional double implicational posets. In the next section we shall then add fusion to get a full assertional residuated p.o. groupoid.

The issue is how to verify 'push and pop'. As we saw above,

$$e \leq a \to a$$

is equivalent to 'push', and so we shall start there. In the representation, the elements e, a will be assigned certain sets $\mathbf{Z}$, $A \subseteq U$, and so we need

$$\mathbf{Z} \subseteq A \to A, \ \textit{i.e., for} \ \zeta \in \mathbf{Z}, \ \textit{if} \ R\alpha\zeta\beta \ \& \ \alpha \in A, \ \textit{then} \ \beta \in A.$$

This suggests that we should define a relation $\alpha \preceq \beta$ *iff* $\exists\zeta \in \mathbf{Z}$, $R\alpha\zeta\beta$, and work only with those subsets A of U that are closed (hereditary) with respect to $\preceq$.

In the canonical frame, we define $\mathbf{Z} = \{C : e \in C\}$, and in particular then $[e) \in \mathbf{Z}$. Recall that in the canonical frame $\sqsubseteq$ is just inclusion between sets. It turns out that $\preceq$ is as well.

Inclusion Lemma. *In the canonical frame,* $C \preceq C'$ *iff* $C \subseteq C'$ *iff* $C \sqsubseteq C'$.

Proof. Recall that in the canonical frame $RC_1C_2C_3$ *iff* $\forall x, y$, if $x \in C_1$ and $x \to y \in C_2$, then $y \in C_3$. Let us suppose that $C \preceq C'$, *i.e.*, $\exists C_z \in \mathbf{Z}$, RCC_zC', *i.e.*, $\forall x, y$, if $x \in C$ & $x \to y \in C_z$, then $y \in C'$. Thus, for any $x \in C$, since $e \in C_z$ and $e \leq x \to x$, we have $x \to x \in C_z$. And so we have $x \in C'$. So $C \subseteq C'$. Conversely, if $C \subseteq C'$, we show $RC[e)\mathbf{Z}C'$, *i.e.*, $\forall x, y$, if $x \in C$ & $x \to y \in [e)$, then $y \in C'$. So we suppose $x \in C$ & $x \to y \in [e)$. This last means $e \leq x \to y$, and so by 'pop' $x \leq y$. But since $x \in C$ (a cone), we have $y \in C \subseteq C'$, and so $y \in C'$ as needed.

The above suggests that the way to define a *concrete assertional implicational poset* is to take a structure $(U, \sqsubseteq, R, \mathbf{Z})$, where $(U, \sqsubseteq, R)$ is an articulated frame (see above), $\mathbf{Z} \subseteq U$ is a member of $\mathcal{P}(U)\uparrow$ and for α, $\beta \in U$,

$$\alpha \sqsubseteq \beta \quad \textit{iff} \quad \exists \zeta \in \mathbf{Z},\ R\alpha\zeta\beta.$$

We then consider the hereditary subsets $\mathcal{P}(U)\uparrow$ and on these define $\to$, $\leftarrow$, $\circ$ as above.

It might be thought that our definition of a *concrete assertional implicational poset* is not yet complete, and that we still need to add a natural condition that assures 'pop':

$$\mathbf{Z} \subseteq A \to B \quad \textit{implies} \quad A \subseteq B.$$

Thus suppose that $\mathbf{Z} \subseteq A \to B$ and $\alpha \in A$. Clearly what is needed to assure that $\alpha \in B$ is that $\exists \zeta \in \mathbf{Z}$, $R\alpha\zeta\alpha$. But this is already assured by the fact that $\sqsubseteq$ is reflexive, which is already built into our definition, since we required that $(U, \sqsubseteq, R)$ be an articulated frame, which then requires $\sqsubseteq$ to be a partial order. There are more requirements imposed on an articulated frame than are strictly needed, but all of this information can be read off of the canonical frame and so there is certainly no harm in requiring it. Thus not only are transitivity and antisymmetry not strictly needed, but not all of the 'tonicity' conditions on R are needed. Indeed, we strictly need only the 'tonicity' condition

$$\textit{if} \quad R\alpha\beta\gamma \ \&\ \beta' \leq \beta, \quad \textit{then} \quad R\alpha\beta'\gamma,$$

which guarantees that $A \to B$ is hereditary given that both A and B are. However, it is convenient to require all of the 'tonicity' conditions (they clearly hold in the canonical frame) just in case we have the operations $\leftarrow$ and $\circ$ around, since the other conditions are needed to verify that hereditary subsets are closed under these operations.

Thus let us now consider the case of an assertional double implicational poset $(S, \leq, \rightarrow, \leftarrow, e)$.

Unfortunately, since we also want

$$\mathbf{Z} \subseteq A \leftarrow A, \textit{ i.e., for } \zeta \in \mathbf{Z}, \alpha \in A, \ \textit{ if } \ R\zeta\alpha\beta, \ \textit{ then } \ \beta \in A,$$

this last suggests that we also define a relation $\alpha \preceq' \beta$ *iff* $\exists\zeta \in \mathbf{Z}$, $R\zeta\alpha\beta$, and that we work with the subsets of U that are hereditary with respect to $\preceq'$ (as well as of course $\preceq$). Things are getting a bit complicated, but the good news is that we can show that in the canonical frame $\preceq$ and $\preceq'$ are in fact identical. We already know from the Inclusion Lemma that $\preceq$ is just $\subseteq$ on cones. The argument can be run through symmetrically to obtain that $\preceq'$ is also just $\subseteq$ on cones. Hence $\preceq = \sqsubseteq = \preceq'$, and so requiring that subsets be hereditary with respect to $\sqsubseteq$ is enough.

The above dictates that the way to define a *concrete assertional double implicational poset* is to take a structure $(U, \sqsubseteq, R, \mathbf{Z})$, where $(U, \sqsubseteq, R)$ is an articulated frame (see above), $\mathbf{Z} \subseteq U$ and for $\alpha, \beta \in U$,

$$\alpha \sqsubseteq \beta \quad \textit{iff} \quad \exists\zeta \in \mathbf{Z}, R\alpha\zeta\beta, \quad \textit{iff} \quad \exists\zeta \in \mathbf{Z}, R\zeta\alpha\beta.$$

We then define $\rightarrow$ and $\leftarrow$ on the hereditary subsets as above.

10 Representations of Assertional Residuated P.O. Groupoids

For a *concrete assertional residuated p.o. groupoid*, we proceed as above, but this time we want to have $\circ$ as well as $\rightarrow$ and $\leftarrow$, and we need to verify

$$\mathbf{Z} \circ A = A, \qquad A \circ \mathbf{Z} = A.$$

Again we use an articulated frame, consider the hereditary subsets $\mathcal{P}(U)\uparrow$ and on these define $\rightarrow, \leftarrow, \circ$ as above.

11 Special Postulates

We next investigate what conditions on the ternary relation R correspond to what postulates on the algebra. Van Benthem has used the label 'correspondence theory' to label the fact that in modal logic a certain property of the accessibility relation will force a certain modal postulate to hold, and *vice versa*. We shall not be so precise here, contenting ourselves on the *vice versa* side to simply verifying that the canonical accessibility relation satisfies the property in question. In Dunn (1986) one can find a list of such correspondences derived from Routley and Meyer, but properties of the relation R are linked to implicational theses. One needs to further link the implicational theses to properties of residuation (as we did above) to get correspondences to properties of fusion.

Thus to illustrate with the simplest example, commutativity ($x \circ y = y \circ x$) corresponds to the commutativity of R in its first two positions ($R\alpha\beta\gamma$ *implies* $R\beta\alpha\gamma$), and Routley-Meyer link this last to assertion. It is easy to see that requiring R to be commutative in its first two places forces $\circ$ as defined on a frame to be commutative, since $A \circ B = \{\chi : \exists\alpha \in A, \beta \in B, R\alpha\beta\chi\} = \{\chi : \exists\beta \in B, \alpha \in A, R\beta\alpha\chi\} = B \circ A$. Conversely, if $\circ$ is commutative, then in the canonical frame $RC_1C_2C_3$ implies $RC_2C_1C_3$. For if $a \in C_2$ and $b \in C_1$, then (since $RC_1C_2C_3$) we have $b \circ a \in C_3$, and so by commutativity of $\circ$ we have that $a \circ b \in C_3$.

Other correspondences are not quite so straightforward. Let us consider right associativity. We need to show that $A \circ (B \circ C) \subseteq (A \circ B) \circ C$. It is easy to see that the requirement that $\chi \in A \circ (B \circ C)$ implies $\chi \in (A \circ B) \circ C$ boils down to $\forall\alpha, \beta, \gamma$,

$$\textit{if } \exists\delta,\ R\alpha\delta\chi \ \&\ R\beta\gamma\delta, \textit{ then } \exists\delta',\ R\alpha\beta\delta' \ \&\ R\delta'\gamma\chi.$$

Using an obvious definition adapted from Meyer and Routley (1972), this can be compactly expressed as

$$R\alpha(\beta\gamma)\chi \quad \textit{implies} \quad R(\alpha\beta)\gamma\chi,$$

which mnemonically echoes the postulate of right associativity.

We show that this condition holds in the canonical frame, given that the underlying $\circ$ itself satisfies right associativity. Let us then assume that we have cones α, β, γ, χ such that for some cone δ, $R\alpha\delta\chi$ and $R\beta\gamma\delta$. We now must show that there exists a cone δ', $R\alpha\beta\delta'$ and $R\delta'\gamma\chi$. Simply define $\delta' = \{d : \exists a \in \alpha, b \in \beta, a \circ b \leq d\}$.[10] Clearly δ' is a cone and $R\alpha\beta\delta'$ holds by definition. $R\delta'\gamma\chi$ is a little trickier. Let us suppose that $d \in \delta'$ and $c \in \gamma$. We must show that $d \circ c \in \chi$. Since $d \in \delta'$ we know that $\exists a \in \alpha$, $b \in \beta$, $a \circ b \leq d$. By right associativity and isotonicity we have,

$$a \circ (b \circ c) \leq (a \circ b) \circ c \leq d \circ c.$$

But $R\beta\gamma\delta$ gives $b \circ c \in \delta$, which together with $R\alpha\delta\chi$ gives $a \circ (b \circ c) \in \chi$, and so the above inequality gives $d \circ c \in \chi$, as needed.

10 δ' is just the cone generated from taking all products of a term from α and a term from β, in a compact notation $[\alpha\beta)$. Note that when we have a semilattice and α and β are filters, δ' so-defined is not necessarily a filter. However, if we simply now understand $[\alpha\beta)$ as the *filter* generated from the product $\alpha\beta$, all works out.

Without further ado we simply list some correspondences, and then make a remark about one of the least straightforward.

$$\begin{array}{rcl@{\qquad}rcl}
 & \circ & & & R & \\
a \circ (b \circ c) & \leq & (a \circ b) \circ c & R\alpha(\beta\gamma)\chi & \textit{implies} & R(\alpha\beta)\gamma\chi \\
(a \circ b) \circ c & \leq & a \circ (b \circ c) & R(\alpha\beta)\gamma\chi & \textit{implies} & R\alpha(\beta\gamma)\chi \\
(x \circ a) \circ b & = & (x \circ b) \circ a & R(\chi\alpha)\beta\gamma & \textit{implies} & R(\chi\beta)\alpha\gamma \\
a \circ (b \circ x) & = & b \circ (a \circ x) & R\alpha(\beta\chi)\gamma & \textit{implies} & R\beta(\alpha\chi)\gamma \\
a \circ b & = & b \circ a & R\alpha\beta\gamma & \textit{implies} & R\beta\alpha\gamma \\
a \circ x & \leq & a \circ (a \circ x) & R\alpha\chi\gamma & \textit{implies} & R a(a\chi)\gamma \\
x \circ a & \leq & (x \circ a) \circ a & R\chi\alpha\gamma & \textit{implies} & R(\chi\alpha)\alpha\gamma \\
a & \leq & a \circ a & & R\alpha\alpha\alpha & \\
a \circ b & \leq & a & R\alpha\beta\chi & \textit{implies} & \alpha \sqsubseteq \chi \\
a \circ b & \leq & b & R\alpha\beta\chi & \textit{implies} & \beta \sqsubseteq \chi
\end{array}$$

Remark 7. The construction from section 5 of C_2 as the complement of $(b]$ works fine as long as we are not concerned about any special properties that canonical R must have. But it must sometimes be modified for certain applications. Also (and not entirely unrelated) for some purposes it is desirable to preserve glbs (when they exist), and for that we have to look at a more complicated definition of C_2.

We thus define $C_2 = \{y : \exists c \in C,\ c \leq a \to y\}$. First we note that $b \notin C_2$, for otherwise $\exists c \in C,\ c \leq a \to b$ and so $a \to b \in C$, contrary to our original supposition. We still have to prove that RC_1CC_2. To this end let us suppose that $x \in C_1$ and $x \to y \in C$, and show that $y \in C_2$. Since $C_1 = [a)$, $a \leq x$, and so $x \to y \leq a \to y$ (suffixing), and so $y \in C_2$.

The important thing about the more complicated construction of C_2 is that it extends nicely to the case where the poset is also a meet semilattice, *i.e.*, a poset closed under binary greatest lower bounds (denoted by $\wedge$, and standardly called *meets*), though we must then postulate $a \to (b \wedge c) = (a \to b) \wedge (a \to c)$. It extends more generally to the case of a complete meet semilattice, *i.e.*, a poset where every subset Y has a glb, postulating $a \to \wedge Y = \wedge\{a \to y : y \in Y\}$.[11] A *filter* can be defined as a cone C that is closed under meets, in symbols if $x, y \in C$, then $x \wedge y \in C$, and a *complete filter* is a cone closed under arbitrary glbs, *i.e.*, if $Y \subseteq C$, then $\wedge Y \in C$.

It is well-known (and easy to see) that mapping an element a into the set of filters of which it is a member preserves meets, and similarly with

11 Actually $a \to (b \wedge c) \leq (a \to b) \wedge (a \to c)$ follows from prefixing, and can replace it as well, and similarly for the complete version. In the presence of $\circ$, distribution over meet can be proven and does not need to be postulated. Cf. Fuchs (1963), Lemma B, p. 90.

complete filters and arbitrary glbs. The important thing about this more complicated definition of C_2 is that for semilattices, if C is a filter, then so is C_2 (and also clearly $C_1 = [a))$, and the analogous fact holds for complete filters in complete semilattices. We argue the latter as the more complicated. Thus suppose $Y \subseteq C_2$, *i.e.*, $\forall y \in Y$, $\exists c \in C$, $c \leq a \to y$. Using the axiom of choice, form the set C' by picking one such c for each y. Clearly $\wedge C' \leq \wedge\{a \to y : y \in Y\} = a \to \wedge Y$, and so $\wedge Y$ satisfies the condition for membership in C_2.

A *downward directed* cone C is a generalization of a filter. Besides being a cone it has the property that if $x_1, x_2 \in C$, then $\exists l \in C$ such that $l \leq x_1, x_2$. For certain applications, where one wants special conditions on the canonical accessibility relation such as $R\alpha\alpha\alpha$, one likes downward directed cones, and one would like to be able to generalize the observation just above to the fact that if C is a downward-directed cone, so is C_2. (Note that trivially $[a)$ is a downward directed cone, since we can set $l = a$.) Unfortunately it appears that this generalization cannot be made unless we also have $\circ$. So if one has either $\wedge$ or $\circ$, one can guarantee certain special properties on the canonical accessibility relation on cones, but otherwise not. This is not true of all special properties, but $R\alpha\alpha\alpha$ is a critical one since it corresponds to square-increasingness and the structural rule Contraction, which distinguish relevant implication from linear implication.

However, if one is willing to give up on cones and complicate the canonical states in various ways, one can still preserve properties such as $R\alpha\alpha\alpha$. Thus, *e.g.*, for an implicational poset satisfying prefixing and permutation, one can restrict states to those cones that are closed under *modus ponens*:

$$\textit{if } a \in C \ \& \ a \to b \in C, \textit{ then } b \in C.$$

And without permutation one has to complicate the construction even more, keeping states as cones closed under *modus ponens* but complicating the definition of the canonical accessibility relation (cf. the definition in Meyer (1982)). Probably the best approach as far as preserving contraction (and other properties) is concerned is to be found in Allwein and Dunn (199+), and to analogize the treatment found there to the case of implicational posets (they have an underlying lattice) one would replace cones with pairs (C, D), where C is a cone, D a dual cone and $C \cap D \neq \emptyset$.

12 Further Generalizations

It is possible to develop a framework where one does not have 'pop', but only 'push', though things get a bit hard to keep track of since then one has both e and e', $\preceq$ and $\preceq'$, and one must look at sets that are hereditary simultaneously under both relations.

When fusion is present this is equivalent to having just

$$e \circ a \leq a, \qquad a \circ e' \leq a,$$

but not the converse of these relations.

There are logics where such a generalization is useful, *e.g.*, 'ticket entailment' (cf. Anderson and Belnap (1975)), which lacks $e \to a \leq a$, and 'non-alethic' modal logics, lacking $\Box a \leq a$.

13 Definition of a Tonoid

In Dunn (1991) a *distributoid* was defined as a structure $\mathfrak{D} = (A, \wedge, \vee, OP)$, where $(A, \wedge, \vee)$ is a distributive lattice and each $f \in OP$ is a (finitary) operation on A that 'distributes' in each of its places over at least one of $\wedge$ and $\vee$, leaving the lattice operation unchanged or switching it with its dual.[12] Since we no longer have the lattice operations to care for, this notion can be simplified so that each $f \in OP$ is merely either isotone or antitone.

More explicitly, let $\mathfrak{T} = (A, \leq, OP)$, where $(A, \leq)$ is a poset, OP is a set of (finitary) operations on A, and each n-ary $f \in OP$ is assigned a 'tonic type' $(\pm_1, \ldots, \pm_j, \ldots, \pm_n)$, where each $\pm_j = +$ or $-$. We then require that in each of its places if $\pm_j = +$, then the function is isotone in that position, *i.e.*, if $b \leq c$, then $f(a_1, \ldots, b, \ldots, a_n) \leq f(a_1, \ldots, c, \ldots, a_n)$, and if $\pm_j = -$, then $f(a_1, \ldots, c, \ldots, a_n) \leq f(a_1, \ldots, b, \ldots, a_n)$. (This can vary from place to place in a given operation.)[13]

There is one more requirement (which generalizes a requirement of Dunn (1991), which in turn generalized a requirement of Jónsson-Tarski). For convenience we require that the lattice be bounded, *i.e.*, there is a least element $\mathbf{0}$ and a greatest element $\mathbf{1}$ (a lattice can always be trivially extended with two additional elements to serve as the bounds).[14] We are

12 This is not quite the complete definition. Inspection of the more precise definition (using 'types') will show that f cannot distribute say over join in one of its positions, changing it to *meet*, and over meet in another of its positions, changing it to *join*. The result would have to be either meet in both cases or else join. Although one can mix and match with the inputs, the outputs must be the same.

13 It is tempting here to avoid the apparatus of tonic types and simply require that the function be either isotone or antitone in each of its positions. But in the degenerate case of a totally unconnected partial order (no point is above any other point), an operation can be both isotone and antitone in the same position. In such a case the formula that we later give for realizing the operation would be ambiguous.

14 Without the assumption of bounds, the following requirement would have to be conditionalized to 'respect' them when they exist.

going to require that the tonoid operations *respect the bounds* in a way that we illustrate using the concrete operations defined on a frame. Given that

$$X \circ Y = \{\chi : \exists\alpha \in X, \exists\beta \in Y, R\alpha\beta\chi\},$$

it is obvious that

$$X \circ \emptyset = \emptyset \circ X = \emptyset.$$

This is the requirement of Jónsson-Tarski. But

$$X \to Y = \{\chi : \forall\alpha, \beta,\ \textit{if}\ \ R\alpha\chi\beta\ \&\ \alpha \in X,\ \ \textit{then}\ \beta \in Y\}$$

also in its own appropriate way respects the bounds, for

$$\emptyset \to X = U, \quad X \to U = U,$$

and so does

$$X \leftarrow Y = \{\chi : \forall\alpha, \beta,\ \textit{if}\ \ R\chi\alpha\beta\ \&\ \alpha \in Y,\ \ \textit{then}\ \beta \in X\}, \quad \text{since}$$
$$X \leftarrow \emptyset = U, \quad U \leftarrow X = U.$$

Algebraically these correspond to the conditions

$$\begin{gathered} x \circ \mathbf{0} = \mathbf{0} \circ x = \mathbf{0}, \\ \mathbf{0} \to x = \mathbf{1}, \qquad x \to \mathbf{1} = \mathbf{1}, \\ x \leftarrow \mathbf{0} = \mathbf{1}, \qquad \mathbf{1} \leftarrow x = \mathbf{1}. \end{gathered}$$

More explicitly, and generalizing to operations of arbitrary finite degree, let us require of the n-ary operation f that

$$\text{when } a_j = \mathbf{0}, \text{ then } \begin{cases} f(a_1, \ldots, a_j, \ldots, a_n) = \mathbf{0} & \text{if } \pm_j = +, \\ f(a_1, \ldots, a_j, \ldots, a_n) = \mathbf{1} & \text{if } \pm_j = -, \end{cases}$$

$$\text{when } a_j = \mathbf{1}, \text{ then } \begin{cases} f(a_1, \ldots, a_j, \ldots, a_n) = \mathbf{1} & \text{if } \pm_j = +, \\ f(a_1, \ldots, a_j, \ldots, a_n) = \mathbf{0} & \text{if } \pm_j = -. \end{cases}$$

Informally, one rotten apple (bound) spoils the barrel, and the formula computes to the same bound or its complement depending on whether the function is isotonic or antitonic in the position to which the bound is input. We must add one more requirement, that the function compute in a uniform way so that its output is always **0** or always **1**.

Summarizing, and also providing a notation and concept that will later be useful in stating an abstract law of residuation, we require of each operation f that it have a *trace*, which formally is simply an $n+1$ tuple of the signs $+$ or $-$, recording whether the function is isotone or antitone in its various argument positions, and also recording which bound is always output when a single argument position is assigned either **1** or **0** depending on whether there is + or − in that position. In notation:

$$(\pm_1, \ldots, \pm_j, \ldots, \pm_n) \longmapsto \pm.$$

Remark 8. The requirement that the operation respect the bounds corresponds to an identically named requirement in Dunn (1991) that was stated using the notion of a 'distribution type', which itself generalized a requirement of Jónsson-Tarski.[15] Since we have at this stage no meets or joins for the operations to distribute over, obviously there can be no such notion. However the present requirement is in effect the 'trace' that remains of the notion of 'distribution type'.

Examples. Besides the examples of Dunn (1991) which required underlying distributive lattices, we can now add (the Lindenbaum algebras of) the implication-fusion fragments of various logics, including classical logic, intuitionistic, relevance logic, linear logic without exponentials, BCK-logic, Łukasiewicz logic. And there is the Lambek calculus, a ternary semantics for which was I once thought a new achievement for partial gaggles, until I read Došen (1992).

14 Representation of Tonoids

We start with a definition of the 'target' structures that will be used in the representation.

Definition. Given a tonoid $\mathfrak{T} = (A, \leq, OP)$, by a *frame* for $\mathfrak{T}$ is meant a structure $(U, \langle R_i \rangle_{i \in I})$ where there is a 1-1 correspondence between OP and I, and for each $f \in OP$, if the degree of f is n, then the corresponding $R_i \subseteq U^{n+1}$.

We will prove the following theorem, the exact statement of which will not be given until after we have some motivating examples.

15 I would like to take this opportunity to make this condition more explicit, to remove a possible ambiguity in Dunn (1991). It should read that if the operation f is of distribution type $(\tau_1, \ldots, \tau_i, \ldots, \tau_n) \longmapsto \tau$, then $f(c_1, \ldots, c_i, \ldots, c_n) = c$, where $c_i = \mathbf{0}$ when $\tau_i = \vee$, and $c_i = \mathbf{1}$ when $\tau_i = \wedge$, and the same for c in relation to τ, *i.e.*, $c = \mathbf{0}$ when $\tau = \vee$, and $c = \mathbf{1}$ when $\tau = \wedge$. In other words, putting in the ith position a bound that 'fits' the input type results in the function computing to a bound that 'fits' the output type, no matter whether the other arguments are bounds or not.

Representation Theorem for Tonoids. *Every tonoid can be represented as a tonoid of sets, with $\leq$ as $\subseteq$ and each operation $f_i \in OP$ interpreted as an operation F_i defined on the power set of U, using R_i according to a pattern to be described after motivation by the two examples below.*

We set out the pattern. Given an n-ary operation f with trace

$$t : (\pm_1, \ldots, \pm_n) \longmapsto \pm,$$

if $\pm = +$, the realization condition for t,

$$\mathcal{C}_t(\alpha_1 \in X_1,\ \ldots,\ \alpha_n \in X_n,\ R(\alpha_1, \ldots, \alpha_n, \beta)),$$

is a universally closed disjunction of

$$\sigma(\alpha_1 \in X_1),\ \ldots,\ \sigma(\alpha_n \in X_n),\ R(\alpha_1, \ldots, \alpha_n, \beta),$$

where each component $\sigma(\alpha_j \in X_j)$ is either $\alpha_j \in X_j$ or $\alpha_j \in \overline{X}_j$ depending on whether $\pm_j$ is $+$ or $-$ respectively. The rule when $\pm = -$ is appropriately dual, *i.e.*, the realization condition is an existentially closed conjunction and the rules reverse for determining whether a component $\sigma(\alpha_j \in X_j)$ is negated.

The reader is advised to consult Dunn (1991) for illustrations of the pattern of representation for 1-placed operations of every trace. In the terms introduced there the representation is done in terms of 'Menu A'. We get three other menus by switching R with each of $\overline{R}$, R^{-1} and $(\overline{R})^{-1}$. We shall not need to consider these other menus until later when we consider how tonoid operations might interact by a generalization of residuation. However, the reader should recognize that the following theorem not only applies to a relation R, but of course applies to the complement of R, or to any relation which is obtained from R by a simple permutation of its terms (as with R^{-1} in the binary case), since these are also relations of the same degree.

We illustrate the concept of a trace by some familiar examples. To each binary operation $*$ of a residuated groupoid ($\circ$ and the two residuals) we assign a trace

$$T(*) = (\pm_1, \pm_2) \longmapsto \pm,$$

where $\pm$ (with or without subscripts) ranges over $+$ and $-$:

$$a \circ b : (-, -) \longmapsto -$$

$$a \leftarrow b : (+, -) \longmapsto + \qquad \textit{one `contrapositive' of } T(\circ),$$

$$a \rightarrow b : (-, +) \longmapsto + \qquad \textit{another `contrapositive'}.$$

Note that it is important that we view the left residual as composed from its arguments in the opposite order from the right residual.

Theorem on Realizing Traces. *Let $(U, \langle R_i \rangle_{i \in I})$ be a frame. Associate with each R_i of degree $n+1$ a trace $t_i : (\pm_1, \ldots, \pm_j, \ldots, \pm_n) \longmapsto \pm$. Define the n-ary operation F_i on subsets of U as follows:*

$$F_i(X_1, \ldots, X_n) = \{\beta \in U : C_t(\alpha_1 \in X_1, \ldots, \alpha_n \in X_n, R_i(\alpha_1, \ldots, \alpha_n, \beta))\}.$$

Then the operation F_i has the trace t_i.

Proof. We content ourselves with observing that the reason that C_{t_i} is required to be a universally closed disjunction when $\pm = +$ is that both universal quantifiers and disjunction preserve implication, and similarly for the case when $\pm = -$. The reason that some atomic sentences $\alpha_j \in X_j$ are negated is that at that place we want to invert the order and also turn a universal set into an empty set, or vice versa.

Representation Theorem for Tonoids. *Every tonoid $(X, \leq, OP)$ is representable as a tonoid defined as above on a frame $(U, \langle R_i \rangle_{i \in I})$.*

Proof. The set U is the set of (proper) cones of A.[16] For each element $a \in X$ we set $h(a)$ to be the set of cones C such that $a \in C$. It is easy to see that h is 1-1 and preserves the inequality both ways. All that remains is to provide for each n-ary operation f_i in OP (with trace t_i) an $n+1$-placed relation R_i defined on cones,

$$h(f_i(a_1, \ldots, a_n)) = F_i(h(a_1), \ldots, h(a_n)),$$

where for sets of cones $X_1, \ldots, X_n$, the set $F_i(X_1, \ldots, X_n)$ is

$$\{C \in U : C_{t_i}(C_1 \in X_1, \ldots, C_n \in X_n, R_i(C_1, \ldots, C_n, C))\}.$$

Before giving the general definition of the relation R_i, we examine 3 of the cases where f_i is a binary operation, there clearly being 8 different trace types. Dunn (1991) contains an exhaustive analysis of the 4 different trace types of unary relations and their 4 different menus.

First let us suppose that we have an operation $\circ$ with trace $(-, -) \longmapsto -$. We need then to show for some appropriate relation R that

$$h(a \circ b) = F_\circ(h(a), h(b)),$$

where $F_\circ$ is the corresponding operation on sets of cones defined using R.

16 A *proper* cone is one that is not the whole poset. It is important for some applications to see that all of the constructions of this paper go through with the restriction to proper cones. One consequence is that the least element $\mathbf{0}$ is then carried into the empty set of cones by h, which is pleasant.

This amounts to showing that for each cone C_3,

$$\begin{array}{rll} C_3 \in h(a \circ b) & \textit{iff} & C_3 \in F_\circ(h(a),\ h(b)), \\ a \circ b \in C_3 & \textit{iff} & \mathcal{C}(C_1 \in h(a), C_2 \in h(b), RC_1C_2C_3), \\ & \textit{iff} & \exists C_1, C_2,\ C_1 \in h(a)\ \&\ C_2 \in h(b)\ \&\ RC_1C_2C_3, \\ & \textit{iff} & \exists C_1, C_2,\ a \in C_1\ \&\ b \in C_2\ \&\ RC_1C_2C_3. \end{array}$$

But this is just the way that $\circ$ was defined above in our representation of residuated p.o. groupoids, and for $RC_1C_2C_3$ we can simply substitute the canonical accessibility relation $R_\circ C_1C_2C_3$, defined as above, meaning $\forall x \in C_1$, $\forall y \in C_2$, $x \circ y \in C_3$.[17] However, for the sake of a later generalization, we shall re-express this as a universally quantified disjunction:

$$\forall x, y,\ x \notin C_1\ \textit{or}\ y \notin C_2\ \textit{or}\ x \circ y \in C_3.$$

Note that the first two components are negated, just as the first two arguments in the trace $(-,-) \longmapsto -$ are negative. We have a universally quantified disjunction 'because' the output of the trace is negative.

The case of $\circ$ was straightforward. Let us now consider a more confusing case and suppose that we have an operation $\rightarrow$ with trace $(-,+) \longmapsto +$. We need then to show for some appropriate relation R that

$$h(a \rightarrow b) = F_\rightarrow(h(a), h(b)),$$

where $F_\rightarrow$ is the corresponding operation on sets of cones defined using R.

This amounts to showing that for each cone C_3,

$$\begin{array}{rll} C_3 \in h(a \rightarrow b) & \textit{iff} & C_3 \in F_\rightarrow(h(a), h(b)), \quad \textit{i.e.}, \\ a \rightarrow b \in C & \textit{iff} & \mathcal{C}(C_1 \in h(a), C_2 \in h(b), RC_1C_2C_3), \\ & \textit{iff} & \forall C_1, C_2,\ C_1 \notin h(a)\ \textit{or}\ C_2 \in h(b)\ \textit{or}\ RC_1C_2C_3, \\ & \textit{iff} & \forall C_1, C_2,\ a \notin C_1\ \textit{or}\ b \in C_2\ \textit{or}\ RC_1C_2C_3, \\ & \textit{iff} & \forall C_1, C_2,\ \textit{if}\ \overline{R}C_1C_2C_3\ \&\ a \in C_1,\ \textit{then}\ b \in C_2. \end{array}$$

This last clearly has the form of the definition for realizing $\rightarrow$ on a frame, except that it has $\overline{R}C_1C_2C_3$ in place of $R_\rightarrow C_1C_3C_2$. But this is easy to fix. Simply define the relation $RC_1C_2C_3$ as $\overline{R_\rightarrow}C_1C_3C_2$, *i.e.*, $R = \overline{R_\rightarrow^{-2}}$, where R^{-j} means 'switch the j-th with the last place'. This has to do with the phenomenon of different 'menus', and will be more fully explained in section 15.

---

17 We can then set $C_1 = [a)$, $C_2 = [b)$. (That $[a)$ and $[b)$ are proper is the whole point of the technicality that operations must respect the bounds.)

We note that $\overline{R_{\to}^{-2}}C_1C_2C_3$ *iff* $\overline{R_{\to}}C_1C_3C_2$ *iff not* $\forall x,y(x \notin C_1$ *or* $x \to y \notin C_3$ *or* $y \in C_2)$ *iff* $\exists x,y(x \in C_1 \;\&\; y \notin C_2 \;\&\; x \to y \in C_3)$. So the canonical relation is definable by way of an existentially quantified conjunction. Note that the pattern of negation on the first two components exactly reverses the pattern of the signs of the first two arguments of the trace $(-,+) \longmapsto +$. We have an existentially quantified conjunction 'because' the output of the trace is positive, and the signs reverse for the same reason.

A similar story occurs when working through the gaggle-theoretic definition of $\leftarrow$ on the propositions of a frame. We let the reader do the manipulations, but its trace $(+,-) \longmapsto +$ yields the realization

$$C_3 \in h(a \leftarrow b) \;\; \textit{iff} \;\; \forall C_1, C_2,\; a \in C_1 \;\; \textit{or} \;\; b \notin C_2 \;\; \textit{or} \;\; RC_1C_2C_3.$$

This time we define $RC_1C_2C_3 = \overline{R_{\leftarrow}^{-1}}C_1C_2C_3$ *iff* $\overline{R_{\leftarrow}}C_3C_2C_1$ *iff not* $\forall x,y(x \leftarrow y \notin C_3$ *or* $y \notin C_2$ *or* $x \in C_1)$ *iff* $\exists x,y(x \notin C_1 \;\&\; y \in C_2 \;\&\; x \leftarrow y \in C_3)$. Again the reason that we have an existentially quantified conjunction is that the output of the trace is positive, and this is also the reason that the pattern of negation reverses.

The general rule for defining $R_f(C_1,\ldots,C_n,C)$ is that if the output of the trace is $-$, the formula is a universally quantified disjunction of $\sigma(x_1 \in C_1),\ldots,\; \sigma(x_n \in C_n),\; f(x_1,\ldots,x_n) \in C$, where the sign σ is positive or negative depending on whether the input to the trace at that component is $+$ or $-$. And if the output of the trace is $+$, the formula is an existentially quantified conjunction of the same components, but where the rule for the sign σ reverses. Finally we note that the 'canonical equivalence for f' can be obtained by taking the equivalence

$$\begin{aligned} &C \in h(f(a_1,\ldots,a_n)) \;\; \textit{iff} \\ &\qquad C_t(C_1 \in h(a_1),\ldots,C_n \in h(a_n), R_f(C_1,\ldots,C_n,C)), \end{aligned}$$

and rewriting it as

$$f(a_1,\ldots,a_n) \in C \;\; \textit{iff} \;\; C_t(a_1 \in C_1,\ldots,a_n \in C_n, R_f(C_1,\ldots,C_n,C)).$$

It is straightforward to see that half of this equivalence is guaranteed by the definition of R_f (left-to-right when the output of the trace of f is $+$, and the other way when it is $-$). The conditional the other way can be argued for by picking principal cones or the complements of principal dual cones. The conditional will have a straightforward existential consequent when the output of the trace is $-$, but one must contrapose it to get an existential consequent when the output of the trace is $+$. Then in each case the consequent is equivalent to an existentially quantified conjunction, and one is set up for a maximalization argument. But one must then be careful to take the complement of the principal ideal $(a_j]$, rather than the filter $[a_j)$, when $a_j \in C_j$ is negated.

15 Definition of a Partial Gaggle

The definition of a *gaggle* in Dunn (1991) presupposes the existence of an underlying distributive lattice. Allwein and Dunn (199+) in effect show that distribution is not essential, and the present paper shows that a lattice is not needed either. I now regret building in a distributive lattice since it leads to such barbarisms as a *partial gaggle*, but it is probably too late to retract terminology. The idea of a *partial gaggle* is that it is a tonoid where the operations are somehow related by a generalization of residuation, so that they form a *family*. The cash value of this is that all of the operations in a family may be realized using the same accessibility relation, but with different definitions in terms of it. The motivating example is of course that of a p.o. residuated groupoid.

Before looking at this example, we first introduce some abbreviations that are useful for our abstract statement of residuation. Thus, where $*$ is a binary operation with trace output $\pm$,

$$S(*, a, b, c) \ \textit{abbreviates} \ \begin{cases} a * b \leq c & \textit{when} \ \ \pm = -, \\ c \leq a * b & \textit{when} \ \ \pm = +. \end{cases}$$

The laws of the left residual and of the right residual can now be stated as follows, where the plus and minus signs underneath are there to aid the reader in discerning the patterns. What happens in each case is that the sign under the operation determines whether the operation applied to the first two terms following it is put on the left or the right side of the inequality. And the term c switches with the term with which it 'contraposes', *i.e.*, for the cases below, it is switched with the term under which there is indicated a change in sign of input to the trace.

$$\begin{array}{llllllllll} (SLR) & S(\ \circ, & a, & b, & c) & \textit{iff} & S(\ \leftarrow, & c, & b, & a) \\ & \ - & - & - & & & \ + & + & - & \end{array}$$

$$a \circ b \leq c \quad \textit{iff} \quad a \leq c \leftarrow b.$$

$$\begin{array}{llllllllll} (SRR) & S(\ \circ, & a, & b, & c) & \textit{iff} & S(\ \rightarrow, & a, & c, & b) \\ & \ - & - & - & & & \ + & - & + & \end{array}$$

$$a \circ b \leq c \quad \textit{iff} \quad b \leq a \rightarrow c.$$

After this motivation, we are at last in a position to define a *partial gaggle*.

Conventions.

(1) Let $\mathfrak{T} = (X, \leq, OP)$ be a tonoid. Recall that every $f \in OP$ has a trace $T(f)$. As a notational convention, let $-+ = -$ and $-- = +$.

(2) We shall say of two n-ary operations f and g that g is a *contrapositive* of f (with respect to the j-th place) when

$$\begin{array}{rl} \textit{if } T(f) = & (\pm_1 \ldots, \pm_j, \ldots, \pm_n) \longmapsto \pm, \\ \textit{then } T(g) = & (\pm_1, \ldots, -\pm, \ldots, \pm_n) \longmapsto - \pm_j\,. \end{array}$$

(3) If $\pm = -$, we write $S(f, a_1, \ldots, a_n, b)$ for $f(a_1, \ldots, a_n) \leq b$, if $\pm = +$, we write $S(f, a_1, \ldots, a_n, b)$ for $b \leq f(a_1, \ldots, a_n)$.

(4) We say that two operations f and g satisfy the *Abstract Law of Residuation* (in their j-th place) when f and g are contrapositives (with respect to their j-th place) and

$$S(f, a_1, \ldots, a_j, \ldots, a_n, b) \;\; \textit{iff} \;\; S(g, a_1, \ldots, b, \ldots, a_n, a_j).$$

(5) Two operations f, $g \in OP$ are *relatives* when they satisfy the Abstract Law of Residuation in some position.

(6) The family of operations OP is *founded* when there is a distinguished operation $f \in OP$ (the *head*) such that any other operation $g \in OP$ is a relative of f.

Definition. A *partial gaggle* is a tonoid $\mathfrak{T} = (X, \leq, OP)$ in which OP is a founded family.

As examples, consider a p.o. residuated groupoid with OP chosen to be any of the following families of operations ($\circ$ is the head of the families of which it is a member): $\{\circ\}$, $\{\circ, \leftarrow\}$, $\{\circ, \rightarrow\}$, $\{\circ, \leftarrow, \rightarrow\}$, $\{\leftarrow\}$, $\{\rightarrow\}$. Note that $\{\rightarrow, \leftarrow\}$ does not formally fall under our definition since the trace of one is not directly the contrapositive of the trace of the other, even though the trace of each is a contrapositive of the trace of $\circ$. However, it would seem not to hurt to further generalize the definition of a *partial gaggle* so as to allow for such indirect relationships of contraposition.

16 Representation of Partial Gaggles

We now have to figure out the realization conditions for the operations on a partial gaggle. Concentrating on p.o. residuated groupoids for concreteness, we know from our results on tonoids how each operation separately can be realized on a frame of cones, but now the question is how to wed the operations together into a single family.

Let us recall that the operations are realized on a frame as follows:

$$\begin{aligned} X \circ Y &= \{\chi : \exists\alpha \in X, \exists\beta \in Y, R\alpha\beta\chi\} \\ X \leftarrow Y &= \{\chi : \forall\alpha, \beta,\ \textit{if}\ R\chi\alpha\beta\ \&\ \alpha \in Y,\ \textit{then}\ \beta \in X\} \\ X \rightarrow Y &= \{\chi : \forall\alpha, \beta,\ \textit{if}\ R\alpha\chi\beta\ \&\ \alpha \in X,\ \textit{then}\ \beta \in Y\}. \end{aligned}$$

Let us also recall that the definitions all use the same accessibility relation R, but that they were not all chosen from the same 'menu'.

Before analyzing the situation further we must define the 'converse' of a relation with more than two terms. Of course there is no such thing really; instead there are just all of the different ways of permuting the terms. But for our purposes here it fortunately turns out that we do not need this full generality. The following suffices:

$$R^{-j}(\alpha_1, \ldots, \alpha_j, \ldots, \alpha_n, \beta)\ \textit{iff}\ R(\alpha_1, \ldots, \beta, \ldots, \alpha_n, \alpha_j).$$

This has to do with the fact that we look at operations whose traces are 'contrapositives' of each other. Note that in the case where R is binary the usual notation R^{-1} results.

If we now look at the concrete example of a residuated groupoid realized on a frame with a 3-placed relation, we see that the definition of $\circ$ used $R\alpha\beta\gamma$, that of $\leftarrow$ used $R^{-1}\alpha\beta\gamma = R\gamma\beta\alpha$ and that of $\rightarrow$ used $R^{-2}\alpha\beta\gamma = R\alpha\gamma\beta$.

The use of R^{-j} does not by itself completely solve the problem of determining the realizing conditions, since, as we observed in section 14, it is not really R^{-1} and R^{-2} that enter as arguments into the formula definitions of $\leftarrow$ and $\rightarrow$, but rather their complements. This is disguised by the fact that we have here expressed the formulas as conditionals, rather than as disjunctions.

We must then address the question as to why R^{-1} and R^{-2} get negated whereas R does not. What is the principle in choosing from different 'menus'?

The answer lies in the residuation laws. The residuals lie on the right side of the inequality, whereas $\circ$ lies on the left. Let us now put this more formally. We assume that we have a founded family of operations OP, where f is the head of the family. Using the trace t of f, let us decide to realize f by an operation F on subsets of U using some relation R and a representing condition in accord with the way of realizing tonoids from section 14.

The only question with respect to g is which 'menu' to use in realizing it as an operation G on subsets of U. We know from our earlier work on tonoids how to choose a definition from a given 'menu' so that the distribution pattern of G will realize that of g. But now we must choose this 'menu' so that f and g satisfy the Abstract Law of Residuation.

Since g is a relative of f, we have that g is a contrapositive of f with respect to some position j. Now we observe that if the output of the trace of g matches that of f, then we choose the definition of G using R^{-j}, whereas if the trace outputs clash, we use $\overline{R^{-j}}$.

We leave to the reader to verify that the Abstract Law of Residuation holds with respect to F and G defined as indicated, and record the above discussion as follows:

Definition. Given two traces t and t' of the same degree, *i.e.*, having the same number of inputs, we say that they *agree* if they have the same output sign, and that they *clash* if they have different output signs.

Theorem on Realizing Contraposed Traces. *Let* $\mathbf{T}$ *be a family of traces* $(\pm_1, \ldots, \pm_j, \ldots, \pm_n) \longmapsto \pm$ *(all of the same degree), where there is some trace* t_0 *in* T *such that every* t' *in* $\mathbf{T}$ *can be obtained from* t_0 *by contraposition. Let* (U, R) *be a frame, where* R *is a relation on* U *of degree* $n+1$*. Define the* n*-ary operation* F_0 *on subsets of* U *as follows:*

$$F_0(X_1, \ldots, X_n) = \{\beta \in U : C_{t_0}(\alpha_1 \in X_1, \ldots, \alpha_n \in X_n, R(\alpha_1, \ldots, \alpha_n, \beta))\}.$$

For a given trace t' *that is a contraposition of* t *in the* j*-th place, define the* n*-ary operation* G *on subsets of* U *as follows:*

$$G(X_1, \ldots, X_n) = \{\beta \in U : C_{t'}(\alpha_1 \in X_1 \ldots, \alpha_n \in X_n, R'(\alpha_1, \ldots, \alpha_n, \beta))\},$$

where R' *is* R^{-j} *or* $(\overline{R})^{-j}$*, depending on whether the trace outputs of* t *and* t' *agree or clash.*

Then the operation F_0 *has trace* t_0*,* G *has trace* t'*, and* F_0 *and* G *satisfy the Abstract Law of Residuation with respect to each other. In short, we have a partial gaggle of operations with traces assigned appropriately.*

Representation Theorem for Partial Gaggles. *Every partial gaggle* $(A, \leq, OP)$ *can be represented using a frame* $(U, \langle R_i \rangle_{i \in I})$*, as for a tonoid, but choosing relations for different operations from different (dual) menus.*

Proof. In the representations of tonoids, a different canonical relation R on the cones was needed for each operation. Here we use the same definition throughout, the definition coming from the trace of the head of the family. The representation otherwise proceeds as for tonoids. But of course we need the fact generalizing that $R_\circ$, $R_\rightarrow$ and $R_\leftarrow$ are all equivalent relations.

17 Modifications for Tonoids and Partial Gaggles with Identities

Consider an n-ary operation $f(x_1, \ldots, x_j, \ldots, x_n)$ whose associated trace output is $\pm$. We shall say that an element e is a *partial-identity element* for f with respect to the j-th place if $f(e, \ldots, x_j, \ldots, e) \leq x_j$ when $\pm = -$, and $x_j \leq f(e, \ldots, x_j, \ldots, e)$ when $\pm = +$. We shall sometimes denote such an element as e_i.

This clearly generalizes the notions of lower left and right identities in a p.o. groupoid, and it is very common for gaggles to have even stronger kinds of identity elements. For example, residuated groupoids often have an element e such that $e \circ x = x \circ e = x$. In this case the single element e serves as an identity element for both places (and note further that we have $=$, and not merely $\leq$). As we saw in section 7, for logical applications identity elements are important, since we need some way to say that a proposition a is 'true', and this may be expressed as $e \leq a$. Of course, in classical logic e can be taken to be the greatest element of the Boolean algebra; in general, as for relevance logic, this is too strong, but e can still be introduced as the 'infinite conjunction' of all of the truths (cf. Dunn (1966)).

In representing partial identities e_j, it is useful to look at *articulated* frames. These have a corresponding set of 'zero' elements $\mathbf{Z}_j$ and the requirement (R is the relation used to represent f):

$$\textit{if} \quad R(\zeta_1, \ldots, \alpha_j, \ldots, \zeta_n, \beta) \ \textit{for some} \ \zeta_1, \ldots, \zeta_n \in \mathbf{Z}_j, \ \textit{then} \ \alpha_j \sqsubseteq \beta.$$

We then require that the representing sets X (the 'propositions') be 'closed' as follows:

$$(\textit{Hereditary Condition}) \qquad \textit{If} \ \alpha \in X \ \& \ \alpha \sqsubseteq \beta, \ \textit{then} \ \beta \in X.$$

The point of the Hereditary Condition is to assure that the sets $\mathbf{Z}_j$ function as partial-identity elements. This seems a conceptual improvement over the corresponding requirement of Dunn (1991), which did not explicitly bring in the additional relation $\sqsubseteq$.

We must also impose some 'tonicity postulates' governing the interaction of $\sqsubseteq$ with the accessibility relation R. Let the head of the family of operations we wish to represent be f, with trace $(\pm_1, \ldots, \pm_j, \ldots, \pm_n) \longmapsto \pm$.[18] We require:

(α) *if* $R\alpha_1 \ldots \alpha_j \ldots \alpha_n\beta$ & $\sigma(\alpha_j \sqsubseteq \alpha'_j)$, *then* $R\alpha_1 \ldots \alpha'_j \ldots \alpha_n\beta$, *where*

$$\text{if } \pm = +, \text{ then } \begin{cases} \sigma(\alpha_j \sqsubseteq \alpha'_j) \text{ is } \alpha_j \sqsubseteq \alpha'_j \text{ when } \pm_j = +, \\ \sigma(\alpha_j \sqsubseteq \alpha'_j) \text{ is } \alpha'_j \sqsubseteq \alpha_j \text{ when } \pm_j = -, \end{cases}$$

and if $\pm = -$, *these reverse*;

(β) *if* $R\alpha_1 \ldots \alpha_j \ldots \alpha_n\beta$ & $\beta \sqsubseteq \beta'$, *then* $R\alpha_1 \ldots \alpha_j \ldots \alpha_n\beta'$.

These somewhat mysterious conditions make sense when one examines the canonical realization conditions which have a similarly structured definition by cases, and sees whether membership conditions in each place are positive or negative.

The purpose of the tonicity postulates is to assure that the operations defined using R carry propositions (hereditary subsets) to propositions. We illustrate this with respect to a residuated groupoid defined on a frame using a 3-placed accessibility relation. In order to verify that $\mathbf{Z} \circ X \subseteq X$, we assume that $\chi \in \mathbf{Z} \circ X$. Then there exist $\zeta \in \mathbf{Z}$ and $\alpha \in X$ such that $R\zeta\alpha\chi$. But then $\alpha \sqsubseteq \beta$, and so by the Hereditary Condition, $\chi \in X$ as needed. One can also verify that $X \subseteq \mathbf{Z} \to X$ and $X \subseteq X \leftarrow \mathbf{Z}$. The head of the family is $\circ$, its trace is $(-,-) \longmapsto -$, and so we need the tonicity postulates:

$$\begin{array}{lll} \textit{if } R\alpha_1\alpha_2\beta \ \&\ \alpha'_1 \sqsubseteq \alpha_1, & \textit{then } R\alpha'_1\alpha_2\beta, \\ \textit{if } R\alpha_1\alpha_2\beta \ \&\ \alpha'_2 \sqsubseteq \alpha_2, & \textit{then } R\alpha_1\alpha'_2\beta, \\ \textit{if } R\alpha_1\alpha_2\beta \ \&\ \beta \sqsubseteq \beta', & \textit{then } R\alpha_1\alpha_2\beta'. \end{array}$$

18 Semilattice-Ordered Gaggles and Lattice-Ordered Gaggles with Involution

We next examine the case of a partial gaggle, where the partial ordering is a semilattice. We shall then examine the case where the partial ordering is a lattice, not necessarily distributive. A good way of looking at this last is to build upon the representation of lattices in Urquhart (1978),

18 In the simpler case where we are working with a tonoid and not a partial gaggle, we let f be any arbitrary operation.

and this is carried out in Allwein and Dunn (199+) for the case of binary operations, with special application to linear logic (without exponentials) and the non-distributive relevance logic **LR**. We say that this is a good way because it does not presuppose the presence of any special operations beyond the lattice operations. But there is a less troublesome opportunity in the presence of a sufficiently strong negation, and this route has been in effect utilized by Goldblatt (1974) with his completeness theorem for orthologic and by Girard (1987) for linear logic.[19] It is this special case that we shall examine here.

Representation Theorem for Semilattice-Ordered Partial Gaggles. *Every semilattice-ordered partial gaggle* $(A, \leq, \wedge, OP)$ *can be represented using a frame* $(U, \langle R_i \rangle_{i \in I})$, *as for a tonoid.*

Proof. The proof is related to that for partial gaggles, where we used principal cones and the complements of dual cones. But the proof must be modified. Principal cones clearly have the property that $\wedge X \in C$ *iff* $\forall x \in X$, $x \in C$, and so sending each element a into the set of cones of which it is a member assure that greatest lower bounds get carried into intersections. But the problem is that the complements of principal dual cones do not have this property. The construction must be modified along the lines of that for C_2 in the Remark of section 11. We shall not do this here, but that Remark in fact suffices for the special applications we have here.

We shall view negation rather abstractly, as a kind of Galois connection (cf. Dunn (199+)). Given two partially ordered sets $(P, \leq)$ and $(Q, \leq')$, a pair of functions $(\neg, -)$ is called a *Galois connection between* P *and* Q (cf. Birkhoff) *iff* $\neg : P \longmapsto Q$, $- : Q \longmapsto P$, *and*

$$x_1 \leq x_2 \textit{ implies } \neg x_2 \leq' \neg x_1, \qquad y_1 \leq' y_2 \textit{ implies } -y_2 \leq -y_1, \qquad (\textit{order inverting, contraposition})$$

$$\textit{for } x \in P,\ x \leq -\neg x, \qquad \textit{for } y \in Q,\ y \leq' \neg - y. \qquad (\textit{Galois double negations})$$

It is easy to see that an equivalent definition would be to require

$$\neg : P \longmapsto Q, \qquad - : Q \longmapsto P,$$

19 We shall not bother to distintinguish here traditional Henkin-style completeness proofs using theories (sets of formulas) and representation theorems using filters. For the Lindenbaum algebra of a logic, filters and theories are the same modulo division by provable equivalence. Cf. Dunn (1975).

and for $x \in P$, $y \in Q$,

$$x \leq -y \quad \textit{iff} \quad y \leq' \neg x. \qquad \textit{(Galois property)}$$

Birkhoff cites J. Schmidt with no proof (cf. Dunn (199+) for a proof).

We prove the following theorem, implicit in Birkhoff (1967):

Theorem. *Let $(S, \leq, \wedge)$ be a semilattice and let $(-, \neg)$ be a Galois connection on the semilattice. For $a, b \in S$, define*

$$\begin{aligned} a \vee b &= -(\neg a \wedge \neg b), \\ L &= \{a \in S : a = -\neg a\} \quad \textit{(the set of } \textbf{\textit{Galois closed elements}}). \end{aligned}$$

Then L is closed under $\wedge$ and $(L, \leq, \wedge, \vee)$ is a lattice.

Proof. We first show that L is closed under $\wedge$. Suppose $a = -\neg a$ and $b = -\neg b$; we need to show that $a \wedge b = -\neg(a \wedge b)$. The inequality from left to right follows from the general property of a Galois connection that $x \leq -\neg x$. The inequality in the other direction is established as follows:

1. $(a \wedge b) \leq a$ *(lower bound)*
2. $\neg a \leq \neg(a \wedge b)$ *(contraposition)*
3. $-\neg(a \wedge b) \leq -\neg a = a$ *(contraposition and Galois closure)*
4. $-\neg(a \wedge b) \leq b$ *(similar)*
5. $-\neg(a \wedge b) \leq a \wedge b$ *(greatest lower bound).*

We next show that $\vee$ is in fact a join. First we show that $a \vee b$ is an upper bound of a (the proof for b is similar):

1. $\neg a \wedge \neg b \leq \neg a$ *(lower bound)*
2. $a = -\neg a \leq -(\neg a \wedge \neg b)$ *(contraposition and Galois closure)*
3. $a \leq a \vee b$ *(definition of $\vee$).*

Finally we show that $a \vee b$ is the least upper bound of a and b:

1. $a \leq x$ *(hypothesis that x is an upper bound of a)*
2. $b \leq x$ *(ditto for b)*
3. $\neg x \leq \neg a$ *(contraposition)*
4. $\neg x \leq \neg b$ *(contraposition)*
5. $\neg x \leq \neg a \wedge \neg b$ *(greatest lower bound)*
6. $-(\neg a \wedge \neg b) \leq -\neg x = x$ *(contraposition and Galois closure)*
7. $a \vee b \leq x$ *(definition of $\vee$).*

Representation Theorem for Galois Connections. *Let $(S, \leq, \wedge)$ be a semilattice and let $(-, \neg)$ be a Galois connection on the semilattice. Then there is some set U with a binary relation $\perp$ on U so that*

$$^{\perp}X = \{\chi : \forall\alpha(\alpha \in X \Rightarrow \chi \perp \alpha)\}$$

$$X^{\perp} \quad = \quad \{\chi : \forall\alpha(\alpha \in X \Rightarrow \alpha \perp \chi)\}$$

define a Galois connection on subsets of U. Further, there is some collection of subsets S' of U closed under intersection and the two unary operations defined above, and a 1-1 map $h : S \longmapsto S'$, so that h is an isomorphism, with $\leq$ carried into $\subseteq$, $\wedge$ into $\cap$, and

$$\begin{aligned} h(-a) &= {}^{\perp}X, \\ h(\neg a) &= X^{\perp}. \end{aligned}$$

Proof. That the operations define a Galois connection is to be found in Birkhoff (1967). Otherwise the theorem is a direct consequence of the above representation for semilattice-ordered partial gaggles. Nonetheless it is worthwhile to examine the essence of its proof.

The idea is to let U be the set of filters and define a relation $\perp$ on filters so that $F \perp G$ *iff*, for some element a, $a \in F$ and $-a \in G$. One could just as well define the 'dual' relation $\perp'$ requiring that $a \in G$ and $\neg a \in F$. We observe that $\perp'$ is just the converse of $\perp$. Thus if $F \perp H$, then by definition for some element a, $a \in F$ and $-a \in G$, but then (by virtue of the relationship $a \leq \neg - a$) for some element b $(= -a)$, $b \in G$ and $\neg b \in F$, *i.e.*, by definition $G \perp F$. The argument is reversible (using the relationship $b \leq -\neg b$).

The key is then to prove that $h(-a) = {}^{\perp}(ha)$, *i.e.*,

$$-a \in F \quad \textit{iff} \quad \forall G(a \in G \quad \textit{implies} \quad F \perp G).$$

Left-to-right is immediate. We prove right-to-left using contraposition, assuming that $-a \notin F$. We then need a filter G such that $a \in G$ and yet not $F \perp G$. Let G be the principle filter $[a)$. Clearly it is not the case that $F \perp G$, for otherwise for some $b \in F$, $a \leq \neg b$. But then $b \leq -a$, and so $-a \in F$. The proof that $h(\neg a) = (ha)^{\perp}$ is obviously symmetric.

We will state below an interesting corollary of the above theorem, once we give a couple of definitions. Let us consider a special case of a Galois connection on a semilattice when $- = \neg$. This amounts to looking at a semilattice $(S, \leq, \wedge, -)$ with so-called 'minimal negation' satisfying 'contraposition' ($a \leq b$ *only if* $-b \leq -a$) and 'constructive double negation' ($a \leq --a$).

Corollary 1. *A semilattice with minimal negation $-$ can be represented as in the Representation Theorem for Galois Connections above using a symmetric relation $\perp$.*

Proof. When $\perp$ is symmetric, $\perp = \perp'$; so $- = \neg$, and so from $(U, \perp)$ we get a semilattice with minimal negation. On the other hand, the relation $\perp$ defined canonically on cones can be seen to be symmetric. Thus $F \perp H$ *iff*, for some element a, $a \in F$ and $-a \in G$. But since $a \leq --a$, we have for some element b, $b\ (= -a) \in G$, $-b\ (= --a) \in F$.

By an *involuted lattice* we mean a structure $(L, \wedge, \vee, -)$ where $-$ is a unary operation on L that is order inverting (contraposition) and of period two ($a = --a$, *i.e.*, 'classical double negation'). It is tempting here to speak of a 'De Morgan lattice' since $-$ can be characterized as satisfying the De Morgan laws (including those with implicit double negation), but this has been reserved in the literature for the case where the lattice is distributive (cf. Dunn (1966)).

Corollary 2. *Let $(L, \wedge, \vee, -)$ be an involuted lattice. This can be represented using a symmetric $\perp$, restricting the representing sets to those which are 'Galois closed', i.e., $X^{\perp\perp} = X$.*

Proof. Let $(U, \perp)$ be a symmetric frame and let $\mathcal{P}(U)^*$ be the Galois closed subsets of U. We know from Corollary 1 that $^{\perp}$ already has all the properties of minimal negation, and the restriction to Galois closed subsets is precisely to assure 'classical double negation'. Incidentally, we also have to observe that the representing function h carries elements into Galois closed sets, but this is obvious since $a = --a$; so $h(a) = h(--a) = (ha)^{\perp\perp}$.

An *ortholattice* is an involuted bounded lattice in which $a \wedge -a = \mathbf{0}$ (and so $a \vee -a = \mathbf{1}$).

Corollary 3 (Goldblatt (1975)). *Every ortholattice can be represented, as in Corollary* 2, *by adding the requirement that $\perp$ is irreflexive.*

Proof. When never $a \perp a$, it is easy to see that $X \cap X^{\perp}$ is empty. In a canonical frame we never have $F \perp F$, for if so, then for some element a we have both $a, -a \in F$, and hence $a \wedge -a = \mathbf{0} \in F$. But then F is no longer proper. Also note that $h(\mathbf{0}) = \emptyset$.

Corollary 4. *Every Girard monoid can be represented, with $\wedge$ and $-$ represented as in Corollary* 2, *and with $\circ$ and $\rightarrow$ represented as with gaggles.*

Proof. For the definition of a *Girard monoid* cf. Allwein and Dunn (199+). The result is related to similar representations given in effect by Girard and by Avron. Here we need to understand the representation theorem as applying to a partial gaggle with a (monoid) identity, but this can be accommodated as in section 15.

19 Other Applications

There are several topics that can be further developed. One has to do with applying distributoids, gaggles and their partial versions to the various logics mentioned in section 2. This has already been carried out in some detail for modal, intuitionistic and relevance logics, and involves 'fiddling with the canonical representation' to get the usual results, which do not always involve an accessibility relation of the same degree as that employed in the gaggle representation. Thus with the intuitionistic logic the usual Kripke semantics uses an accessibility relation of degree two, whereas our representation uses a relation of degree three.

In this section we assume that there is an underlying distributive lattice, although as a later footnote makes clear this is not really necessary for the observations regarding intuitionistic implication and negation. We illustrate this with the definition of implication from the Kripke semantics for intuitionistic logic, which uses a binary accessibility relation that we denote by $\sqsubseteq$:

$$A \to_k B = \{\chi : \forall\beta,\ \textit{if}\ \ \chi \sqsubseteq \beta,\ \textit{then either}\ \beta \notin A\ \textit{or}\ \beta \in B\}.$$

The Kripke semantics also assumes the *Intuitionistic Hereditary Condition*: *if* $\alpha \in A$ & $\alpha \sqsubseteq \beta$, *then* $\beta \in A$.

The definition that comes out of the gaggle approach is rather

$$A \to_g B = \{\chi : \forall\alpha, \beta,\ \ \textit{if}\ R\alpha\chi\beta\ \&\ \alpha \in A,\ \textit{then}\ \beta \in B\}.$$

But in the canonical model $R\alpha\chi\beta$ means

$$\forall a, x,\ \ \textit{if}\ \ a \in \alpha\ \&\ x \in \chi,\ \textit{then}\ a \circ x \in \beta.$$

Since in intuitionistic logic,

$$a \wedge b \leq c\ \ \textit{iff}\ \ \ b \leq a \to c,$$

we can let $\circ = \wedge$. It is then easy to argue that in the canonical frame, $R\alpha\chi\beta$ *iff* $\alpha \subseteq \beta$ & $\chi \subseteq \beta$.

For left-to-right, assume that $R\alpha\chi\beta$, *i.e.*, $\forall a, x$, *if* $a \in \alpha$ & $x \in \chi$, *then* $a \wedge x \in \beta$. We show $\alpha \subseteq \beta$ by assuming that $a \in \beta$. Let $x = \mathbf{1}$ (the top element, which always exists in a Heyting lattice setting $\mathbf{1} = a \to a$). We then obtain $a \in \alpha$ & $1 \in \chi$ *only if* $a \wedge \mathbf{1} \in \beta$. But $\mathbf{1} \in \chi$ and $a \wedge \mathbf{1} = a$. The argument that $\chi \subseteq \beta$ is parallel, letting $x = \mathbf{1}$. For right-to-left, assume that $\alpha \subseteq \beta$ & $\chi \subseteq \beta$. We show that $R\alpha\chi\beta$ by assuming that $a \in \alpha$ & $x \in \chi$

and showing $a \wedge x \in \beta$. Our assumptions tell us that $a \in \beta$ and $x \in \beta$, and so, since β is a filter,[20] $a \wedge x \in \beta$.

We can now substitute $\alpha \subseteq \beta$ & $\chi \subseteq \beta$ for $R\alpha\chi\beta$ in the Kripke definition of implication, obtaining

$$\chi \in A \to_k B \ \textit{ iff } \ \forall\alpha, \beta, \ \textit{if } \alpha \sqsubseteq \beta \ \& \ \chi \sqsubseteq \beta \ \& \ \alpha \in A, \textit{ then } \beta \in B.$$

Now more fiddling is required. Let us assume the right-hand side,

$$(*) \ \forall\alpha, \beta, \ \textit{if } \alpha \sqsubseteq \beta \ \& \ \chi \sqsubseteq \beta \ \& \ \alpha \in A, \textit{ then } \beta \in B,$$

and show that

$$(**) \ \forall\beta, \ \textit{if } \chi \sqsubseteq \beta \ \& \ \beta \in A, \textit{ then } \beta \in B.$$

Given the reflexiveness of $\sqsubseteq$, this follows immediately, when we put β for α.

The converse is only slightly trickier. Let us assume (**) and the antecedent of (*), $\alpha \sqsubseteq \beta$ & $\chi \sqsubseteq \beta$ & $\alpha \in A$. We must now show that $\beta \in B$. Since $\alpha \in A$ and $\alpha \sqsubseteq \beta$, it follows from the Intuitionistic Hereditary Condition that $\beta \in A$. But since the antecedent of (*) also contains $\chi \sqsubseteq \beta$, we can apply (**) to obtain $\beta \in B$ as desired.

The reader by now is likely wondering about the relation of our gaggle-theoretic treatment of intuitionistic negation to the standard Kripke definition. This can be handled similarly by way of a 'fiddle' on the canonical model. Cf. Dunn (199+) for details. The reader is also referred to that paper for a comparison of the gaggle-theoretic treatment of De Morgan negation using a binary relation to that of the Białynicki-Birula/Rasiowa/Routleys definition using a unary function.

An especially cute 'fiddle' occurs with the case of Boolean algebras (classical logic) viewing Boolean complement as a distributoid operation since it distributes over join, changing it to meet (it also does the dual, but one is enough). One can get the properties of Boolean complement by postulating that the relation R is irreflexive ($X \cap X^{\perp} = \mathbf{0}$), symmetric (so $X^{\perp} = {}^{\perp}X$, and the properties of the negation of minimal logic result), and that the propositions X are 'closed' in the sense that if $\alpha \in X$ and $\alpha\overline{R}\chi$, then $\chi \in X$ ($X \cup X^{\perp} = \mathbf{1}$). We verify that all of these properties hold in the canonical representation in terms of prime filters. One finally observes that the use of the relation R is otiose: prime (proper) filters in Boolean algebras are maximal filters, and one can show that they are

20 It greatly simplifies things here to have the assumption that we have at least an underlying semilattice so that our states can be assumed to be filters and not just cones. If we have merely an underlying partial order our states have to be cones that are closed under *modus ponens*, as in the construction alluded to in the remark of section 11.

both consistent and complete with respect to Boolean complement, *i.e.*, $x \in F$ *iff* $x^{\perp} \notin F$. Since FRG is just $\exists x(x \in F \mathbin{\&} x^{\perp} \in G)$, it follows that FRG *iff* $F \neq G$. But then $G \in X^{\perp}$ *iff* FRG for every $F \in X$, *iff* $F \neq G$ for every $F \in X$, *iff* $G \notin X$. But then $X^{\perp}$ becomes ordinary set-theoretic complement (relative to the class of prime filters), and so we get the ordinary (Stone) representation for Boolean algebras out of the one delivered by the representation for distributoids.

References

G. Allwein and J. M. Dunn (199+), Kripke models for linear logic, *The Journal of Symbolic Logic* (forthcoming).

A. R. Anderson, N. D. Belnap *et al.* (1975), *Entailment: The Logic of Relevance and Necessity, Vol. 1*, Princeton (Princeton University Press).

G. Birkhoff (1940, 1948, 1967), *Lattice Theory*, Providence (American Mathematical Society).

T. S. Blyth and M. F. Janowitz (1972), *Residuation Theory*, New York (Pergamon Press).

J. Certaine (1943), *Lattice-Ordered Groupoids and Some Related Problems*, Doctoral Dissertation, Harvard.

H. B. Curry (1963), *Foundations of Mathematical Logic*, New York (McGraw-Hill).

K. Došen (1988–1989), Sequent-systems and groupoid models, *Studia Logica 47–48*, 353–385, 41–65.

——— (1992), A brief survey of frames for the Lambek calculus, *Zeitschrift für mathematische Logik und Grundlagen der Mathematik 38*, 179–187.

J. M. Dunn (1966), *The Algebra of Intensional Logics.* Doctoral Dissertation, University of Pittsburgh, Ann Arbor (University Microfilms). Portions relevant to this paper are reprinted in Anderson, Belnap, *et al.* as §8 and §28.2.

——— (1986), Relevance logic and entailment, in *Handbook of Philosophical Logic, Vol. III*, eds D. Gabbay and F. Guenthner, Dordrecht (D. Reidel Publishing Company), 117–224.

——— (1991), Gaggle theory: An Abstraction of Galois connections and residuation with applications to negation and various logical operations, *Logics in AI, Proceedings European Workshop JELIA 1990*, LNCS 478, Berlin (Springer).

——— (199+), Perp and star: Two treatments of negation, *Philosophical Perspectives* (*Philosophy of Language and Logic*) *8*, ed. J. Tomberlin (forthcoming).

C. J. Everett (1944), Closure operators and Galois theory in lattices, *Transactions of the American Mathematical Society 55*, 514–525.

L. Fuchs (1963), *Partially Ordered Algebraic Systems*, New York (Pergamon Press).

J.-Y. Girard (1987), Linear logic, *Theoretical Computer Science 50*, 1–102.

R. I. Goldblatt (1974), Semantic analysis of orthologic, *Journal of Philosophical Logic 3*, 19–35.

B. Jónsson and A. Tarski (1951–52), Boolean algebras with operators, *American Journal of Mathematics 73–74*, 891–939, 127–162.

S. A. Kripke (1963), Semantical analysis of intuitionistic logic I, in *Formal Systems and Recursive Functions*, eds J. Crossley and M. Dummett, Amsterdam (North-Holland), 92-130.

J. Lambek (1958), The mathematics of sentence structure, *American Mathematical Monthly 65*, 153–172.

——— (1981), The influence of Heraclitus on modern mathematics, in *Scientific Philosophy Today*, eds J. Agassi and R. S. Cohen, Dordrecht (Reidel).

E. J. Lemmon (1966), Algebraic semantics for modal logics, *The Journal of Symbolic Logic 32*, 46–65, 191–218.

S. Mac Lane (1971), *Categories for the Working Mathematician*, New York (Springer).

R. K. Meyer (1982), The pure calculus of entailment is the pure calculus of entailment, in Routley with Plumwood, Meyer and Brady (1982), 425–429 (ms. approximately 1972).

R. K. Meyer and R. Routley (1972), Algebraic analysis of entailment, I, *Logique et Analyse N.S. 15*, 407–428.

H. Ono and Y. Komori (1985), Logics without the contraction rule, *The Journal of Symbolic Logic 50*, 169–201.

O. Ore (1944), Galois connexions, *Transactions of the American Mathematical Society 55*, 493–513.

R. Routley and R. K. Meyer (1972–73), The semantics of entailment, I, in *Truth, Syntax, Modality*, ed. H. Leblanc, Amsterdam (North-Holland), 199–243.

R. Routley with V. Plumwood, R. K. Meyer, and R. Brady (1982), *Relevant Logic and its Rivals, Part 1, The Basic Philosophical and Semantical Theory*, Atascadero, California (Ridgeview Publishing Company).

M. H. Stone (1936), The theory of representations for Boolean algebras, *Transactions of the American Mathematical Society 40*, 37–111.

——— (1937), Topological representations of distributive lattices and Brouwerian logics, *Časopis pro pěstování matematiky a fysiky 67*, 1–25.

A. Urquhart (1978), A topological representation theorem for lattices, *Algebra Universalis 8*, 45–58.

M. Ward and R. P. Dilworth (1939), Residuated lattices, *Transactions of the American Mathematical Society 45*, 335–54.

A General Theory of Structured Consequence Relations

Dov M. Gabbay

Department of Computing
Imperial College of Science, Technology and Medicine, London[1]

1. Introduction

We want to give an answer to the following question:

What is a logical system?

Obviously our starting point should be a language which allows us to express declarative statements and we need to give a formal definition of when we accept a reasoning system based on the language as a logical system.

The traditional answer to that (say in 1975) was that to present a logical system we need to define the well formed formulas of the language, say **L**, and further define a consequence relation for **L**. This is a relation between finite sets of formulas Δ, Γ of **L**, written $\Delta \vdash \Gamma$, satisfying the properties of Reflexivity, Monotonicity and Cut.

Reflexivity: $\Delta \vdash \Gamma$ if $\Delta \cap \Gamma \neq \varnothing$

Monotonicity: $\Delta \vdash \Gamma$ implies $\Delta, \Delta' \vdash \Gamma, \Gamma'$

Cut[2]**:** $\Delta \vdash A; \Delta, A \vdash \Gamma$ imply $\Delta \vdash \Gamma$

---

1 This paper is part 3 in a series of papers entitled 'Theoretical Foundations for Non-monotonic Reasoning', the other parts are [9] and [12]. The research reported here was supported by SERC project *Syntactical Foundations for Non-monotonic Reasoning*, GR/G46671. I am grateful to Seiki Akama, Jon Barwise, Heinrich Wansing and the anonymous referees for their critical reading of earlier drafts of this manuscript.

2 There are several versions of the **Cut Rule** in the literature, they are all equivalent for the cases of classical and intuitionistic logic but are not equivalent in the context of

The consequence relation may be defined in various ways. Either through an algorithmic system $\mathbf{S}_{\vdash}$,or implicitly by postulates on the properties of $\vdash$.

Thus a logic is obtained by specifying $\mathbf{L}$ and $\vdash$. Two algorithmic systems $\mathbf{S}_1$ and $\mathbf{S}_2$ which give rise to the same $\vdash$ are considered the same logic.

It soon became apparent, through the applications of logic in the analysis of language and in Artificial Intelligence, that there are many reasoning systems which we would intuitively like to admit into the family of logics, which are not covered by the above definition. These include a variety of non-monotonic systems, families of resource logics and a large class of probabilistic and network systems.

These reasoning systems are mostly presented algorithmically, as proof systems $\mathbf{S}$, and the majority of them give rise to a consequence relation which does not satisfy the postulates above. Further, there are many different proof systems for the *same* consequence relation, which are of such different motivation, mathematical presentation and intuitions, that one does not feel it is correct to identify them as the *same* logical systems.

It was obvious that we needed to modify our concept of a logical system in response to current reasoning practices.

The first attempt in this direction was in 1985, in [9], where we proposed that a consequence relation should only be required to satisfy *Restricted Monotonicity* and not full monotonicity, namely:

Restricted Monotonicity $\Delta \vdash A, \Delta \vdash \Gamma$ imply $\Delta, A \vdash \Gamma$.

The above weakening on the conditions of the notions of a consequence relation allows one to accept more reasoning systems as logical systems.

This idea was further developed in the literature by several authors, thus creating the new area of *axiomatic non-monotonic reasoning*.

Makinson [23] proposed a semantics for the above basic logic (i.e. the smallest consequence relation satisfying the above three conditions) and proved a completeness result. Kraus, Lehmann and Magidor [19] studied a variety of extensions of the basic system, providing them with preferential

this paper. The version in the main text we call **Transitivity** (**Lemma Generation**). Another version is:

$$\frac{\Gamma \mid\!\sim A \quad \Delta, A \mid\!\sim B}{\Delta, \Gamma \mid\!\sim B}.$$

This version implies **monotonicity**, when added to **Reflexivity**. Another version we call **Internal Cut**:

$$\frac{\Delta, A \mid\!\sim \Gamma \quad \Delta \mid\!\sim A, \Gamma}{\Delta \mid\!\sim \Gamma}.$$

semantics in the spirit of Shoham [28, 29]. Several other authors continued to develop this area, among them Wójcicki [35, 36], Lehmann [21], Lehmann and Magidor [22], Besnard [6] and Akama [2]. A general survey is available in Makinson [24].

The above notion of consequence, although general enough to cover many non-monotonic systems such as circumscription and negation by failure, does exclude, unfortunately, such non-monotonic systems as probabilistic systems, resource logics such as linear logic and other such systems such as inheritance networks. Obviously a more general notion is required, a notion which can unify the probabilistic approach with the rule based approach; to name one important need.

On the algorithmic front, we have no alternative but to accept that different proof systems should be considered as different logics, even though they give rise to the same consequence relation. This view was put forward in 1983, and supported at length in [10, 11, 13].

Thus the situation in 1988–89 was that although we were able to accept more systems into the family of logics we still had to exclude many well known reasoning systems.

Furthermore, there was no unifying framework and a satisfactory general answer to what is a logical system. In 1989, we turned to proof theory [10] and introduced the notion of *Labelled Deductive Systems* (*LDS*).

This is a systematic framework where the logical units are labelled formulas (of the form $t : A$, where t is a term in some algebra) and where the logical rules are rules for manipulating both formulas and their labels. The intuition behind *LDS* is that in $t : A$, A carries declarative information and the label t, represents further information of a different nature which we do not want to 'code' into A. It turns out that many monotonic, non-monotonic and probabilistic systems can be presented as *LDS* systems. An *LDS* database Δ is a constellation of labelled formulas with additional structure on the labels. For example Δ may be an algebra τ with a function $\delta(x)$, $x \in \tau$, assigning formulas $A_x = \delta(x)$, for each $x \in \tau$.

Proof rules are given to define the notion $\Delta \vdash s : B$. This approach is developed in detail in [10]. It is possible to have $\Delta \vdash s_i : B$ for several labels s_i, for example s_i may show from which part of Δ the formula B is proved. In many applications one is interested in the notion of $\Delta \vdash B$, where we want to ignore the label and just give a yes or no answer to the query 'Does B follow from Δ'. This can be done in several ways. The process of extracting (through some proof method) a decision of the form $\Delta \vdash B$ or $\Delta \vdash \sim B$ or neither, from the set $\{t \mid \Delta \vdash t : B\} \cup \{t \mid \Delta \vdash t : \sim B\}$ is called *Flattening* (see example 1.4). Obviously the flattening process is functionally dependent on the structure of Δ (as induced by the labels). The flattening process is actually *hiding* the labels and all it really uses is a structured database Δ proving or not proving a formula B. We can in this case define axiomatically the properties of the consequence $\Delta \vdash B$. We thus end up with a notion of consequence between Δ and B, where Δ is structured and the labels in Δ are hidden, and used only in the proof system. It turns out that this simplified notion is very useful and can be developed on its own as a direct extension of the notion of monotonic consequence relation. We call our studies the area of *structured consequence relations*. It can be developed independently of *LDS*, as a direct contribution to the abstract area of *axiomatic non-monotonic reasoning*.

The purpose of this paper is to develop the notion of structured consequence relation and further refine the notion of a non-monotonic consequence relation to relate to non monotonic systems arising from resource boundedness, and controlled inference. The refinement is in the direction of structuring the assumptions. Some pioneering work in the direction of looking at structural connectives can already be found in [5].

Many non-monotonic databases are naturally structured. The database may be, for example, a list of assumptions, or a partially ordered set of assumptions, and consequences are proved from the database using an inferential (logical) discipline which takes into account such a structure. Non-monotonicity may arise because of these 'resource' considerations and inferential restrictions.

We begin our study with several motivating examples.

Example 1.1. (Concatenation Logic) A database in this logic is a sequence of formulas $(A_1, \ldots, A_n)$. The inference rule is modus ponens. To derive B from the database we must use all the assumptions in the order shown. Further, when we use modus ponens $A, A \to B \vdash B$, the implication $A \to B$ must be supported by assumptions earlier in the database sequence than all those which support the minor premise A. Thus, for example

(a_1) $A \to (B \to C)$

(a_2) B

(a_3) A

does not prove C because we have to start by using (a_1) and (a_3) 'jumping over' (a_2). However,

(b_1) $A \to (B \to C)$

(b_2) A

(b_3) B

does prove C because (b_1) and (b_2) give $(B \to C)$ then (b_3) is used and we get C.

Another example is:

(c_1) $A \to (A \to B)$

(c_2) A

This database does not prove B because (c_2) needs to be used twice. However, the database:

(d_1) $A \to (A \to B)$

(d_2) A

(d_3) A

does prove B.

Another example is:

(e_1) $A \to (A \to B)$

(e_2) A

(e_3) A

(e_4) C

This database does not prove B because (e_4) is not used. Even if (e_4) were B (i.e. $C = B$) we still could not derive B because if we use (e_4) we are not using all the assumptions and we are not starting our proof at the first one. The alternative, using (e_1) - (e_3) is also not valid because (e_4) is not used.

Concatenation logic has applications in any area where the database is perceived as a list, either for conceptual reasons or for computational reasons.

As an example we consider its applications to Prolog. Consider the database

1. $q \to q$
2. q

and the query $?q$.

Prolog implementations will store the database as a list and compute the query from the top of the list. This query will therefore loop. In concatenation logic, the body of clause 1 needs to be asked from clause 2. Thus concatenation logic gives a discipline for the Prolog pointer.

We continue and present two more examples.

Example 1.2. (Ordered Logic) This area in non monotonic reasoning was started by D. Vermeir. An ordered logic database is an ordered set of clauses. The following diagram (Figure 1) describes an ordered logic database:

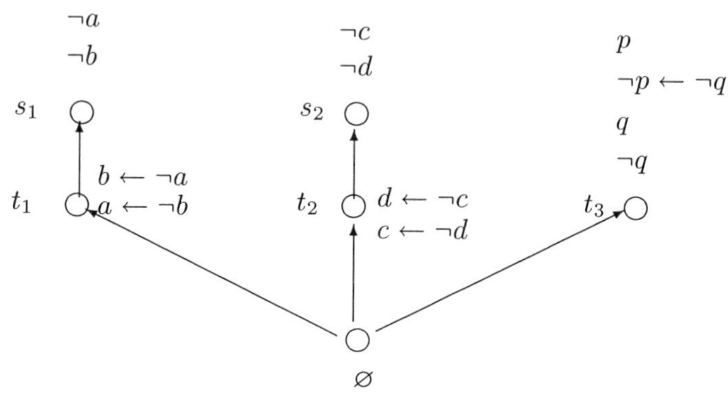

Figure 1

The nodes are t_1, t_2, t_3, s_1, s_2 and $\varnothing$ and are partially ordered as in the figure. This induces ordering on the clauses.

To motivate such databases, consider an ordinary logic program $C_1 = \{p \leftarrow \neg q\}$. The computation of a logic program assumes that, since q is not a head of any clause, $\neg q$ is part of the data, (this is the *closed world assumption*). Suppose we relinquish this principle and adopt the principle of asking an *advisor* what to do with $\neg q$. The advisor might say that $\neg q$ succeeds or might say that $\neg q$ fails. The advisor might have his own program C_2, to consult. He might run the goal q (or $\neg q$), look at what he gets and then advise. To make the situation symmetrical and general we must allow for Logic programs to have rules with both q and $\neg q$ (i.e. literals) in heads and bodies and have any number of negotiating advisors. Thus we can have $C_2 = \{\neg q\}, C_1 = \{p \leftarrow \neg q\}$ and C_1 depends on C_2. Ordered logic develops and studies various aspects of such an advisor system which is modelled as a partially ordered set of theories. Such a logic is useful, e.g. for multi-expert systems where we want to represent the knowledge of several experts in a single system. Experts may then be ordered according to an 'advisory' or a relative preference relation.

A problem to consider is what happens when we have several advisors that are in conflict. For example, C_1 depends on C_2 and C_1 depends on C_3. The two advisers, C_2 and C_3, may be in conflict. One may advise $\neg q$, the other q. How do we decide? There are several options:

1. We can accept q if all advisors say yes to q.
2. We can accept q if at least one advisor says yes to q.
3. We can apply some non-monotonic or probabilistic mechanism to decide.

If we choose options (1) or (2) we are essentially in modal logic. To have a node t and to have $?q$ refer to advisors $t_1, \ldots, t_n$ with $t < t_i$, $i = 1, \ldots, n$ is like considering $?\Box q$ at t in modal logic with $t_1, \ldots, t_n$ possible worlds in option (1) and like considering $\Diamond q$ at t in option (2). Option (3) is more general, and our new approach is most useful. We see from this advisors examples an application area where the structure arises naturally and usefully. The area of ordered logic is surveyed in [32].

Example 1.3. (Defeasible Logic) This important approach to non-monotonic reasoning was introduced by Nute [26]. The idea is that rules can prove either an atom q or its negation $\neg q$. If two rules are in conflict, one proving q and one proving $\neg q$, the deduction that is stronger is from a rule whose antecedent is logically more specific. Thus the database:

$$\begin{array}{l} \text{Bird } (x) \rightarrow \text{ Fly } (x) \\ \text{Big } (x) \wedge \text{ Bird } (x) \rightarrow \neg \text{ Fly } (x) \\ \text{Big } (a) \\ \text{Bird } (a) \end{array}$$

will entail $\neg$ Fly (a) because the second rule is more specific.

We can view the above as a structured database Δ in which the ordering of the rules is done by the logical strength of their antecedents relative to some background theory Θ (which can be a subtheory of Δ of some form). Deduction pays attention to the strength of rules.

Example 1.4. (Application to Planning) The notion of a structured database allows one to give a logical modelling to actions. To give an idea of how it is done, consider a blocks world example. Assume three blocks a, b and c and the initial state (model) Δ_1 to be {on (a, c), on $(table, b)$, on $(table, a)$}. We move c from the top of a by our first action α and put it on the table and then move it again and put it on b by action β. The question is can we model the pair $(\Delta_1, \alpha + \beta)$? Clearly after executing α we get Δ_2 = { on $(table, a)$, on $(table, b)$, on $(table, c)$} and after executing β we get Δ_3 = {on $(table, a)$, on $(table, b)$, on (b, c)}. If we want one database to model $(\Delta_1, \alpha + \beta)$, we can use the following structured database, where $0, \alpha, \beta$ are labels.

0. on (a, c)

0. on $(table, b)$

0. on $(table, a)$

α: on $(table, c)$

β: on (b, c)
$0 < \alpha < \beta$.

If we ask the query ?(on (x, c) we get three answers with three labels: 0: on (a, c); α: on $(table, c)$, , $c)$; β: on (b, c), $0 < \alpha < \beta$.

The situtation is no different from the previous example where we get Fly(a) and $\neg$Fly(a) from different parts of the database. The relative strength of the data will determine which answer is accepted.

To sum up, we are saying we can model actions by adding their results to the data to form a structured database.

Example 1.5. (Probabilistic Nets) A typical example is a network of dependencies of the form

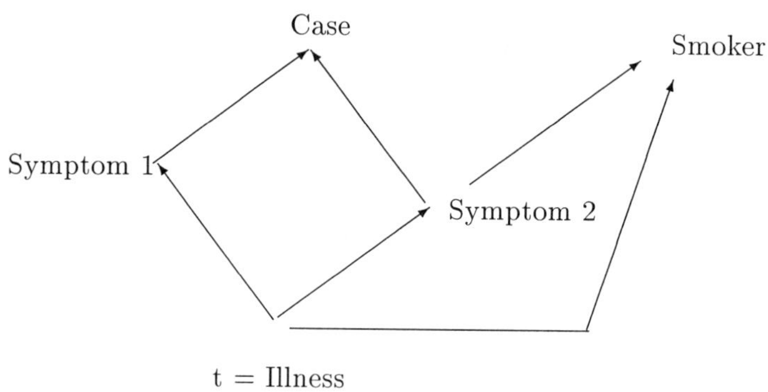

There are various probabilistic dependencies in the network and one can put in numerical values at the nodes. If some of the values are known, the network can predict values at the remaining nodes. Thus

$$\text{network} \vdash \text{value } x\text{: node } t$$

means that the value x is predicted at the node t.

Suppose we put in a value y at the node Smoker. On the basis of that and the other values we get a value x at node t. Suppose the value y: Smoker was obtained from another network, for example

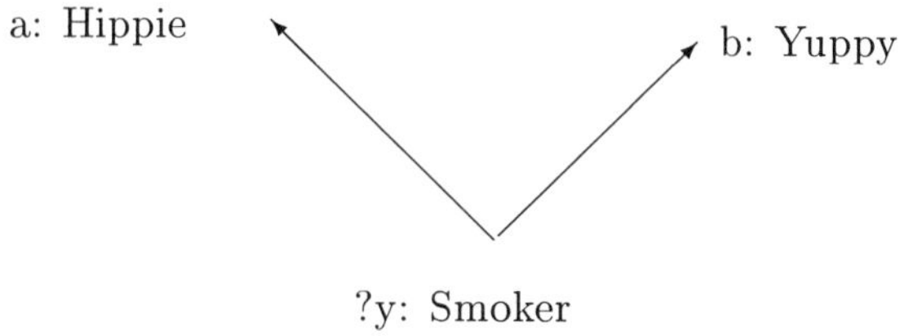

on the grounds that hippies smoke because they are hippies and yuppies smoke because they are stressed.

The Surgical Cut rule means that if we replace in the first network the node 'Smoker' by the second diagram to obtain a new more complex network, then the complex network will give the same answer at t.

A notion of proper replacement (or Substitution) needs to be defined, which will include adjustments of the probability dependencies between nodes. If such adjustment is possible, then **Surgical Cut** is satisfied. There are working systems of this kind satisfying **Surgical Cut**. One such a commercial system is HUGIN, based on [20].

We now introduce the new notion of a non-monotonic consequence relation, which we call S-consequence relation (Structured Consequence Relation). A precise definition will be given in a later section. We need three auxiliary notions for our definition:

1. We have already indicated that we need to deal with structured databases. We thus have to specify precisely what structures are allowed as databases, for the consequence relation to be defined. These must include the one formula database (A) and the empty database $\varnothing$.

2. The notion of a structured database must also include the concept of *structured addition of data*. By this we mean the notion of how to combine two databases together. The traditional way is to take their union but when the databases are already structured, the union has to be structured as well. We denote the structured addition of the databases Δ and Γ by $\Delta + \Gamma$. This binary operation, '+', has to be defined as part of the concepts underlying the consequence relation. It usually satisfies associativity

$$\Delta_1 + (\Delta_1 + \Delta_3) = (\Delta_1 + \Delta_2) + \Delta_2$$

and the empty database must satisfy

$$\varnothing + \Delta = \Delta + \varnothing = \Delta.$$

Commutativity is not required to hold.

3. We also need a concept of substitution of one structured database Δ inside another, Γ, achieved by replacing one formula A in Γ by Δ. Formally we need to make sense of the symbol $\Gamma[A]$, reading Γ is a structured database which contains A somewhere inside, and the symbol $\Gamma[\Delta]$, which denotes the database resulting from substituting Δ for A inside Γ. These notions are needed to formulate the Cut Rule.

We can now define the three rules which we call **Identity**, **Surgical Cut** (or **Substitutional Cut**) and **Directional Monotonicity** as follows:

Identity

$$(A) \mathrel{|\!\sim} A$$

Surgical (Substitutional) Cut

$$\frac{\Delta \mathrel{|\!\sim} A \qquad \Gamma[A] \mathrel{|\!\sim} B}{\Gamma[\Delta] \mathrel{|\!\sim} B.}$$

Directional Monotonicity (right hand side)

$$\frac{\Delta \mathrel{|\!\sim} A}{\Delta + \Gamma \mathrel{|\!\sim} A}$$

Directional Monotonicity (left hand side)

$$\frac{\Delta \mathrel{|\!\sim} A}{\Gamma + \Delta \mathrel{|\!\sim} A}$$

Definition 1.6. (Tentative Definition of Structured Consequence Relation) Let **L** be a language. Let $\mathcal{L}$ be a family of order types of the form $(\tau, \leq)$. Assume that the empty set and one element set types are in $\mathcal{L}$. Assume that the notion of substitution of one order type τ for a point x in another order type $\tau'(x)$ is well defined (as in (3) above).

We say that a relation $\Delta \mathrel{|\!\sim} A$, (between ordered multisets Δ of wffs of an order type τ and a single wff A) is a *S-consequence relation* iff it satisfies **Identity** and **Surgical Cut**.

Example 1.7. (Consequence presentation of CL) Consider a language with $\to$. Consider data structures in the form of lists $\Delta = (A_1, \ldots, A_n)$. An S-consequence relation on pairs (Δ, A), written as $\Delta \Vdash A$, is any relation satisfying the following two conditions.

1. **Identity**

$$(A) \Vdash A$$

2. **Surgical (Substitutional) Cut for (the data structure) Lists**

$$(A_1, \ldots, A_n) \Vdash A$$

$$\frac{(C_1, \ldots, C_m, A, D_1, \ldots, D_k) \Vdash B}{(C_1, \ldots, C_m, A_1, \ldots, A_n, D_1, \ldots, D_k) \Vdash B}$$

Let $\Vdash_{\mathbf{CL}}$ be the smallest S-consequence relation satisfying the *right hand side deduction theorem* for the data structure namely:

Right-hand side Deduction Theorem:

$$(A_1, \ldots, A_n) \Vdash A \to B \text{ iff } (A_1, \ldots, A_n, A) \Vdash B$$

Note that we need to show that such a smallest $\Vdash$ exists. This is simple because the family of S-consequence relations for lists which satisfy **Identity**, **Surgical Cut** and **Right hand side Deduction Theorem** is closed under intersections.

The above example illustrates two points:

1. How to define a logic by conditions on its consequence relation and its data structures.

2. The notions of *a* Cut Rule and *a* Deduction Theorem depend on the data structures.

To further illustrate our ideas, note that the well-known intuitionistic logic can be defined in a similar manner, as shown next.

1. *Intuitionistic Consequence*
   Data structures are sets of wffs. $\Delta \Vdash A$ is the smallest consequence relation on the data structure satisfying Reflexivity, Monotonicity, Cut and the Deduction Theorem.

2. *R-mingle Consequence*
   Data structures are sets of wffs. $\Delta \Vdash A$ is the smallest consequence relation satisfying Identity, Cut and the Deduction Theorem.

3. *Linear Consequence* (**LL**)
   Data structures are multisets of formulas. $\Delta \Vdash A$ is the smallest Consequence Relation satisfying Identity, Cut and the Deduction Theorem.

Notice that the difference between R-mingle and linear consequence is only in the data structures.

Example 1.8. (Beamtenlogik) This is 'civil servant' logic. The answer to a query from a file is yes only if the query is explictly listed in the file. Formally the datastructures are sets of formulas and we define

$$\Delta \Vdash A \quad \text{iff} \quad A \in \Delta.$$

Example 1.9.

1. Let **L** be the language of classical or intuitionistic logic and let the data structures be sets. Define a consequence relation $\mid\!\sim_{H^0}$ by $\{A_1, \ldots, A_n\} \mid\!\sim_{H^0} B$ iff $B \vdash_H \bigvee_{i=1}^n A_i$ where $\vdash_H$ is provability in classiccal logic ($H = C$) or intuitionistic logic ($H = I$). We show that $\mid\!\sim_{H^0}$ is an S-consequence relation.

   It is obvious that it satisfies **Identity**, we check the **Surgical Cut** rule.

   Assume, for example, that $\Delta, A \mid\!\sim_{H^0} B$ and $\Gamma \mid\!\sim_{H^0} A$. We must show that $\Delta, \Gamma \mid\!\sim_{H^0} B$. From our assumptions we have $B \vdash_H A \vee \bigvee \Delta$ and $A \vdash_H \bigvee \Gamma$.

   We get therefore

   $$B \vdash_H \bigvee \Gamma \vee \bigvee \Delta$$

   hence by definition

   $$\Delta, \Gamma \mid\!\sim_{H^0} B.$$

   Note that the following holds for all A, B

   $$A \vdash_H B \text{ iff } B \mid\!\sim_{H^0} A.$$

2. Define another consequence relation $\mid\!\sim_{H^1}$ by $\{A_1, \ldots, A_n\} \mid\!\sim_{H^1} B$ iff for some A_i, $B \vdash_H A_i$. We show that $\mid\!\sim_{H^1}$ is a consequence relation. It is obvious that it satifies **Identity**. We check the **Surgical Cut**.

   Assume $\Delta, A \mid\!\sim_{H^1} B$ and $\Gamma \mid\!\sim_{H^1} A$. We must show that $\Delta, \Gamma \mid\!\sim_{H^1} B$. From our assumptions, $B \vdash_H X$, for some $X \in \Delta \cup \{A\}$ and $A \vdash_H Y$ for some $Y \in \Gamma$. We need to show that $B \vdash_H Z$ for some $Z \in \Delta \cup \Gamma$. Such Z exists because:

   If $X \in \Delta$ let $Z = X$
   If $X = A$ let $Z = Y$.

   Clearly for all A, B,

   $A \vdash_H B$ iff $B \mid\!\sim_{H^1} A$.

2. Structured Consequence Relations: A General Notion of a Logical System

We have seen that the general notion of a database is a configuration of formulas. The proof discipline from such databases would rely on the configuration in defining the allowable proof moves.

This section develops the general notions needed for S-consequence relation. We begin with definitions and examples.

Definition 2.1. 1. Let **L** be a language for well formed formulas. **L** will contain atomic propositions and connectives and the notion of a wff is defined inductively in the traditional way.

2. Let $\mathcal{L}$ be a theory in first or higher order logic and let $\mathcal{M}$ be a class of classical models of $\mathcal{L}$. We assume that the structure with one point and empty relations is in $\mathcal{M}$. This structure is denoted by $\ll t \gg$. Assume that two model theoretic operations are defined on $\mathcal{M}$ and assume that $\mathcal{M}$ is closed under these operations. The operations are the following:

   (a) If τ_1 and τ_2 are two structures in $\mathcal{M}$ then an operation $+$ is available for constructing the structure $\tau = \tau_1 + \tau_2$. '+' usually satisfies associativity, but not necessarily in the general case. If $\varnothing$ is the empty set then $+$ is defined and $\varnothing + \tau = \tau + \varnothing = \tau$.

   (b) Let τ_1 and τ_2 be two disjoint structures in $\mathcal{M}$ and let $x \in \tau_1$ be a point in τ_1. Then a binary substitution function $Sub^x_{\tau_2}\, \tau_1(x)$ is defined which yields the result of substituting τ_2 in $\tau_1(x)$ for x (i.e. $\lambda x \tau_1(x)[\tau_2]$ is well defined). The result of the substitution is also in $\mathcal{M}$. *Sub* is associative, and behaves like substitution if done properly (no clash of variables).

   (c) For some purposes (see section on structural connectives) we need to allow for the notion of *deletion*. That is we must know how to delete the point x from $\tau(x)$. We can regard deletion as substituting the empty structure $\varnothing$ for x, i.e. $\lambda x \tau(x)[\varnothing]$.

   Note that (a)–(c) above are very general, almost saying nothing. For each particular case, '+' and '*Sub*' must be specifically defined. For our purpose, i.e. the study of general structured consequence relations, we do not need to know more. See [15] for a general algebraic example with a semantical interpretation.

3. A database is any pair (τ, δ) where $\tau \in \mathcal{M}$ and δ is a function assigning for each $x \in \tau$ a wff $\delta(x)$.

4. A consequence relation for the pair $(\mathcal{M}, \mathbf{L})$ is any relation $\mid\!\sim$ between databases $\Delta = (\tau, \delta)$ and single wffs A, written in the form $\Delta \mid\!\sim A$, satisfying the following:

   (a) **Identity**

   $$(A) \mid\!\sim A$$

   (b) **Surgical Cut for $\mathcal{M}$.** If

   $$\Delta = (\tau_1, \delta_1),\ \Gamma = (\tau_2, \delta_2),\ \tau_1 \cap \tau_2 = \varnothing,\ x \in \tau_2 \text{ and } \delta_2(x) = A$$

   and if further

   $$\Delta \mid\!\sim A$$
   $$\Gamma \mid\!\sim B$$

   then

   $$\Gamma[\Delta] = (\tau_3, \delta_3) \mid\!\sim B$$

   where

   $$\tau_3 = Sub^{x}_{\tau_1}\, \tau_2(x)$$

   and

   $$\delta_3(y) = \begin{cases} \delta_1(y) & \text{if } y \in \tau_1 \\ \delta_2(y) & \text{if } y \in \tau_2. \end{cases}$$

Example 2.2. 1. Let $(T, <_T)$ be an ordering, and let $t \in T$. Let $(S, <_S)$ be another ordering with $T \cap S = \varnothing$. We define the substitution $[(T, <_T) \mid t \mid (S, <_S)]$ to be the ordering with $T \cup S - \{t\}$ as the set of points and order $<$ defined by

$$x < y \quad \begin{array}{l} \text{if } x, y \in T \text{ and } x <_T y \\ \text{or } x, y \in S \text{ and } x <_S y \\ \text{or } x \in T \text{ and } y \in S \text{ and } x < t \\ \text{or } x \in S \text{ and } y \in T \text{ and } t < y. \end{array}$$

The idea of the definition is to take the ordering $(T, <_T)$ and to take a point $t \in T$ and take another ordering $(S, <_S)$ and to replace t by the entire set S. We now have an ordering with S in the middle. Thus all points of T which were before (after) t are now before (after) all the points of S. Inside S the ordering is the same as before, i.e. $<_S$.

2. Define $(T, <_T) + (S, <_S)$ to be the ordering with all elements of T coming before all elements of S.

3. Let **L** be a language. Assume that part of the definition of **L** is the definition of allowable databases of the form $\{t : A_t \mid t \in T\}$, where $(T, <_T)$ is an order configuration. Assume that the order configurations of **L** are closed under the substitutions defined in the above. Let $\Vdash$ be a relation defined in **L** between data structures and formulas. Then the Cut Rule for $\Vdash$ is defined to be the following:

   Surgical Cut Rule

   If $\{t : A_t \mid t \in T\} \Vdash B$
   and $\{s : B_s \mid s \in S\} \Vdash C$
   and $x \in S$ and $B_x = B$
   then $\{r : A_r \mid r \in S \cup T - \{x\}\} \Vdash y : C$

   where the configuration on $S \cup T - \{x\}$ is $[(S, <_S) \mid x \mid (T, <_T)]$ defined above.

Example 2.3. Consider modal logic. A configuration is an ordered set of possible worlds. Thus if for example we have the configurations in Figures 2 and 3 then the application of the cut rule will yield the configuration of Figure 4

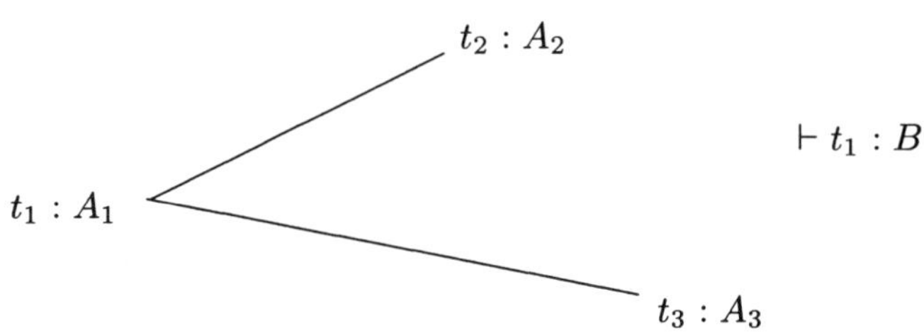

Figure 2

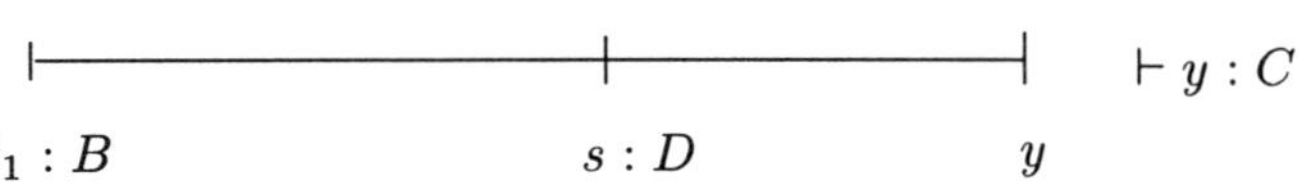

Figure 3

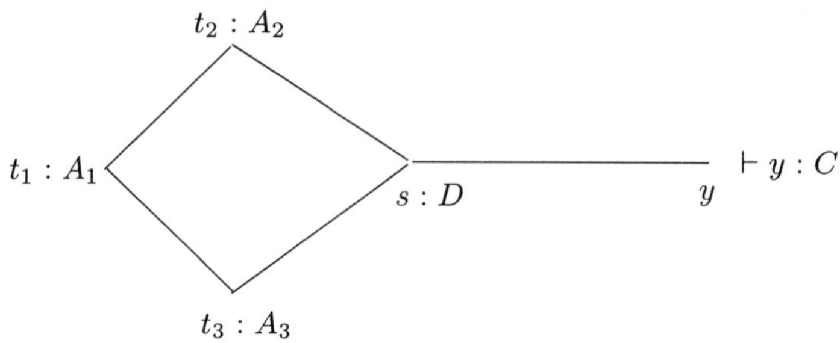

Figure 4

We now want to propose a form for the deduction theorem. The discipline must tell us how a database Δ can prove an implicational formula of the form $A \to B$.

Let us understand $A \to B$ as a sort of deductive 'functor' which 'given' A can 'yield' us B. Thus $\Delta \vdash_{\mathbf{CL}} A \to B$ means that if we 'add' A to Δ i.e. 'give' A to Δ, we can 'get' or 'yield' B from Δ 'and' A.

In general application areas, Δ is a structured database; some configuration of formulas. Thus the notion of adding A to Δ needs to be defined. We need to specify where to fit A in the configuration of Δ. In the general case precise definitions must be given. Since there may be more than one way of 'adding' A to Δ, we may have several possible versions of the deduction theorem as well as several possible implications.

Example 2.4. Consider **CL** (or examples 1.1 and 1.7) and its data structures of lists of wffs. Let $\Delta = (A_1, \ldots, A_n)$ and consider A of examples 1.1 and 1.7. This new data item can be 'added' to Δ in several ways:

1. Add A to the end of the sequence to form

$$(A_1, \ldots, A_n, A)$$

2. Add A to the beginning of the sequence to form

$$(A, A_1, \ldots, A_n)$$

3. Add A to the middle of the sequence to form

$$(A_1, \ldots, A_k, A, A_{k+1}, \ldots, A_n)$$

4. Of course we can have combinations, for example, add A to the middle or the beginning, etc

The **CL** deduction theorem allows for A to be added to the end of the sequence. We can thus write $A \to_e B$ for the **CL** implication to indicate that given A it can give B provided A is added to the end of the data sequence. The next connective to consider is the implication $A \to_b B$ where 'b' means 'beginning'. Here we expect a deduction theorem of the form

$$(A, A_1, \ldots, A_n) \vdash B,$$

iff

$$(A_1, \ldots, A_n) \vdash A \to_b B$$

For example, since

$$A \to_b B \vdash A \to_b B$$

we will get in this logic

$$A, A \to_b B \vdash B.$$

In concatenation logic the implication must come earlier than the minor premise. In this logic, the minor premise comes first. The set of theorems of this logic, with $\to_b$, is unchanged; we get **CL** again, except that in the semantics we now concatenate to the left. A different system is obtained if we have both implications (see [33]). This is actually the Lambek Calculus. The Lambek Calculus can be defined as the smallest S-consequence relation with two implications $\to_e$ and $\to_b$ satisfying the appropriate Deduction Theorems for the list data structures.

For example, since

$$A \to_e B, A \vdash B$$

we get that

$$A \vdash (A \to_e B) \to_b B$$

The stipulation that

$$A \to_e B \text{ iff } A \to_b B$$

gives us linear logic, with $\to$ being either of $\to_b$ or $\to_e$.

Both $\to_e$ and $\to_b$ satisfy a Deduction Theorem.

Having illustrated our concepts, we can now generalise the notion of the deduction theorem to arbitrary data constellations. To achieve that we also need to generalise the notion of implication.

Definition 2.5. 1. Let $\mathcal{M} = \{\tau_i\}$ be a class of structures τ_i, as in Definition 2.1. Let φ be a wff in some language $\mathcal{L}$ capable of expressing properties of the structures in $\mathcal{M}$.

Assume that for every τ in $\mathcal{M}$, φ defines a unique element (or possibly a unique set of elements) in τ.

The formula φ defines a contraction mapping on structures τ. Let τ' be the structure obtained from τ by taking out the points identified by φ. Thus, for example, for lists structures τ, if φ identifies the last element in the list then τ' is the list without its last element. We can denote the contraction mapping function by '$\| \varphi$' and write $\tau' = \tau \| \varphi$.

Thus, given two structures τ_1, τ_2 we may say that $\tau_2 = \tau_1 \| \varphi$ iff τ_2 is obtained by dropping out of τ_1 the elements identified by φ. We write $\tau[\|\varphi]^n$ to mean $\tau\|\varphi \ldots \|\varphi$, n times. For example, for lists τ and φ which identifies the last element of the list, we have $(t_1, \ldots, t_n) = (t_1, \ldots, t_n, x)\|\varphi$ and $\varnothing = (t_1, \ldots, t_n)[\|\varphi]^n$.

2. Let a database discipline be given, involving a family of structures $\mathcal{M}$. A database Δ in this discipline is any pair (τ, δ) where τ is a finite element of $\mathcal{M}$ (e.g. a list) and δ associates with each $t \in \tau$ a formula $\delta(t) = A_t$. We also write $\Delta = \Delta'\|\varphi$, just in case $\Delta = (\tau, \delta), \Delta' = (\tau', \delta')$ and $\tau = \tau'\|\varphi$ and $\delta = \delta' \upharpoonright \tau$, i.e. δ is the restriction of δ' to τ.

Example 2.6. Let $\mathcal{M}$ be the class of all finite lists including the empty list. Define $+$ as concatenation. Let $\mathcal{L}$ be a language with '$+$' and '$<$'. φ can be a formula in $\mathcal{L}$ which defines uniquely the last element of any finite list. Thus for $\tau = (a_1, \ldots, a_n, b)$, we have $\tau\|\varphi = (a_1, \ldots, a_n)$. These lists are the data structures of **CL**.

Definition 2.7. (The Notion of the Deduction Theorem) Let Δ, Δ' be two databases and assume that $\tau = \tau'\|\varphi$. Assume further that we have

$$\Delta\|\varphi \mathrel{|\!\sim} A \to B \text{ iff } \Delta' \mathrel{|\!\sim} B$$

Consider a special connective denoted by $\to_\varphi$. We say that $\to_\varphi$ satisfies the *Deduction Theorem with respect to* φ iff whenever $\Delta \mathrel{|\!\sim} B$ and $\Delta = (\tau, \delta)$ and τ has n elements then

$$\Delta[\|\varphi]^n \mathrel{|\!\sim} \varnothing : \delta(t_1) \to_\varphi (\ldots \to_\varphi (\delta(t_n) \to_\varphi B) \ldots)$$

where $\ll t_i \gg \; = \tau[\|\varphi]^i - \tau[\|\varphi]^{i+1}$ i.e. t_i is the element identified by φ in $\tau[\|\varphi]^i$.

Our discussions so far presented the consequence relation in the form of $\Delta \mid\!\sim A$, where A is a single formula on the right hand side. This is a Tarski type consequence relation. The general form of the consequence relation should be $\Delta \mid\!\sim \Gamma$, where both Δ and Γ are structured databases. This is a Scott type consequence relation. In fact, there is no reason at all to assume that the same structures can serve on both sides of the turnstile. If we refer to the left hand side as *the data* and the right hand side as the goal, then the general Scott type S-consequence relation has the form $\Delta \mid\!\sim \Gamma$, where Δ is a data structure and Γ is a goal structure. For example, Δ may be a set of formulas while Γ may be a list of formulas. This is indeed the case, for example, in Logic Programming, where we have

$$\Delta \mid\!\sim (A_1, \ldots, A_n) \text{ iff } \Delta \mid\!\sim A_1 \text{ and } \ldots \text{ and } \Delta \mid\!\sim A_n$$

If Δ is a sequence, as in the case of **CL**, or a multiset, as in the case of linear logic, then we can have

$$\begin{array}{l}\Delta \mid\!\sim (A_1, \ldots, A_n) \\ \text{iff for some } \Delta_i,\ \Delta = \Delta_1 + \ldots + \Delta_n \text{ and } \Delta_i \mid\!\sim A_i,\ i = 1, \ldots, n\end{array}$$

where $+$ is concatenation or multiset union, respectively.

Another example is the case of Gentzen systems for classical logic, where both Δ and Γ are sets and where we have $\Delta \mid\!\sim \{A_1, \ldots, A_n\}$ iff $\Delta \mid\!\sim A_1 \vee \ldots \vee A_n$.

In the above cases the Scott consequence relation $\Delta \mid\!\sim \Gamma$, for Γ a complex structure, can be reduced to its Tarski part $\Delta \mid\!\sim A$, where A a single formula. It is not clear whether we should insist on this property to hold in the general case.

To get an idea of our options, let our starting point be two structured Tarski type consequence relations $\mid\!\sim_1$ and $\mid\!\sim_2$ on the same language **L** with $\mathcal{L}_1$, $\mathcal{L}_2$ and $\mathcal{M}_1$ and $\mathcal{M}_2$ the two structural languages satisfying the conditions of definition 2.1. We want to 'extend' $\mid\!\sim_1$ and $\mid\!\sim_2$ to a Scott type consequence relation $\|\!\sim$, allowing for $\Delta\|\!\sim\Gamma$ where Δ is an $\mathcal{M}_1$ structured database and Γ is an $\mathcal{M}_2$ structured goal. Thus the allowable structures for $\|\!\sim$ are $\mathcal{M}_1$ for the data and $\mathcal{M}_2$ for goals. Clearly it must satisfy the rules of **Identity** and **Surgical Cut** in the following form:

$$A\|\!\sim A$$

and

$$\frac{\Delta[A]\|\!\sim\Gamma, \Delta'\|\!\sim A}{\Delta[\Delta']\|\!\sim\Gamma}.$$

We may wish to rename the Cut Rule as **Left hand side Surgical Cut**, to stress that cut is done on the left.

A **Right hand side Surgical Cut** would be

$$\frac{\Delta\|\!\!\sim\Delta[B] \quad B\|\!\!\sim\Gamma'}{\Delta\|\!\!\sim\Gamma[\Gamma']}$$

The question is, what other requirements do we have for $\|\!\!\sim$?

We obviously want to make use of both $|\!\!\sim_1$ and $|\!\!\sim_2$. The new consequence $\|\!\!\sim$ should extend them both in some sense. Since both consequence relations are general, the only structures they share for certain are the unit formulae structures A. Given Δ, Γ and A, we know what $\Delta \mid\!\!\sim_1 A$ is and what $\Gamma \mid\!\!\sim_2 A$ is, but no more.

The next two definitions tell us what is the general notion of a Scott type S-consequence relation and how to form a Scott type $\|\!\!\sim$ out of $|\!\!\sim_1$ and $|\!\!\sim_2$.

Definition 2.8. Let $\mathbf{L}$ be a language for wffs and let $(\mathcal{M}_1, \mathcal{L}_1)$ and $(\mathcal{M}_2, \mathcal{L}_2)$ be two languages for data and for goal structures respectively. Note that $\mathcal{L}_1$ and $\mathcal{L}_2$ come with their own respective notions of structrual substitutions and structural additions $+_1$ and $+_2$ respectively.

A relation $\Delta\|\!\!\sim\Gamma$ is said to be a (Scott type) S-consequence relation on the data and goal structures iff it satisfies the following two conditions:

Identity $A\|\!\!\sim A$

Surgical Cut $\dfrac{\Delta[A]\|\!\!\sim\Gamma[B] \quad \Delta'\|\!\!\sim A \quad B\|\!\!\sim\Gamma'}{\Delta[\Delta']\|\!\!\sim\Gamma[\Gamma'].}$

Note that some of the traditional formulations of the Cut Rule may not hold, for example

Internal Cut $\dfrac{\Delta + A \mid\!\!\sim \Gamma \quad \Delta \mid\!\!\sim A + \Gamma}{\Delta \mid\!\!\sim \Gamma}$

Transitivity $\dfrac{\Delta \mid\!\!\sim A \quad \Delta + A \mid\!\!\sim B}{\Delta \mid\!\!\sim B}$ may not hold.

Further note that **Identity** can hold only for single formulas. I do not know whether the rule of identity formulated for single formulas A would imply the rule of identity for all common structures Δ.

Definition 2.9. Let $|\!\!\sim_1$ and $|\!\!\sim_2$ be two Tarski type S-consequence relations on structured databases with structures $\mathcal{M}_1$ and $\mathcal{M}_2$ respectively, based on the same language $\mathbf{L}$, as in Definition 2.1. Assume that $|\!\!\sim_1$ and $|\!\!\sim_2$ are *compatible*, namely they satisfy for all A, B.

$$A \mid\!\!\sim_1 B \quad \text{iff} \quad B \mid\!\!\sim_2 A.$$

Define $|\!\!\sim_{1,2}$ to be the following Scott type S-consequence relation, called the *Minimal Amalgamation* of $|\!\!\sim_1$ and $|\!\!\sim_2$:

1. The data structures of $\mid\!\sim_{1,2}$ are those of $\mathcal{M}_1$. The goal structures of $\mid\!\sim_{1,2}$ are those of $\mathcal{M}_2$
2. Let $\Delta \mid\!\sim_{1,2} \Gamma$ hold iff (definition) for some wff A we have that both $\Delta \mid\!\sim_1 A$ and $\Gamma \mid\!\sim_2 A$ hold.

Lemma 2.10. *Let $\mid\!\sim_1$ and $\mid\!\sim_2$ be two compatible Tarski type consequence relations. Let $\Vdash\!\sim$ be their minimal amalgamation. Then $\Vdash\!\sim$ is a Scott type consequence relation.*

Proof. It is clear that **Identity** holds. We check the rule of **Surgical Cut**.

Assume

$$\Delta[A] \Vdash\!\sim \Gamma[B],\ \Delta' \Vdash\!\sim A \text{ and } B \Vdash\!\sim \Gamma',$$

We show

$$\Delta[\Delta'] \Vdash\!\sim \Gamma[\Gamma'].$$

By definition there exist C_1, C_2, C_3 such that

$$\Delta[A] \mid\!\sim_1 C_1,\ \ \Gamma[B] \mid\!\sim_2 C_1,\ \ \Delta' \mid\!\sim_1 C_2,\ \ A \mid\!\sim_2 C_2,\ \ B \mid\!\sim_1 C_3,\ \ \Gamma' \mid\!\sim_2 C_3$$

all hold. From compatibility we get

$$C_2 \mid\!\sim_1 A \quad \text{and} \quad C_3 \mid\!\sim_2 B.$$

By **Surgical Cut** we get

$$\Delta' \mid\!\sim_1 A \quad \text{and} \quad \Gamma' \mid\!\sim_2 B.$$

Hence by **Surgical Cut** again we get

$$\Delta[\Delta'] \mid\!\sim_1 C_1 \quad \text{and} \quad \Gamma[\Gamma'] \mid\!\sim_2 C_1.$$

Hence by definition we get

$$\Delta[\Delta' \Vdash\!\sim \Gamma[\Gamma'].$$

■

Example 2.11. Let $\Delta \vdash_H A$ be the classical (for $H = C$) or intuitionistic (for $H = I$) consequence relation, for the data structures of single formulas. Let $\mid\!\sim_{H^0}$ be the consequence relation of Example 1.9(1). Form $\mid\!\sim_{H,H^0}$ as in the preceding definition. We show that what we get is classical (or intuitionistic) Scott type relation $\Vdash_H$ for sets as structures on both sides of the turnstile.

We have $\Delta \mid\!\sim_{H,H^0} \Gamma$ iff for some $A, \Delta \vdash_H A$ and $\Gamma \mid\!\sim_{H^0} A$. But $\Gamma \mid\!\sim_{H^0} A$ iff $A \vdash_H \bigvee \Gamma$. Hence $\Delta \mid\!\sim_{H,H^0} \Gamma$ iff for some A, $\Delta \vdash_H A$ and $A \vdash_H \bigvee \Gamma$ and in classical or intuitionistic logic we can continue, iff $\bigwedge \Delta \vdash_H \bigvee \Gamma$, because we can take $A = \bigvee \Gamma$.

Example 2.12. Let **L** be the language of intutionistic implication $\to$ and conjunciton $\wedge$. Let $\Delta \vdash_I A$ be the intuitionistic consequence relation for this language, with the goal structure of single formulas on the right. Let $\mid\!\sim_{H^1}$ be the consequence relation for **L**, defined in example 1.9(2).

Then $\mid\!\sim_{I,H^1}$ is the amalgamated consequence relation satisfying

$$\Delta \vdash_{I,H^1} \Gamma \quad \text{iff} \quad \text{for some } A \in \Gamma \text{ we have } \Delta \vdash_I A.$$

In my book [8] in Chapter 1 it was shown that $\mid\!\sim_{I,H^1}$ is the consequence relation arising from the *Beth Semantics* for intuitionistic logic. It was further shown that there are two Scott type consequence relations which agree (extend) the Tarski type $\Delta \mid\!\sim_I A$, for A a single wff. The maximal one, arising from the Kripke semantics, defined as $\Delta \vdash_I A_1 \vee \ldots \vee A_n$, and the minimal one, defined as in this example.

Definition 2.13. Let **L** be a language and let $(\mathcal{M}_1, \mathcal{L}_1)$ and $(\mathcal{M}_2, \mathcal{L}_2)$ be structures for the data and goal respectively. Let $\mid\!\sim_i$, $i = 1, 2$, be Tarski type S-consequence relations on the data structures $\mathcal{M}_i$ respectively with the goal structure being single wffs and let $\Vdash\!\sim$ be a Scott type S-consequence on both data structures $\mathcal{M}_1$ and goal structures $\mathcal{M}_2$.

(a) We say that $\Vdash\!\sim$ *agrees* with $\mid\!\sim_1$ in the data (respectively $\Vdash\!\sim$ *agrees* with $\mid\!\sim_2$ in the goal) iff for all Δ, Γ, A, (1) holds (respectively (2) holds).

1. $\Delta \Vdash\!\sim A$ iff $\Delta \mid\!\sim_1 A$
2. $A \Vdash\!\sim \Gamma$ iff $\Gamma \mid\!\sim_2 A$.

(b) We say that $\Vdash\!\sim$ agrees with $(\mid\!\sim_1, \mid\!\sim_2)$ iff $\Vdash\!\sim$ agrees with $\mid\!\sim_1$ in the data and with $\mid\!\sim_2$ in the goal.

Theorem 2.14. *Let $\mid\!\sim_1$ and $\mid\!\sim_2$ be two compatible Tarski type S-consequence relations, i.e. satifying for all A, B, $A \mid\!\sim_1 B$ iff $B \mid\!\sim_2 A$. Then there exist two Scott-type consequence relations, a maximal $\Vdash\!\sim^+$ and a minimal $\Vdash\!\sim^-$ which agree with $(\mid\!\sim_1, \mid\!\sim_2)$. In other words, for any $\Vdash\!\sim$ which agrees with $(\mid\!\sim_1, \mid\!\sim_2)$ we have*

1. *for all Δ, Γ: $\Delta \Vdash\!\sim^- \Gamma$ implies $\Delta \Vdash\!\sim \Gamma$;*
2. *for all Δ, Γ: $\Delta \Vdash\!\sim \Gamma$ implies $\Delta \Vdash\!\sim^+ \Gamma$.*

Proof. Let $\Vdash\!\sim^-$ be the minimal amalgamation of $\mid\!\sim_1$ and $\mid\!\sim_2$ of Definition 3.9.

We show that $\parallel\!\sim^-$ agrees with $(\mid\!\sim_1, \mid\!\sim_2)$. Assume $\Delta \parallel\!\sim^- A$, then by definition for some B, $\Delta \mid\!\sim_1 B$ and $A \mid\!\sim_2 B$. Hence by compatibility $B \mid\!\sim_1 A$ and by the Cut Rule, $\Delta \mid\!\sim_1 A$.

Similarly assume $A \parallel\!\sim^- \Gamma$. Then for some B, $A \mid\!\sim_1 B$ and $\Gamma \mid\!\sim_2 B$. By compatibility $B \mid\!\sim_2 A$ and by Cut, $\Gamma \mid\!\sim_2 A$.

We now show minimality. Assume that $\parallel\!\sim$ agrees with $(\mid\!\sim_1, \mid\!\sim_2)$. Assume that $\Delta \parallel\!\sim^- \Gamma$, we show that $\Delta \parallel\!\sim \Gamma$.

From our assumption we have that for some A

$$\Delta \mid\!\sim_1 A \quad \text{and} \quad \Gamma \mid\!\sim_2 A$$

Hence from agreement $\Delta \parallel\!\sim A$ and $A \parallel\!\sim \Gamma$ and hence by Cut, $\Delta \parallel\!\sim \Gamma$.

We now have to prove the existence of $\parallel\!\sim^+$. To this end we show that for any finite set $\{\parallel\!\sim_i\}$ of consequence relations which agree with $\mid\!\sim_1$ and $\mid\!\sim_2$ one can define a consequence relation $\parallel\!\sim$ agreeing with $(\mid\!\sim_1, \mid\!\sim_2)$, with the property that for all i $\parallel\!\sim_i \subseteq \parallel\!\sim$.

An application of Zorn's Lemma will now yield the existence of $\parallel\!\sim^+$.

Let $\parallel\!\sim_i$ be given for $i = 1, \ldots, m$. We define the desired $\parallel\!\sim$. Define $\Delta \parallel\!\sim \Gamma$ to hold iff there exists a sequence of pairs (Δ_n, Γ_n), $n = 1, \ldots, k$ such that $\Delta_k = \Delta$, $\Gamma_k = \Gamma$ and each element (Δ_n, Γ_n) in the sequence satisfies one of the following conditions:

1. For some $\parallel\!\sim_i$, $\Delta_n \parallel\!\sim_i \Gamma_n$ (i.e. (Δ_n, Γ_n) is an 'axiom').
2. There exist earlier n_1, n_2, n_3 in the sequence such that:
   - $(\Delta_{n_1}, \Gamma_{n_1}) = (\Delta', A)$, A a wff.
   - $(\Delta_{n_2}, \Gamma_{n_2}) = (B, \Gamma')$, B a wff.
   - $(\Delta_{n_3}, \Gamma_{n_3})$ has the form $(\Delta_{n_3}[A], \Gamma_{n_3}[B])$ and $(\Delta_n, \Gamma_n) = (\Delta_{n_3}[A/\Delta'], \Gamma_{n_3}[B/\Gamma'])$.

Intuitively $\parallel\!\sim$ is essentially the sequent system with $\Delta \parallel\!\sim_i \Gamma$ as axioms, for $i = 1, \ldots, m$ and where the surgical cut is the inference rule.

It is clear from the definition of $\parallel\!\sim$ that **Identity** and **Surgical Cut** are satisfied and that $\parallel\!\sim_i \subseteq \parallel\!\sim$.

We now have to show that $\parallel\!\sim$ agrees with $(\mid\!\sim_1, \mid\!\sim_2)$. i.e. we have to show that for any Δ, Γ, A (1) and (2) hold:

1. $\Delta \parallel\!\sim A$ iff $\Delta \mid\!\sim_1 A$
2. $A \parallel\!\sim \Gamma$ iff $\Gamma \mid\!\sim_2 A$

We prove this by induction on the length of the sequence defining $\parallel\!\sim$, and on the assumption that $\parallel\!\sim_i$, $i = 1, \ldots, m$ all agree with $(\mid\!\sim_1, \mid\!\sim_2)$.

For sequences of length 1 the claim follows form the assumption.

Consider a sequence of length k which justifies $\Delta \| \!\sim A$. Then we have the following situation from earlier members of the sequence:

- $\Delta' \| \!\sim A'$ and hence by the induction hypothesis $\Delta' \mid\!\sim_1 A'$
- $B \| \!\sim A$ and hence by the induction hypothesis $A \mid\!\sim_2 B$.
- $\Delta[A'] \| \!\sim B$ and hence $\Delta(A') \mid\!\sim_1 B$.

We also know that $\Delta[\Delta'] \| \!\sim A$ and we want to show that it also agrees with $(\mid\!\sim_1, \mid\!\sim_2)$, i.e. that $\Delta[\Delta'] \mid\!\sim_1 A$. From compatibility we get $B \mid\!\sim_1 A$ and from Cut we get $\Delta[\Delta'] \mid\!\sim_1 A$.

Similarly consider a sequence of length k which justifies $A \| \!\sim \Gamma$. This means that earlier elements in the sequence are

- $A' \| \!\sim \Gamma'$ hence $\Gamma' \mid\!\sim_2 A'$
- $A \| \!\sim B$ hence $A \mid\!\sim_1 B$
- $B \| \!\sim \Gamma[A']$ and hence $\Gamma[A'] \mid\!\sim_2 B$

We also know that $A \| \!\sim \Gamma[\Gamma']$ and we want to show that $\Gamma[[\Gamma'] \mid\!\sim_1 A$. By compatibility $B \mid\!\sim_2 A$ and by Cut, $\Gamma[\Gamma'] \mid\!\sim_2 A$. ■

Example 2.15. (Symmetric Amalgamation) Given a Scott type S-consequence relation $\mid\!\sim$, the two Tarski consequence relations derived from it, namely

$$\Delta \mid\!\sim_1 A \quad \text{iff} \quad \Delta \mid\!\sim A$$
$$\Gamma \mid\!\sim_2 A \quad \text{iff} \quad A \mid\!\sim \Gamma$$

may not be identical or isomorphic, even in the case where Δ and Γ are based on the same data structures. We say that the Scott type $\mid\!\sim$ is a symmetrical amalgamation of the Tarski type $\mid\!\sim_0$ iff the two derived consequence relations $\mid\!\sim_1$ and $\mid\!\sim_2$ are the same and equal to $\mid\!\sim_0$. For a given $\mid\!\sim_0$, it is always possible to construct the symmetrical amalgamation. We cannot take the naive identity $\mid\!\sim_1 = \mid\!\sim_2 = \mid\!\sim_0$ because then the definition of amalgamation will yield

$$\Delta \mid\!\sim \Gamma \quad \text{iff} \quad \text{for some } A, \Delta \mid\!\sim_1 A \quad \text{and}$$
$$\Gamma \mid\!\sim_2 A \quad \text{iff} \quad \text{for some } A, \Delta \mid\!\sim_0 A \text{ and } \Gamma \mid\!\sim_0 A.$$

If we let A = truth we get $\Delta \mid\!\sim \Gamma$ for all Δ, Γ.

What is needed is an isomorphism function $*$ such that

$$A^{**} = A \text{ and } \Gamma \mid\!\sim_2 A \quad \text{iff} \quad \Gamma^* \mid\!\sim_1 A^*.$$

We thus get

$$\Delta \mid\!\sim \Gamma \quad \text{iff} \quad \text{for some } A, \Delta \mid\!\sim_1 A \text{ and } \Gamma^* \mid\!\sim_1 A^*$$

now since we also have

$$A \mid\!\sim_1 B \quad \text{iff} \quad B \mid\!\sim_2 A$$

we get that $*$ must satisfy

$$A \mid\!\sim_1 B \quad \text{iff} \quad B^* \mid\!\sim_1 A^*.$$

In case of classical logic the mapping $*$ is negation

$$A \vdash B \quad \text{iff} \quad \neg B \vdash \neg A.$$

The above examples have shown how to extend S-consequence relations to ones with a more complex goal structure by minimal amalgamation. When the original consequence relation $\mid\!\sim$ satisfies the Deduction Theorem, it is possible to extend $\mid\!\sim$ by making use of this fact. The Deduction Theorem is similar to a left to right shift operator (studied in a later section). Thus by using shift operators we can give meaning to $\Delta \mid\!\sim \Gamma$ for Γ more complex, in terms of shifting Γ to the left and reducing the problem to a known one.

Thus for example one may write $\Delta \mid\!\sim \Gamma$ iff some boolean combination $\Delta_i \mid\!\sim \Gamma_i \ldots$ holds, where each Γ_i has less elements than Γ.

Classical logic can be viewed to be a special case of reducing $\{A_i\} \vdash_C \{B_j, C\}$ to $\{A_i, \neg C\} \vdash_C \{B_j\}$.

The Intuitionistic consequence, arising from the Kripke Semantics (see 2.12), cannot be viewed in this manner. We have no means of shifting wffs from the right hand side to the left hand side.

Let $\mid\!\sim'$ be a consequence relation with $\mathcal{M}_1'$ and $\mathcal{M}_2'$ being the structures for the data and goals respectively. Let $\mid\!\sim''$ be another consequence relation with structures $\mathcal{M}_1''$ and $\mathcal{M}_2''$ respectively. We want to define the composition $\mid\!\sim^*$ of $\mid\!\sim'$ and $\mid\!\sim''$. The data structures (goal structures) of $\mathcal{M}_1^*(\mathcal{M}_2^*)$ are obtained from $\mathcal{M}_1'$ and $\mathcal{M}_1''$ (respectively $\mathcal{M}_2'$ and $\mathcal{M}_2''$) by the repeated substitution of one data structure inside the other.

If the structures in $\mathcal{M}_1'$ are lists of points $(x_1, \ldots, x_n)$ and the structures in $\mathcal{M}_1''$ are sets then the structures in $\mathcal{M}_1^*$ are hereditary lists of non-empty sets of points $(X_1, \ldots, X_n)$. If $\mathcal{M}_1'$ contains multisets of points then $\mathcal{M}_1^*$ is a hereditary set of multisets of non-empty sets of points.

Definition 2.16. (Composition of two S-consequence relations) Let $\mid\!\sim'$ and $\mid\!\sim''$ be two consequence relations with structures $(\mathcal{M}'_1, \mathcal{M}'_2)$ and $(\mathcal{M}''_1, \mathcal{M}''_2)$ respectively. We assume further that we have available a notion of substitution of a τ' structure inside a τ'' structure and vice versa as needed below. We define the composition $\mid\!\sim^*$ of $\mid\!\sim'$ and $\mid\!\sim''$ as follows:

1. The structures of $\mathcal{M}^*_i$ of $\mid\!\sim^*, i = 1, 2$ are defined by induction.

   (a) Let $\mathcal{M}^*_i(0) = \mathcal{M}'_i \cup \mathcal{M}''_i$.

   (b) Assume $\mathcal{M}^*_i(k)$, for $k < n$ are all defined. Let $\tau \in \mathcal{M}^*_i(k)$ for some $k < n$ and $\tau'(x)$ (respectively $\tau''(x)$) is in $\mathcal{M}'_i$ (respectively $\mathcal{M}''_i$). Then $\tau'(\tau)$ (resp $\tau''(\tau)$) is in $\mathcal{M}^*_i(n)$.

   (c) Let $\mathcal{M}^*_i = \bigcup_n \mathcal{M}^*_i(n)$.
   Note: It is easy to show that if $\tau_1(x) \in \mathcal{M}^*_i$ and $\tau_2 \in \mathcal{M}^*_i$ then $\tau_1(\tau_2) \in \mathcal{M}^*_i$.

2. Let $\Delta = (\tau_1, \delta_1)$ and $\Gamma = (\tau_2, \delta_2)$ be two databases for $\mid\!\sim^*$.

   We can assume by definition that for some $\alpha_1, \beta_1, \alpha_2, \beta_2, \varepsilon, \eta$:

$$\begin{aligned}
&\tau_1 = \beta_1(\alpha_1) \text{ hence } \Delta = \Delta_0(\varepsilon)\\
&\tau_2 = \beta_2(\alpha_2) \text{ hence } \Gamma = \Gamma_0(\eta), \text{ where}\\
&\Delta_0(x) = (\beta_1(x), \delta_1), \Gamma_0(x) = (\beta_2(x), \delta_2)\\
&\epsilon = (\alpha_1, \delta_1), \eta = (\alpha_2, \delta_2)
\end{aligned}$$

   where β_1, β_2 are either in $\mathcal{M}'_1$ or in $\mathcal{M}''_i$ for $i = 1, 2$ respectively.

   Note that δ_i do not give a value to x.

   We say that $\Delta \mid\!\sim^* \Gamma$ iff for some C, D

$$\begin{aligned}
&\varepsilon = (\alpha_1, \delta_1) \mid\!\sim^* C\\
&\eta = (\alpha_2, \delta_2) \mid\!\sim^* D \text{ and}\\
&\Delta_0[C] \mid\!\sim^* \Gamma_0[D]
\end{aligned}$$

Let us summarise our current view:

- A logical language **L** is defined by introducing the notion of well formed formulas and by further stating a family of data structures and defining the notions of substitution on the data structure, and of adding (combining) data structures.

- Minimal conditions for a relation between data structures and formulas to be a consequence relation are those of **Identity** and **Surgical Cut**. The smallest consequence relation (if it exists) for a language **L** is denoted by **S**(**L**).
- The smallest structured consequence relation for the language with $\to$ only, can be extended either in the direction of non-monotonic consequence relations or in the direction of Substructural Logics or both.
- We need and can develop adequate semantics for **S**(**L**) in the spirit of [19], see [14].

3. Structural Connectives

Let $\mid\!\sim$ be a Scott type consequence relation on a language **L**. The consequence relation involves structures for the data and structures for goal. The purpose of this section is to show that given these structures, there are some natural connectives, called structural connectives, which one may wish to add into the language **L**, with porperties dictated by the structure. In fact, the entire consequence relation may be in some cases (e.g. linear logic, see next section) no more and no less than a reflection (via the structural connectives) of the properties of the data structures involved.

Let $\mid\!\sim$ be any Scott type consequence relation. Let $\mathcal{M}_1$ be structures for the data and let $\mathcal{M}_2$ be the structures for the goals. Let $+_1$ be structural addition for the data and $+_2$ for the goal. We can thus write

$$\Delta +_1 A \mid\!\sim B +_2 \Gamma$$

which may or may not hold.

We assume nothing special about $\mid\!\sim$, only that it satisfies the mimimal properties of **Identity** and **Surgical Cut**. We do not assume any special connectives in the language **L** of $\mid\!\sim$, and we aim to introduce several structural connectives.

It might be useful to think in terms of two examples, namely concatenation logic, where the structures are lists, and linear logic, where the structures are multisets. Both cases would have $\to$ in the language **L**. To be even more specific, think of the Tarski type consequence relations to be the relations $\Vdash_{\mathbf{CL}}$ and $\Vdash_{\mathbf{LL}}$ for **CL** and for **LL** of Example 1.7 respectively, and assume that $\mid\!\sim$ is some Scott extension of one of them which agrees with it, the goal structures being the same as the data structures. Note further that there may be more than one such $\mid\!\sim$ agreeing with the Tarski type $\Vdash_{\mathbf{CL}}$ or $\Vdash_{\mathbf{LL}}$ of Example 1.7, and that in the sequel we may choose one such $\mid\!\sim$ with special symmetry conditions.

Let us begin with a general Scott type consequence $\mid\!\sim$.

The first thing to bear in mind is that in the general case, the structural operators '$+_1$' and '$+_2$' which come with $\mid\!\sim$ are in the metalevel, on the structures of $\mathcal{M}_1$ and $\mathcal{M}_2$ respectively, and not in the language **L** of $\mid\!\sim$ itself. For the special case of **CL**, '$+_i$' is concatenation and for the case of **LL**, '$+_i$' is multiset union. Let us now add another metalevel operator, a sort of notational shift operator 'o', having the following properties:

- $\Delta +_1 A^o \mid\!\sim \Gamma$ means the same as $\Delta \mid\!\sim A +_2 \Gamma$
- $\Delta \mid\!\sim B^o +_2 \Gamma$ means the same as $\Delta +_1 B \mid\!\sim \Gamma$

In fact, $\Delta +_1 A^o \mid\!\sim B^o +_2 \Gamma$ should be the same as $\Delta +_1 B \mid\!\sim A +_2 \Gamma$.

Thus the operator 'o' in 'A^o' is used to indicate, in the metalevel, where the formula A should be. It has the status of a metalevel annotation, indicating that although the formula A appears in one place in the structure, it should be in another place. Thus $\Delta +_1 A^o \mid\!\sim \Gamma$ really says through the annotation 'o' on A, that A is really on the right hand side, namely $\Delta \vdash A +_2 \Gamma$. For general structures the annotation may be problematic. If $\Delta[x]$ indicates that x stands somewhere in the structure Δ, then we know what $\Delta[A]$ means. We also have to say what $\Delta[A^o]$ means. We have to specify the following:

1. Since A^o in $\Delta[A^o]$ means that although A^o shows up in $\Delta[x]$ at the place x (i.e. we have $\Delta[A^o/x]$), A should really be somewhere else, we need to specify where A should be.
2. If A^o is not at place x in $\Delta[x]$, then there is *nothing* at x. Hence our notion of substitution should also allow for empty substitution $\Delta[\varnothing/x]$, which is really deletion.

If both Δ and Γ are multisets, as in the case of linear logic, then we have no such problem, otherwise a precise definition is required. Even in the case of **CL**, this is not simple, and requires care.

In **CL**, the data structures are lists $(A_1, \ldots, A_n)$. Of course, the list $(A_1, \ldots, A_n, B)$ is generally not the same data structure as $(B, A_1, \ldots, A_n)$. We can mark B as B^o to mean that B should be *shifted* to the other end of the list. Thus

$$(A_1, \ldots, A_n, B^o) =_{\text{def}} (B, A_1, \ldots, A_n)$$
$$(B^o, A_1, \ldots, A_n) =_{\text{def}} (A_1, \ldots, A_n, B).$$

Note that $\Delta_1 +_2 \Delta_2$ (and similarly $\Gamma_1 +_1 \Gamma_2$) is not meaningful, since $+_2$ (resp $+_1$) is a goal (data) structural addition and cannot be applied to the data (goal).

So far we have introduced several meta operations. We now want to put them in the object level. We have seen in connection with the Deduction Theorem in the previous section that it is possible to take a meta operation and add a connective for it in the object language. We now do just that for '$+_1$', '$+_2$' and 'o' of the metalevel. We denote the object level connectives by '$\oplus_1$', '$\oplus_2$' and '$\odot$', respectively.

Our purpose is to study their obvious properties. The connection between $\{A + B\}$ and $\{A \oplus B\}$ is that they must be declaratively identical. This means that $A \oplus B$ should behave essentially like $A + B$ in all $\mid\!\sim$ contexts. We thus have the following rules:

Identity

$$A +_1 B \mid\!\sim A \oplus_1 B$$

$$A \oplus_2 B \mid\!\sim A +_2 B$$

The other direction of the identity of $A \oplus_i B$ with $A +_i B$, cannot be written directly but can be characterised by a Cut rule:

Surgical Cut

$$\frac{\Delta_1 +_1 (A +_1 B) +_1 \Delta_2 \mid\!\sim \Gamma_1 +_2 (C +_2 D) +_2 \Gamma_2}{\Delta_1 +_1 (A \oplus_1 B) +_1 \Delta_2 \mid\!\sim \Gamma_1 +_2 (C \oplus_2 D) +_2 \Gamma_2}$$

We believe that the object level counterparts of the structural connectives of any system can always be conservatively added to the system. We may have to assume some mild and very reasonable sufficient conditions.

Example 3.1. In the case of **CL**, if we add the structural connective $\oplus$ to the language, we get the following system **CL**($\oplus$), with the following two rules:

$$(A, B) \mid\!\sim A \oplus B$$

which we get from **Identity**, and we should also have the following from the **Cut Rule**

$$\frac{(A_1, \ldots, A_n, A, B, B_1, \ldots, B_m) \mid\!\sim C}{(A_1, \ldots, A_n, A \oplus B, B_1, \ldots, B_m) \mid\!\sim C}$$

These we call **Structural Rules** for the structural connective $\oplus$, because they arise from the fact that $\oplus$ is no more than an object level reflection of the structured database operation $+$.

In classical logic the data structures are sets and conjunction '$\wedge$' serves as the structural connective. The goal structures are also sets and '$\vee$' serves as the structural connectives.

In case of **CL** we also have a **Right hand side Deduction Theorem**, which, when combined with the structural connective $\oplus$ will yield rules like

$$A \oplus B \mid\!\sim C \quad \text{iff} \quad A \mid\!\sim B \to C$$

and

$$\varnothing \mid\!\sim A \to (B \to A \oplus B).$$

We now have to check which rules for $\odot$ follow from the properties of shift. We know that $\Delta +_1 A^{\odot} \mid\!\sim B^{\odot} +_2 \Gamma$ should hold iff $\Delta +_1 B \mid\!\sim A +_2 \Gamma$. We immediately get

- $A^{\odot\odot} \mid\!\sim A$
- $A \mid\!\sim A^{\odot\odot}$

Consider

$$\Delta +_1 C^o +_1 D^o \mid\!\sim \Gamma$$

which is the same as

$$\Delta \mid\!\sim (C \oplus_2 D) +_2 \Gamma$$

which is the same as

$$\Delta +_1 (C \oplus_2 D)^o \mid\!\sim \Gamma$$

We thus have that $(C \oplus_2 D)^o$ must be the same as $C^o \oplus_1 D^o$. Similarly $(C \oplus_1 D)^o$ must behave the same as $C^o \oplus_2 D^o$. We thus get

- $(C \oplus_2 D)^{\odot} \mid\!\sim C^{\odot} \oplus_1 D^{\odot}$
- $(C \oplus_1 D)^{\odot} \mid\!\sim C^{\odot} \oplus_2 D^{\odot}$

We still have a problem of what to do (what meaning we can prove or have) to $A \oplus_1 B$ when it appears on the right hand side and $C \oplus_2 D$ when it appears on the left hand side. The typical expression is:

$$C \oplus_2 D \mid\!\sim A \oplus_1 B$$

clearly it should be the same as

$$(A \oplus_1 B)^{\odot} \mid\!\sim (C \oplus_2 D)^{\odot}$$

which is the same as

$$A^{\odot} \oplus_2 B^{\odot} \mid\!\sim C^{\odot} \oplus_1 D^{\odot}$$

which does not help, as again we get the wrong '$\oplus$' on the wrong side.

In general, there is nothing to be done. However, if we have the object level '$\rightarrow_r$' with the Deduction Theorem holding, a few more equivalences can be derived.

Consider

$$A +_1 B \mathrel{|\!\sim} C$$

this holds iff

$$A \mathrel{|\!\sim} B \rightarrow_r C$$

but also iff

$$A \mathrel{|\!\sim} B^o +_2 C.$$

Hence

$$B \rightarrow_r C \text{ is equivalent to } B^o \oplus_2 C.$$

The difference between the shift and the Deduction theorem is that the connective '$\rightarrow_r$' is in the object language **L**, while our '$+_2$' and 'o' are in the metalevel. $\Delta \mathrel{|\!\sim} B^o +_2 C$ is just another way of writing, in the metalevel, $\Delta +_1 B \mathrel{|\!\sim} C$.

However, if we do have the object level connectives, '$\oplus_1$', and '$\oplus_2$', we can write down equivalences. We therefore get that

- $A \oplus_2 B$ is equivalent to $A^{\odot} \rightarrow_r B$
- $A \oplus_1 B$ is equivalent to $(A \rightarrow_r B^{\odot})^{\odot}$.

Thus '$\oplus_1$', '$\oplus_2$' are reducible, in a system containing '$\rightarrow_r$', to the connectives '$\rightarrow_r$' and '$\odot$'.

Let us now pause and summarise what we have learnt so far.

Suppose we start with a consequence relation $\mathrel{|\!\sim}$ with $+_1$ for data and $+_2$ for goals. Suppose we introduce a shift operation which shifts from the right hand side of the data to the left hand side of the goals and vice versae, and suppose that we have an object level connectives $\odot$ for the shift and $\rightarrow_r$ satisfying right hand side Deduction Theorem. Then we can define in the object level the two operators '$\oplus_1$' and '$\oplus_2$' using '$\rightarrow_r$' and the object level '$\odot$' as follows:

$$A \oplus_2 B =_{\text{def}} A^{\odot} \rightarrow_r B$$
$$A \oplus_1 B =_{\text{def}} (A \rightarrow_r B^{\odot})^{\odot}.$$

The above discussion showed how to add a structural connective $\oplus$ to reflect the notion $+$, which is part of the defintion of any consequence relation. Any specific consequence relation allows for a family $\mathcal{M}$ of structures. Any of these structures $\tau \in \mathcal{M}$ can be reflected in the object level by a special connective $\textcircled{\tau}$. The way it is done is described in the next definition.

Definition 3.2. (Left Structural Connectives and Left Structural Rules) Let $\mathcal{M}$ be a class of structures and let $\mid\!\sim$ be a consequence relation structured on $\mathcal{M}$, as defined in 2.1. Let τ be a fixed structure in $\mathcal{M}$ and assume τ has n elements $a_1, \ldots, a_n$. Indicate this fact by writing $\tau(a_1, \ldots, a_n)$. Enrich the language of $\mid\!\sim$ with an additional n-place connective $\textcircled{\tau}(A_1, \ldots, A_n)$ with the following *Left Structural Rules for* $\textcircled{\tau}$.
$\textcircled{\tau}$ *Identity*

$$(\tau(a_1, \ldots, a_n), \delta) \mid\!\sim \textcircled{\tau}(\delta(a_1), \ldots, \delta(a_n)).$$

$\textcircled{\tau}$ *Surgical Cut*
6pt Let $\Delta_1 = (\tau_1, \delta_1)$ and $\Delta = (\tau, \delta)$. Let $x \in \tau_1$ and let $\delta_1(x) = B$. Let $\Delta_2 = (\tau_2, \delta_2)$ be the result of structural substitution of Δ in Δ_1, i.e. $\Delta_2 = \Delta_1[B/\Delta]$.

This means that $\tau_2 = \lambda x \tau_1(x)[\tau]$.

Consider $\Delta_1' = (\tau_1(x), \delta_1')$ which is just like Δ_1 except that $\delta_1'(x) = \textcircled{\tau}(\delta(a_1), \ldots, \delta(a_n))$. Thus $\Delta_1' = \Delta_1[B/\textcircled{\tau}(\delta(a_1), \ldots, \delta(a_n))]$.

The **Surgical Cut Rule** is therefore

$$\frac{\Delta_1[B/\Delta] \mid\!\sim A}{\Delta_1[B/\textcircled{\tau}(\delta(a_1), \ldots, \delta(a_n)] \mid\!\sim A}$$

or if formulated with a slight abuse of notation:

$$\frac{\lambda x \tau_1(x)[\tau] \mid\!\sim A}{\lambda x \tau_1(x)[\textcircled{\tau}(\delta(a_1), \ldots, \delta(a_n))] \mid\!\sim A}$$

The cut rule really ensures the other direction of the identity, by forcing $\textcircled{\tau}$ to behave properly.

We are now ready to explain what a **Structural Shift Connective** is and what **Structural Shift Rules** are.

Definition 3.3. (Structural Shift Connectives) Let $\mid\!\sim$ be a Scott type S-consequence relation and let $\mathcal{M}_i$ and $\mathcal{L}_i$ be as in definition 2.8. Let φ_i be the wff of $\mathcal{L}_i$ each identifying a unique point in each $\tau_i \in \mathcal{M}_i$. Further assume that for each $\tau(a_1, \ldots, a_n)$ there exist a unique $\tau_2(a_1, \ldots, a_n, z)$ such that $\tau = \tau_2 \| \varphi_2$. Assume $\tau' = \tau_1 \| \varphi_1$. We can write $\tau' = f_1(\tau_1)$, and $\tau_2 = f_2(\tau)$ where f_i are function symbols for the functional dependence which exists.

Let $\odot_{\varphi_2}^{\varphi_1}$ be a new unary connective to the language of $\mid\!\sim$. Enrich $\mid\!\sim$ with this connective and the following structural shift rule:

(φ_1, φ_2) Structural Shift Rule

Let $\Delta_1 = (\tau_1(x), \delta_1)$ be the data structure with x the point identified in τ_1 by φ_1, and let $\Delta_2 = (\tau_2(y), \delta_2)$ be the goal structure with y the point identified by φ_2. Assume that $\delta_1(x)$ and $\delta_2(y)$ satisfy that either $\delta_1(x) = \odot^{\varphi_1}_{\varphi_2} A$ and $\delta_2(y) = A$ or $\delta_1(x) = A$ and $\delta_2(y) = \odot^{\varphi_1}_{\varphi_2} A$. For each of the above the following holds:

$$\Delta_1 \mid\!\sim \Delta_2 \parallel \varphi_2 \quad \text{iff} \quad \Delta_1 \parallel \varphi_1 \mid\!\sim \Delta_2$$

The functions induced by φ_1, φ_2 define the two places, and the shift rule tells us that $\odot^{\varphi_1}_{\varphi_2} A$ shifts A from one place to the other place.

It is worth noting that with a bit of abuse of notation, the shift operation can be written as

1. $\Delta_1[\odot A] \mid\!\sim \Delta_2[\varnothing]$ iff $\Delta_1[\varnothing] \mid\!\sim \Delta_2[A]$
2. $\Delta_1[A] \mid\!\sim \Delta_2[\varnothing]$ iff $\Delta_1[\varnothing] \mid\!\sim \Delta_2[\odot A]$

where $\varnothing$ is the empty set and its substitution at point x in $\Delta[x]$ means deletion.

Example 3.4. Consider **CL** with its list structures. Let φ_1 define the last element of a list and let φ_2 define the second element of a list. Then if $\odot$ denotes $\odot^{\varphi_1}_{\varphi_2}$ we get to the following two shift rules:

1. $(A_1, \ldots, A_m, \odot B) \mid\!\sim (C_1 \ldots C_n)$
   iff $(A_1 \ldots, A_m) \mid\!\sim (C_1, B, C_2, \ldots, C_n)$
2. $(A_1, \ldots, A_n) \mid\!\sim (C_1, \odot B, C_2, \ldots, C_n)$
   iff $(A_1, \ldots, A_n, B) \mid\!\sim (C_1, \ldots, C_n)$.

The shift operation suggest a new structural connective $\textcircled{!}A$. Its meaning can be motivated by the following observation. Given $\Delta_1(x)$ and $\Delta_2(y)$ the shift rule, stated intuitively, says for example that:

$$\Delta_1[A] \mid\!\sim \Delta_2(\varnothing) \quad \text{iff} \quad \Delta_1[\varnothing] \mid\!\sim \Delta_2[\odot A].$$

We understand the operation as shifting A from Δ_1 into Δ_2. After the shift, A is no longer present in Δ_1 but is recorded in Δ_2. That is why we write $\Delta_1[\varnothing]$. If we do not take A out Δ_1, we get a different rule namely

$$\Delta_1[\textcircled{!}A] \mid\!\sim \Delta_2[\varnothing] \quad \text{iff} \quad \Delta_1[\textcircled{!}A] \mid\!\sim \Delta_2[\odot A].$$

This means that although A is supposed to 'shift' to the right hand side, it stays while 'duplicating' itself. The connective '$\textcircled{!}$' indicates the duplications property.

Example 3.5. Consider the previous example where we had

$$\begin{aligned}&(A_1 \ldots A_m, B) \mid\!\sim (C_1, \ldots, C_n)\\ &\text{iff} \quad (A_1, \ldots, A_n) \mid\!\sim (C_1^{\odot}, B, C_2, \ldots, C_n)\end{aligned}$$

If the formula B does not delete itslef when shifting, but duplicates itself we indicate this by writing $\textcircled{!}B$ instead of B.

$$\begin{aligned}&(A_1, \ldots, A_n \textcircled{!}B) \mid\!\sim (C_1, \ldots, C_n)\\ &\text{iff } (A_1, \ldots, A_m, \textcircled{!}B) \mid\!\sim (C_1 \odot B, C_2, \ldots, C_n)\end{aligned}$$

Thus $\textcircled{!}B$ at the last place on the left simply indicates an infinite number of B's. Of course we can shift the entire $\textcircled{!}B$ to the right and get

$$\begin{aligned}&(A_1, \ldots, A_m, \textcircled{!}B) \mid\!\sim (C_1, \ldots, C_n)\\ &\text{iff } (A_1, \ldots, A_m) \mid\!\sim (C_1 \odot \textcircled{!}B, C_2, \ldots, C_n)\end{aligned}$$

Some properties of $\textcircled{!}$ and $\odot$ can be obtained from the above meaning, especially if other connectives are present, for example

$$\textcircled{!}B + B \equiv \textcircled{!}B$$

Another structural connective is the proof-net like connective, which identifies copies. Consider again $\Delta_1[x]$ and $\Delta_2[y]$. Using a shift connective, we can either put $x = A$ and $y = \varnothing$ or put $x = \varnothing$ and $y = \odot A$. In either case we get A on the left ($\Delta_1[A]$) and nothing on the right.

Another possibility is to put A on the right and nothing on the left, ie $\Delta_1[\varnothing], \Delta_2[A]$. We can write this possibility also as $\Delta_1[\odot A]$, $\Delta_2[\varnothing]$.

What we should *not* write is

$$\Delta_1[\odot A] \mid\!\sim \Delta_2[A]$$

because here we are duplicating A twice on the right. We could of course add an additional marker say $\dagger\dagger$, to indicate that these two copies of A should really be *one* copy. So we can write:

$$\Delta_1[\odot \dagger_A \dagger] \mid\!\sim \Delta_2[\dagger_A \dagger]$$

where $\dagger\dagger$ marks the fact that there is only one copy. Thus $(\odot \dagger_A \dagger, \dagger_A \dagger)$ is not the same as $(\dagger_A \dagger, \odot \dagger_A \dagger)$ becuase the first corresponds to A on the right and the second corresponds to A on the left.

Example 3.6. To continue the previous example,

$$(A_1, \ldots, A_n, \odot \dagger_A \dagger) \mid\!\sim (C_1, \dagger_A \dagger, C_2, \ldots, C_n)$$

is the same as

$$(A_1, \ldots, A_m) \mid\!\sim C_1, A, C_2, \ldots, C_n)$$

while

$$(A_1, \ldots, A_m, \dagger_A \dagger) \mid\!\sim (C_1, \odot \dagger_A \dagger, C_2, \ldots, C_m)$$

is the same as

$$(A_1, \ldots, A_m, A) \mid\!\sim (C_1, \ldots, C_n).$$[3]

We now turn out attention to the so called 'additive' connectives. As before, we start with a general consequence relation $\mid\!\sim$.

Consider now the case where we have that both $\Delta \mid\!\sim \Gamma[A_1]$ and $\Delta \mid\!\sim \Gamma[A_2]$ hold. We can abbreviate the above situation by writing in the metalevel $\Delta \mid\!\sim \lambda x \Gamma(x)[A_1 \wedge A_2]$ or just $\Delta \mid\!\sim \Gamma[A_1 \wedge A_2]$ where '$\wedge$' is a metaleval symbol, just like '$+_1$', '$+_2$' and 'o'. Similarly if we have $\Delta[A_1] \mid\!\sim \Gamma$ and $\Delta[A_2] \mid\!\sim \Gamma$ we can abbreviate it using '$\vee$' and write $\Delta[A_1 \vee A_2] \mid\!\sim \Gamma$. Note that '$\wedge$' and '$\vee$' only abbreviate other notation and are not operations on the data. We could modify the data structures if we want to accommodate the respective operations but that is not necessary.

Note that $\Delta[A_1 \wedge A_2] \mid\!\sim \Gamma$ is not meaningful, as we have not said what it means when '$\wedge$' appears on the left. Similarly for $\Delta \mid\!\sim \Gamma[A_1 \vee A_2]$.

The following hold by definition

$$\Delta \mid\!\sim A \wedge B \quad \text{iff} \quad \Delta \mid\!\sim A \text{ and } \Delta \mid\!\sim B$$

$$A \vee B \mid\!\sim \Gamma \quad \text{iff} \quad A \mid\!\sim \Gamma \text{ and } B \mid\!\sim \Gamma.$$

Example 3.7. Suppose we add the structural connectives Ⓥ and Ⓐ to the language; what are the obvious properties it must satisfy? In the general case we cannot say much beyond the obvious, so let us see what we get in the special case of linear logic. Here the Deduction Theorem comes into play:

---

3 The perceptive reader may ask why we are doing all of this shifting. Instead of writing plainly $(C, A) \mid\!\sim B$, for example, we code it as $C \mid\!\sim (B, \odot A)$, putting A on the right and then *marking* it by $\odot$ to indicate that it should really be on the left. Then to confuse matters even further, we write A on the left anyway and mark it so that the duplication is cancelled, i.e. $(C, \dagger_A \dagger \mid\!\sim (B, \odot \dagger_A \dagger)$.

The answer can be found in understanding the annotation as indicating *resource*. A structured database presents formulas to be used in the deduction in a way compatible with their structured layout. For linear logic the layout is a multiset and so each formula is to be used exactly once. In the course of the proof the formulas may be scattered about and/or duplicated. We need annotations to keep track of what is happening. This is why these connectives are useful.

1. $\Delta \mid\!\sim A \to B \circledwedge C$ iff $\Delta, A \mid\!\sim B \circledwedge C$
   iff $\Delta, A \mid\!\sim B$ and $\Delta, A \mid\!\sim C$
   iff $\Delta \mid\!\sim A \to B$ and $\Delta \mid\!\sim A \to C$
   iff $\Delta \mid\!\sim (A \to B) \circledwedge (A \to C)$.

2. $\Delta \mid\!\sim A \circledvee B \to C$
   iff $\Delta, A \circledvee B \mid\!\sim C$
   iff $\Delta, A \mid\!\sim C$ and $\Delta, B \mid\!\sim C$
   iff $\Delta \mid\!\sim A \to C$ and $\Delta \mid\!\sim B \to C$
   iff $\Delta \mid\!\sim (A \to C) \circledwedge (B \to C)$.

3. From the Identity Rule we get, since $A \circledvee B \mid\!\sim A \circledvee B$ that
   $A \mid\!\sim A \circledvee B$
   $B \mid\!\sim A \circledvee B$

4. Similarly since by the Identity Rule $A \circledwedge B \mid\!\sim A \circledwedge B$ we get that:
   $A \circledwedge B \mid\!\sim A$
   $A \circledwedge B \mid\!\sim B$

5. Since $A \mid\!\sim A \circledvee B$ we get $(A \circledvee B)^{\odot} \mid\!\sim A^{\odot}$ and in a similar way $(A \circledvee B)^{\odot} \mid\!\sim B^{\odot}$. Hence we get $(A \circledvee B)^{\odot} \mid\!\sim A^{\odot} \circledwedge B^{\odot}$.

6. Since $A \circledwedge B \mid\!\sim A$ we get $A^{\odot} \mid\!\sim (A \circledwedge B)^{\odot}$. Similarly $B^{\odot} \mid\!\sim (A \circledwedge B)^{\odot}$ and therefore $A^{\odot} \circledvee B^{\odot} \mid\!\sim (A \circledwedge B)^{\odot}$.

7. Since $(A \to C) \circledwedge (B \to C) \mid\!\sim A \to C$ and $A \oplus_1 (A \to C) \mid\!\sim C$ we get by **Surgical Cut** that $A \oplus_1 ((A \to C) \circledwedge (B \to C)) \mid\!\sim C$ and similarly $B \oplus_1 ((A \to C) \circledwedge (B \to C)) \mid\!\sim C$ hence $(A \circledvee B) \oplus_1 ((A \to C) \circledwedge (B \to C)) \mid\!\sim C$ and therefore $(A \to C) \circledwedge (B \to C) \mid\!\sim A \circledvee B \to C$.

So far our study of structural connectives and shift connectives used no special properties of $\mid\!\sim$ beyond the existence of $\to_r$ in the object language. We shall see that we are going to need some symmetry assumptions on $\mid\!\sim$. Suppose we start with the Tarski $\mid\!\sim_{\mathbf{LL}}$ of Example 1.7 and assume that $\mid\!\sim_1$ and $\mid\!\sim_2$ are both Scott type and are compatible with $\mid\!\sim_{\mathbf{LL}}$. Everything we have said and done so far would apply to both $\mid\!\sim_1$ and $\mid\!\sim_2$ equally. We can thus add the structural connective '$\odot$' and get all the equivalences mentioned earlier to hold. However there might be a difference between the case of $\mid\!\sim_1$ and that of $\mid\!\sim_2$, depending on their properties.

Example 3.8. Consider the consequence relation $\mid\!\sim_{LL^1}$ defined by

$$\Delta \mid\!\sim_{LL^1} A \quad \text{iff} \quad \text{for some } B \in \Delta,\ A \mid\!\sim_{\mathbf{LL}} B,$$

where Δ is a multiset.

This is a consequence relation by 1.9(2).

Let $\mid\!\sim_1$ be the minimal amalgamation of $\mid\!\sim_{\mathbf{LL}}$ and $\mid\!\sim_{LL^1}$. Clearly $\Delta \mid\!\sim_1 \Gamma$ iff for some $A \in \Gamma, \Delta \mid\!\sim_{\mathbf{LL}} A$ holds. This means that $\mid\!\sim_1$ is monotonic in Γ.

Let us add to the $\mid\!\sim_1$ the shift '$\odot$' and consider the following chain: $A \mid\!\sim_1 B, C$ iff $A \mid\!\sim_1 B^{\odot} \to C$ iff $A, B^{\odot} \mid\!\sim_1 C$. If we continue the chain using the fact that $\mid\!\sim_1$ agrees with $\mid\!\sim_{\mathbf{LL}}$ we get that the above holds iff $A, B^{\odot} \mid\!\sim_{\mathbf{LL}} C$. We can now get that $\mid\!\sim_{\mathbf{LL}}$ is monotonic in $B^{\odot}$, i.e. $\Delta \mid\!\sim_{\mathbf{LL}} C$ implies $\Delta, B^{\odot} \mid\!\sim_{\mathbf{LL}} C$. This is so because $\Delta \mid\!\sim_{\mathbf{LL}} C$ implies $\Delta \mid\!\sim_1 B, C$ and hence $\Delta, B^{\odot} \mid\!\sim_{\mathbf{LL}} C$ from the equivalence chain. Although $\mid\!\sim_{\mathbf{LL}}$ itself is not monotonic, there is nothing wrong with monotonicity in $B^{\odot}$, because we know that $\Delta, B^{\odot} \mid\!\sim_{\mathbf{LL}} C$ really means $\Delta \mid\!\sim_1 B, C$. However, it does force us syntactically to note which wff is of the form B and which is of the form $B^{\odot}$, and excludes writing $B^{\odot\odot}$ instead of B. Clearly this is not desirable. We therefore need to assume further properties on the $\mid\!\sim$ which agrees with $\mid\!\sim_{\mathbf{LL}}$, namely the property that it is *symmetrical*, namely $A \mid\!\sim \Gamma$ iff $\Gamma^{\odot} \mid\!\sim_{\mathbf{LL}} A^{\odot}$.

In terms of Definition 2.9 we want $\mid\!\sim$ to be the amalgamation of $\mid\!\sim_{\mathbf{LL}}$ with itself.

We now proceed with a series of definitions leading to the notion of self amalgamation.

Definition 3.9. (The dual of a consequence relation) Let $\mid\!\sim$ be a consequence relation in language $\mathbf{L}$. Consider a dual language $\mathbf{L}^*$ defined as follows:

1. The atoms of $\mathbf{L}^*$ are all atoms of the form q^*, where q is an atom of $\mathbf{L}$.
2. The connectives of $\mathbf{L}^*$ are all connectives of the form $\sharp^*$, where $\sharp$ is a connective of $\mathbf{L}$.

Let $*$ be a mapping from $\mathbf{L}$ to $\mathbf{L}^*$ defined by

3. $(q)^* = q^*$, for q atomic
4. $(\sharp(A_1, \ldots, A_n))^* = \sharp^*(A_1^*, \ldots, A_n^*)$, for a connective $\sharp$.
5. Define $\Delta^* \mid\!\sim^* A^*$ iff def $\Delta \mid\!\sim A$.

Example 3.10. In classical logic let $\wedge$ and $\vee$ be duals and let $A^* = \neg A$.

For further study of the ideas of this section, see [34].

4. A Case Study: What is the Logic of Classical Linear Logic

A lot has been said about linear logic. From our point of view, the system is very simple; it is the logic based on the data structures of multisets, satisfying the Deduction Theorem and fortified with additional structural and shift connectives.

It is our purpose in this section to present linear logic from this point of view.[4]

We start with the data structures of multisets for both the data and goal. Consider the language with $\rightarrow$ only and consider the smallest Scott type consequence relation $\mid\!\sim$ satisfying the deduction theorem for $\rightarrow$.

We know from example 1.7 that the smallest Tarski type consequence relation $\mid\!\sim_1$ for multisets satisfying the deduction theorem does characterise implicational intuitionistic linear logic. Thus it is clear that

$$\Delta \mid\!\sim_1 A \text{ iff } \Delta \mid\!\sim \{A\}.$$

What is not clear is the nature of the consequence $\Delta \mid\!\sim_2 A$ defined by

$$\Delta \mid\!\sim_2 A \quad \text{iff} \quad \{A\} \mid\!\sim \Delta.$$

We shall address this problem later.

Given $\mid\!\sim$, let us add to the language the structural connectives $\oplus_1, \oplus_2$ and the shift connective $\odot$, as we did in the previous section, when we were discussing such connectives for an arbitrary consequence relation. We get for our case that the following holds, where $X \equiv Y$ means both $X \mid\!\sim Y$ and $Y \mid\!\sim X$:

1. $A^{\odot\odot} \equiv A$
2. $A \oplus_2 B \equiv A^{\odot} \rightarrow B$
3. $A \oplus_1 B \equiv (A \rightarrow B^{\odot})^{\odot}$
4. $(A \oplus_1 B)^{\odot} \equiv A^{\odot} \oplus_2 B^{\odot}$
5. $(A \oplus_2 B)^{\odot} \equiv A^{\odot} \oplus_1 B^{\odot}$

4 The reader is cautioned that the phrase 'from our point of view' is important. Intuitionistic logic, from our point of view, is essentially the smallest monotonic consequence relation with the Deduction Theorem for $\rightarrow$, with sets as structures and the structural and shift connectives. However, intuitionistic logic arose in a context of far greater importance and motivation in the historical development of logic. Similarly, linear logic has its own historical context and to understand its role one has to go back as early as 1976 [17]. For these reasons, Avron [4] should be viewed in context.

We can also add the 'duplication' connective ⓣ and markers. Let us agree that $\dagger_n\dagger$ is the corresponding connective on the right. We thus have:

6. $A \oplus_1 ⓣ_1 \equiv ⓣ_1 A$
7. $A \oplus_2 ⓣ_2 A \equiv ⓣ_2 A$
8. $A \oplus_1 B^{\dagger_n\dagger} \mathrel{|\!\sim} C \oplus_2 (B^{\odot})^{\dagger_n\dagger}$ is the same as $A \oplus_1 B \mathrel{|\!\sim} C$
9. $A \oplus_1 B^{\odot\dagger_n\dagger} \mathrel{|\!\sim} C \oplus_2 B^{\dagger_n\dagger}$ is the same as $A \oplus_1 B^{\odot} \mathrel{|\!\sim} C$.

The moral of the story is that the properties of all of these connectives are determined by the structure and their geometric meaning.

The additives Ⓐ and Ⓥ can be added to linear logic as in the Example 3.7. When added to the language alongside $\oplus_1$ and $\oplus_2$, one can form formulas with arbitrary nestings of Ⓐ, Ⓥ, $\oplus_1, \oplus_2$ within themselves. Given the meaning of these connectives, the corresponding meatalevel structures are the composition of two consequence relations, one for the additives over sets and one for the multiplicatives over multisets, as in 2.16.

In the case of the additives of linear logic, the structures τ^* were multisets of sets (not hereditary). These can be regarded as a family of structures $\{\tau\}$ of $\mathcal{M}'$ (i.e. multisets) obtained by choosing all possible points from the sets of points in τ^*. For example if τ^* is $(\{a,b\},\{c,d\},\{e\})$ then the following set of τ's are associated

$$(a,c,e),\ (a,d,e),\ (b,c,e) \text{ and } (b,d,e).$$

We write formally $\tau \in \tau^*$ to indicate the connection. We can define the function δ^* to give values (wff) to each point. Thus databases and goals become lists of sets of formulas or multisets of sets of formulas.

For example, if Δ^* is $(\{A,B\}, \{C,D\}, \{E\})$ we get the following associated databases

$$(A,C,E),\ (A,D,E),\ (B,C,E) \text{ and } (B,D,E).$$

We write again $\Delta \in \Delta^*$ to indicate the association.

Our goal is to define a consequence relation $\Delta^* \mathrel{|\!\sim}^* \Gamma^*$ by stipulating: $\Delta^* \mathrel{|\!\sim}^* \Gamma^*$ iff$_{\text{def}}$ for all $\Delta \in \Delta^*$, $\Gamma \in \Gamma^*$, $\Delta \mathrel{|\!\sim} \Gamma$.

We need to show that we get an S-consequence relation and for that we need to define the *Surgical Cut Rule*; we need to define substitution of one structure in another. We refer back to the general definition of the *composition* of two S-consequence relations.

We now present a theorem about the Hilbert formulation of linear logic with $\to$ and $\neg$ which shows that $\odot$ can be taken as $\neg$ and that $\neg$ can be mapped into the implicational fragment.

Theorem 4.1. *Let* **LL**$(\neg)$ *be the extension of* **LL** *of Example* 1.7 *with the unary symbol* $\neg$ *and the following axioms*:

$$\Vdash \neg\neg A \leftrightarrow A$$
$$\Vdash (\neg(A \to \neg B) \to C) \leftrightarrow (A \to (B \to C))$$
$$\Vdash (A \to B) \to (\neg B \to \neg A)$$

Let $\bot$ *be an arbitrary atom of* **LL**. *Let* $\varphi^{\bot}(A)$ *be a translation from* **LL**$(\neg)$ *into* **LL**$(\bot)$ *defined as follows*:

$$\varphi^{\bot}(q) = (q \to \bot) \to \bot, \quad q \textit{ atomic}$$
$$\varphi^{\bot}(A \to B) = \varphi^{\bot}(A) \to \varphi^{\bot}(B)$$
$$\varphi^{\bot}(\neg A) = \varphi^{\bot}(A) \to \bot$$

Then the following holds:

$$\mathbf{LL}(\neg) \Vdash A \quad \textit{iff} \quad \mathbf{LL} \Vdash \varphi^{\bot}(A).$$

Proof. In [10]. ■

Remark 4.2. The previous theorem yields the following:

1. Semantics for **LL**$(\neg)$, through the semantics for **LL**.
2. It shows that $\neg, \oplus_1$ and $\oplus_2$ (i.e. all the structural connectives of linear logic) are already definable from the impliction $\to$, via the 'internal' translation $\varphi^{\bot}$, for a fixed atom $\bot$.
3. It shows that $\neg$ can also be regarded as a negation, and not merely as a shift operator.

Example 4.3. (Cut Free Formulation of Linear Logic) In this section, we started with a consequence relation $\vdash_{\mathbf{LL}}$ for multisets as data, added the obvious structural and shift connectives together with the additives, defined the symmetric amalgamation and got a conseqeunce relation $\mid\!\sim$. Certain properties of $\mid\!\sim$ were listed in the previous example. It is possible to list enough properties of $\mid\!\sim$ to enable one to derive the *Cut Rule.* When this is done one gets a cut free formulation of the sytem $\mid\!\sim$. Although we have been referring to $\mid\!\sim$ as linear logic, we have not proved that $\mid\!\sim$ is indeed the system known in the literature as linear logic. To show that, all we need to do is to take a known formulation of linear logic, say $\vdash_1$, see e.g. [18], and prove that all $\vdash_1$ rules are valid in $\mid\!\sim$. This will show that $\vdash_1 \subseteq \mid\!\sim$. If we prove the Cut Elimination for $\vdash_1$ this will show that $\mid\!\sim \subseteq \vdash_1$. Thus essentially Cut Elimination for $\vdash_1$ establishes that $\mid\!\sim$ is indeed linear logic.

References

1. S. Abramsky, Computational interpretation of linear logic, *Theoretical Computer Science 111*, 1993, 3–57.
2. S. Akama, Mathematical foundations of non-monotonic logics, *Journal of Automated Reasoning* (to appear).
3. C. E. Alchourrón, P. Gärdenfors and D. Makinson, On the logic of theory change: Partial meet contraction and revision functions, *Journal of Symbolic Logic 50*, 1985, 510–530.
4. A. Avron, The semantics and proof theory of linear logic, *Theoretical Computer Science 57*, 1988, 161–184.
5. N. D. Belnap, Display logic, *Journal of Philosophical Logic 11*, 1982, 375–417.
6. P. Besnard, Axiomatisation in the metatheory of non-monotonic inference systems, *Proceedings of COCSI '88*, Edmonton 1988, 117–124.
7. G. Crocco, L. Farinas del Cerro, Structural Logics, in *What is a Logical System?*, D. M. Gabbay (Ed.), Oxford University Press (to appear).
8. D. M. Gabbay, *Semantical Investigations in Heyting's Intuitionistic Logic*, Reidel, Dordrecht, 1981.
9. ———, Theoretical foundations for non-monotonic reasoning in expert systems, in *Proceedings of the NATO Advanced Study Institute on Logics and Models of Concurrent Systems, La Colle-sur-Loup, France*, K. R. Apt (Ed.), Springer, Berlin, 1985, 439–457.
10. ———, *Labelled Deductive Systems*, 1st draft September 1989, 6th draft February 1991, Oxford University Press (to appear).
11. ———, Algorithmic proof with diminishing resource, I, in *Proceedings of CSL '90*, E. Börger, H. Kleine Büning, M. M. Richter and W. Schönfeld (Eds.), Springer, Berlin, 1991 (LNCS 533), 156–173.
12. ———, Theoretical foundations for non-monotonic reasoning, Part 2: Structured non-monotonic theories. In *SCAI '91 - Proceedings of the Third Scandinavian Conference on AI*, IOS Press, Amsterdam, 1991, 19–40.
13. ———, Theory of algorithmic proof, in *Handbook of Logic in Theoretical Computer Science*, Vol. 2, S. Abramsky, D. M. Gabbay and T. Maibaum (Eds.), Oxford University Press, 1992, 311–413.

14. ———, *Fibred Semantics and the Weaving of Logics*, Manuscript, Imperial College, September 1992.

15. ———, How to construct a logic for your application, in *GWAI-92: Advances in Artificial Intelligence*, H. J. Ohlbach (Ed.), Springer, Berlin, 1992 (LNAI 671), 1–30.

16. P. Gärdenfors, *Knowledge in Flux: Modelling the Dynamics of Epistemic States*, MIT Press, Cambridge MA, 1988.

17. J-Y. Girard, Three-valued logic and cut-elimination: The actual meaning of Takeuti's conjecture, *Dissertationes Mathematicae 136*, Warsaw, 1976.

18. ———, Linear logic, *Theoretical Computer Science 50*, 1987, 1–102.

19. S. Kraus, D. Lehmann and M. Magidor, Non-monotonic reasoning, preferential models and cumulative logics, *Artificial Intelligence 44*, 1990, 167–207.

20. S. L. Lauritzen and D. J. Spiegelhalter, Local computation with probabilities on graphical structures and their application to expert systems (with discussion). *Journal of the Royal Statistical Society (B) 50*, 157–224.

21. D. Lehmann, What does a conditional knowledge base entail?, in *Proceedings of the 1st International Conference on Principles of Knowledge Representation and Reasoning - KR '89, Toronto, 1989*, Morgan Kaufmann.

22. D. Lehmann and M. Magidor, Rational logics and their models: a study in cumulative logics (submitted for publication).

23. D. Makinson, General theory of cumulative inference, in *Proceedings of the 2nd International Workshop on Non-monotonic Reasoning*, Reinfrank, M. (Ed.), Springer, Berlin, 1989 (LNAI 346), 1–18.

24. ———, General patterns in nonmonotonic reasoning, in *Handbook of Logic in Artificial Intelligence and Logic Programming, Vol. 3*, D. M. Gabbay, C. J. Hogger and J. A. Robinson (Eds.), Oxford University Press (to appear).

25. D. Makinson and P. Gärdenfors, Relations between the logic of theory change and nonmonotonic logic, in *The logic of theory change*, A. Fuhrmann and M. Morreau (Eds.), Springer, Berlin, 1990 (LNAI 465), 185–205.

26. D. Nute, *LDR – A Logic for Defeasible Reasoning*, ACMC Research Report 01-0013, 1986.

27. M. Ryan, Defaults and revision in structured theories, *Proceedings of the 6th Annual IEEE Symposium on Logic in Computer Science (Amsterdam 1991)*, IEEE Computer Society Press, Los Alamitos CA, 1991, 362–373.

28. Y. Shoham, A semantical approach to non-monotonic logics, in *Proceedings of the 2nd Annual Symposium on Logic in Computer Science (Ithaca NY 1987)*, IEEE Computer Society Press, Los Alamitos CA, 1987, 275–279.

29. ———, *Reasoning about Change: Time and Causation from the Standpoint of Artificial Intelligence*, MIT Press, Cambridge MA, 1988.

30. A. Urquhart, Semantics for relevant logics, *Journal of Symbolic Logic 37*, 1972, 159–169.

31. J. van Benthem, The Lambek Calculus, in *Categorial Grammars and Natural Language Structures*, R. Oehrle, E. Bach and D. Wheeler (Eds.), Reidel, Dordrecht 1987, 35–68.

32. D. Vermeir and E. Laenens, An overview of ordered logic, in *Abstracts of the Third Logical Biennial*, Varna, Bulgaria, 1990.

33. H. Wansing, *Formulas as types for a hierarchy of sublogics of intuitionistic propositional logic*, Technical report no 9/90, Free University of Berlin, 1990, 1–29.

34. ———, *Tarskian structured consequence relations* (to appear).

35. R. Wójcicki, An axiomatic treatment of non-monotonic arguments, *Studia Logica* (to appear).

36. ———, Heuristic rules of inference in non-monotonic arguments, *Studia Logica* (to appear).

Decidability and Interpolation for a First-Order Relevance Logic

Aleksandar Kron

Faculty of Philosophy, University of Belgrade[1]

1 Introduction

Grishin (cf. [7], [8], [9], and [10]) investigated the logic obtained from the classical first-order calculus by deleting the contraction axiom-scheme

$$A \to (A \to B) \to . A \to B.$$

He showed that this logic was decidable (cf. [7] and [9]); the result has been obtained by using a suitable Gentzen-style formulation and by proving a cut-elimination theorem. Another decidable fragment of contractionless predicate logic was investigated by Ketonen and Weyhrauch (cf. [13]).

In both contractionless logics mentioned above, the distribution laws

$$A\&(B \vee C) \to (A\&B) \vee (A\&C)$$

and

$$(A \vee B)\&(A \vee C) \to A \vee (B\&C)$$

fail to be theorems.

In this paper we are concerned with a subsystem **GQRW** of the quantificational contractionless relevance logic **RWQ**. The above forms of distribution are included in **RWQ**; they will be accepted in **GQRW** as well.

Until recently there was no known Gentzen-style formulation even for propositional contractionless relevance logics. The first Gentzen-style formulations of the positive fragment of the propositional relevance logic **R**

1 This research has been supported by the Science Fund of Serbia (grant number 0401A) through the Mathematical Institute in Belgrade.

were obtained by Kripke (cf. [15]), Dunn (cf. [1] and [5]) and Mints (cf. [17]). Urquhart has shown in [22] that this fragment, as well as **R** itself, is undecidable. However, Giambrone proved that the contractionless positive fragment of **R** was decidable, by using Dunn's style of gentzenization via extensional and intensional sequences. Eventually, Brady (cf. [3]) succeeded to combine the results of Dunn, Giambrone, Belnap and Bull (cf. [6], [2], and [4]) to obtain a cut-free Gentzen-style formulation of **RW** (contractionless **R**), since negation in contractionless systems enjoys some properties of constructible falsity investigated by Zaslavskiĭ in [23]. Note, however, that in most of these relevance systems the deducibility problem is undecidable (cf. [22]).

An obvious reading of [7], [6] and [3] shows that to prove the axioms of distributionless **RW** in a Gentzen-style one does not need extensional sequences. To have distribution laws, we may introduce appropriate rules; this is done in our paper. However, closure under the unrestricted Cut rule becomes unprovable. This entails the underivability of some theorems of **RWQ** even though all the axioms of **RWQ** are derivable.

The system **GQRW** is a Gentzen-style system, a subsystem of first-order **RWQ**. All rules of **GQRW** are derived in **RWQ** (for $\neg\rightarrow$ cf. [16]). We show that it has a property of propositional intuitionistic logic. Let antecedent and consequent occurrences of subformulas be introduced as in [1]; then the property can be stated as follows: (a) if $A \rightarrow (B \vee C)$ is provable and there is no antecedent occurrence of either $\vee$ or $\exists$ in A, then either $A \rightarrow B$ or $A \rightarrow C$ is provable, and, in the first-order case, (b) if $A \rightarrow \exists x B$ is provable and there is no antecedent occurrence of either $\vee$ or $\exists$ in A, then there is a term t such that $A \rightarrow B(t)$ is provable. The property has been investigated in intuitionistic logic by Harrop in [11] and [12], by Rasiowa and Sikorski in [18] and by Kleene in [14], and used many times (cf. [19] and [20] for example) for theories based on intuitionistic logic. We prove here that **GQRW** has the same property, although the formula $\neg\neg A \rightarrow A$ is a theorem of **GQRW** (fortunately enough, the formula $p \vee \neg p$ is not). Moreover, the property in **GQRW** can be extended to & and $\forall$: (a) if $B \& C \rightarrow A$ is provable in **GQRW** and there is neither an antecedent occurrence of $\vee$ or $\exists$ nor a consequent occurrence of & or $\forall$ in A, then either $B \rightarrow A$ or $C \rightarrow A$ is provable in **GQRW**, and (b) if $\forall x B \rightarrow A$ is provable in **GQRW** and there is neither an antecedent occurrence of $\vee$ or $\exists$ nor a consequent occurrence of & or $\forall$ in A, then there is a term t such that $B(t) \rightarrow A$ is provable in **GQRW**. This property, a consequence of the presence of $\neg\neg A \rightarrow A$ and De Morgan laws, enjoyed by **GQRW**, is used in an essential way in the proof of closure under a restricted Cut rule as well as in the proof of a weak Herbrand's theorem.

We summarize the difference between systems where intensional and extensional sequences are used and **GQRW**. Having intensional and extensional sequences allows us to accept a contraction rule for extensional sequences. This enables the derivation of distribution axioms as well as closure under Cut, which makes a standard formulation of modus ponens admissible in propositional logic. Hence, there is a proper Gentzen-style formulation of **RW**. However, such a contraction rule blocks Grishin's proof of decidability of the corresponding first-order logic. In addition, Urquhart has proved that the interpolation theorem fails for **E**, **T** and **R**.[2]

In **GQRW** there is no contraction rule at all. The rules that are necessary for the derivation of distribution theorems make the proof of closure under Cut possible only under some restrictions. Hence, **GQRW** is not a Gentzen-style formulation of **RWQ**. Even worse, neither is the propositional subsystem of **GQRW** a Gentzen-style formulation of **RW** nor does **GQRW** contain the whole of **TW**. On the other hand, **GQRW** is a decidable first-order logic with the Craig-Lyndon interpolation property.

2 The System GQRW

GQRW is supposed to be given in a first-order language. We shall write $\rightarrow$, $\&$, $\vee$, $\neg$, $\forall$ and $\exists$ for (relevant) implication, conjunction, disjunction, negation, universal and existential quantifier, respectively. We allow vacuous occurrences of quantifiers. We shall write $A(\substack{x \\ t})$ for the result of substituting the term t for all free occurrences of the variable x in A.

Let X, Y, Z, ... denote finite (possibly empty) sequences of formulas. If X and Y are such sequences, then X,Y is the sequence obtained by writing down all members of X and then all members of Y, retaining their internal order in X and Y, respectively. A *sequent* (*consecution*) is such a nonempty sequence X of formulas.

An *axiom* is any sequent of the form

$$p, \neg p$$

where p is an atomic formula.

The rules of **GQRW** are the following:

$$\frac{X, A, B, Y}{X, B, A, Y} \qquad \frac{X, A, Y}{X, \neg\neg A, Y}$$

$$\frac{X, A, Y}{X, A\&B, Y} \qquad \frac{X, B, Y}{X, A\&B, Y} \qquad \frac{X, \neg A, Y \quad X, \neg B, Y}{X, \neg(A\&B), Y}$$

$$\frac{X, A, Y \quad X, B, Y}{X, A \vee B, Y} \qquad \frac{X, \neg A, Y}{X, \neg(A \vee B), Y} \qquad \frac{X, \neg B, Y}{X, \neg(A \vee B), Y}$$

2 According to oral communications by Kosta Došen and Ross T. Brady.

$$\frac{X, A\&B, Y \quad X, A\&C, Y}{X, A\&(B \vee C), Y} \qquad \frac{X, B\&A, Y \quad X, C\&A, Y}{X, (B \vee C)\&A, Y}$$

$$\frac{X, \neg(A \vee B), Y \quad X, \neg(A \vee C), Y}{X, \neg(A \vee (B\&C)), Y} \qquad \frac{X, \neg(B \vee A), Y \quad X, \neg(C \vee A), Y}{X, \neg((B\&C) \vee A), Y}$$

$$\frac{X, \neg A, Y \quad U, B, V}{X, U, A \to B, Y, V} \qquad \frac{X, A, \neg B, Y}{X, \neg(A \to B), Y}$$

$$\frac{X, A(\tbinom{x}{t}), Y}{X, \forall x A, Y} \quad \frac{X, \neg A(\tbinom{x}{y}), Y}{X, \neg\forall x A, Y} \quad \frac{X, \neg(A \vee B(\tbinom{x}{y})), Y}{X, \neg(A \vee \forall x B), Y} \quad \frac{X, \neg(B(\tbinom{x}{y}) \vee A), Y}{X, \neg(\forall x B \vee A), Y}$$

$$\frac{X, A(\tbinom{x}{y}), Y}{X, \exists x A, Y} \quad \frac{X, \neg A(\tbinom{x}{t}), Y}{X, \neg\exists x A, Y} \quad \frac{X, A\&B(\tbinom{x}{y}), Y}{X, A\&\exists x B, Y} \quad \frac{X, B(\tbinom{x}{y})\&A, Y}{X, \exists x B\&A, Y}$$

The rules are named according to the structure of the member introduced by the given rule. For example, the first one is called *permutation* and denoted by *Perm*. The third and the fourth are & rules, the fifth is $\neg\&$, and the last two are $\&\exists$ and $\exists\&$.

We assume that in $\forall$ and $\neg\exists$ the term t (called *discharged*) is free for x in A, and that in $\neg\forall$, $\exists$, $\neg \vee \forall$, $\neg\forall\vee$, $\&\exists$ and $\exists\&$ the variable y (the *eigenvariable* of the rule) is not free in the conclusion of these rules.

Perm is the only *structural* rule; the remaining rules are *logical.* In the sequel the use of *Perm* will not be mentioned. The last eight rules are called *quantificational*, in contrast to the remaining rules, which are *propositional.*

By a *derivation* of a sequent X in **GQRW** we understand a tree, branching upward, with usual properties. The notion of *immediately above* (*below*) is taken as primitive, and that of *above* (*below*) is its transitive closure. For a node Y of such a tree the *weight* of Y is the number of nodes above Y, on all branches containing Y. We say that a sequent is *derivable with weight* w iff there is a derivation $\mathcal{T}$ of X and the weight of X in $\mathcal{T}$ is w. In the sequel 'derivable' means 'derivable in **GQRW**'.

A formula A is a *theorem* of **GQRW** iff the sequent $\neg A$ is derivable. All theorems of **GQRW** are theorems of quantificational **RW**. To see this, it suffices to establish the following fact by using induction on weight.

Let X be the sequence

$$A_1, \ldots, A_{m-1}, A_m, A_{m+1}, \ldots, A_n,$$

and let A be any formula

$$A_1 \to \ldots \to . A_{m-1} \to . A_{m+1} \to \ldots \to . A_n \to \neg A_m;$$

if X is derivable in **GQRW**, then A is provable in **RWQ**.

By an *application* *App*(ρ) of a rule ρ in a derivation $\mathcal{T}$ we mean nodes X' (and X'') of $\mathcal{T}$ such that (1) X is the conclusion of ρ, and X' (and X'')

is (are) premiss(es) of ρ, and (2) X' (and X'') is (are) in $\mathcal{T}$ immediately above X. In an application $App(\rho)$ of a rule ρ, the premiss(es) and the conclusion of ρ are called *the premiss(es)* and *the conclusion* of $App(\rho)$, respectively.

By the *principal* member of the conclusion X, in an application $App(\rho)$ of a logical rule ρ, we mean the member of the conclusion of $App(\rho)$ which occurs in no premiss of $App(\rho)$. An occurrence α of a connective or quantifier is *principal* if α is the principal occurrence in the principal member. For example, in & and in &∨ the displayed occurrence of & is principal. By a *sub-principal* formula in the conclusion of $App(\rho)$ we understand an occurrence of the proper subformula (if there is any) of the principal member of $App(\rho)$ which occurs in no premiss of $App(\rho)$. For example, B and $B \vee C$ in & and in &∨ are sub-principal members of $App(\&)$ and $App(\&\vee)$, respectively.

If a member of the conclusion is not principal, it is called a *parametric member* (pm). If $\mathcal{T}$ is a derivation of X,A,Y and A is a parametric member, then there are nodes (1) $X_1,A,Y_1,\ldots$, (n) X_n,A,Y_n of $\mathcal{T}$ above X,A,Y such that for any i, $1 \leq i \leq n$, either A is principal in (i) or (i) is an axiom.

A *constituent* member in a premiss, in an application $App(\rho)$ of a logical rule, is the member of a premiss of $App(\rho)$ which is not a member in the conclusion of $App(\rho)$. For example, A and $A\&B$ are constituent members of $App(\&)$ and $App(\&\vee)$, respectively.

We shall write $\alpha(B(A))$ (and $\alpha(X(A))$) to denote B and a particular occurrence of A in B (X and a particular occurrence of A in a member of X).

By the *degree* of a formula A we understand the number $d(A)$ of occurrences of connectives or quantifiers in A.

Following [21], we say that a derivation $\mathcal{T}$ is *regular* iff (1) all eigenvariables in $\mathcal{T}$ are distinct and (2) no eigenvariable is free in a sequent below the node in which it is an eigenvariable. Without loss of generality we shall assume that all derivations are regular.

Let Y and X be a premiss and the conclusion in an application $App(\rho)$ of a rule ρ, respectively; it may happen that for a particular occurrence $\alpha(A)$ of a subformula A in Y (in X) there is a corresponding occurrence $\beta(A')$ of A' in X (in Y), where either A' is A or A and A' are $B(\frac{x}{t})$ and QxB, respectively, for some B, Q being a quantifier and t free for x in B. The official definition of the corresponding occurrence of a subformula A in Y (in X) is left to the reader.

Let Y be a node in a derivation $\mathcal{T}$ of a sequent X; we say that an occurrence $\alpha(A)$ of a formula A is *introduced in* $\mathcal{T}$ *via* Y if either Y is the axiom $A, \neg A$, or A is the principal member of Y, or there is a sub-principal member B of Y such that A is a subformula of B and there is no occurrence

$\beta(A')$ of A' in a constituent member of a premiss of Y corresponding to $\alpha(A)$.

GQRW has some classical properties. For example, the connectives & and $\vee$ are associative, commutative, absorptive, idempotent and mutually distributive, in the sense that the corresponding formulas are theorems of **GQRW**.

However, in absence of contraction of any kind, there are theorems of **RW** such that the corresponding sequents are underivable in **GQRW**. Consider Giambrone's example:

$$(A \to A)\&(A \to (B \vee C)) \to .((A\&B \to D)\&(A\&C \to D)) \to .A \to D.$$

One of the corresponding sequents in **GQRW** would be

$$(A \to A)\&(A \to (B \vee C)), (A\&B \to D)\&(A\&C \to D), A, \neg D.$$

This sequent is underivable in **GQRW**. We give another example:

$$((I \to .H \to G)\&((A \to C)\&F) \to G) \to .(A \to B)\&I\&(L \to M) \to \\ .(M \to F)\&(N \to .(B \to C) \vee H) \to .L\&N \to G.$$

The corresponding sequent

$$((I \to .H \to G)\&((A \to C)\&F) \to G), (A \to B)\&I\&(L \to M), \\ (M \to F)\&(N \to .(B \to C) \vee H), L\&N, \neg G.$$

is underivable in **GQRW**. [3]

3 Antecedent and Consequent Occurrences

In this section we develop the main method to be used in the proof that **GQRW** is closed under a restricted Cut rule.

The *depth* $\delta(\alpha(X(A)))$ of an occurrence $\alpha(X(A))$ is defined as follows: if $\alpha(X(A))$ is a member of X, then $\delta(\alpha(X(A))) = 0$; if $\alpha(X(A))$ is an occurrence of a proper subformula in a member of X, then either $\delta(\alpha(X(A))) = \delta(\alpha(X(\neg(A))))+1$, or else $\delta(\alpha(X(A))) = \delta(X(\alpha(B))) = \delta(\alpha(X(A*B)))+1$, where $*$ is a binary connective, or else $\delta(\alpha(X(A))) = \delta(\alpha(X(QxA))) + 1$, where Q is a quantifier.

For any sequent X we define *antecedent* and *consequent* occurrences of subformulas in X, by induction on the depth of the given occurrence.

3 I wish to thank R. K. Meyer, S. Giambrone and R. T. Brady for showing that this formula is a theorem of $\mathbf{TW}_+$.

1. Any member A of X is an antecedent occurrence in X of the formula A;

2. If $A \to B$ is an antecedent (consequent) occurrence, then the displayed occurrence of A is a consequent (antecedent) occurrence and the displayed occurrence of B is an antecedent (consequent) occurrence.

3. If $A * B$ is an antecedent (consequent) occurrence, then the displayed occurrences of both A and B are antecedent (consequent) occurrences, where $*$ is either & or $\vee$.

4. If $\neg A$ is an antecedent (consequent) occurrence, then the displayed occurrence of A is a consequent (antecedent) occurrence.

5. If QxA is an antecedent (consequent) occurrence, where Q is a quantifier, then the displayed occurrence of A is an antecedent (consequent) occurrence.

By an antecedent (consequent) occurrence in X of a binary connective $*$, or else of $\neg$, or else of a quantifier, we mean an antecedent (consequent) occurrence in X of a formula $A * B$, or $\neg A$, or $\forall xA$ or $\exists xA$, respectively.

The above definition of antecedent and consequent occurrences is the same as that in [1]. The apparent difference disappears after the following consideration: suppose that we had sequents with nonempty antecedent and consequent parts; any member of the antecedent (consequent) part of the sequent would be an antecedent (consequent) occurrence of that member; since in our sequents the consequent part is empty, we have the same definition as in [1].

Lemma 1. (a) *Let $\alpha(A)$ be an antecedent occurrence of a subformula A in a sequence X; if $\beta(B)$ is an antecedent (consequent) occurrence of a subformula B in the sequence A (consisting of the formula A alone), then $\beta(B)$ is an antecedent (consequent) occurrence of the subformula B in X.*

(b) *Let $\alpha(A)$ be a consequent occurrence of a subformula A in a sequence X; if $\beta(B)$ is an antecedent (consequent) occurrence of a subformula B in the sequence A, then $\beta(B)$ is a consequent (antecedent) occurrence of the subformula B in X.*

Proof. By induction on the depth of $\beta(B)$ in A.

The next lemma states an important property of $\alpha(C)$ and $\beta(C')$.

Lemma 2. (a) *If an occurrence $\alpha(C)$ of a subformula C of the constituent member A of App(ρ) of a logical rule ρ is an antecedent (consequent) occurrence of C in a premiss Y of App(ρ), then the corresponding occurrence $\beta(C')$ of C' in the principal member B of App(ρ) is an antecedent (consequent) occurrence of C' in the conclusion X of App(ρ).*

(b) *If an occurrence $\alpha(C)$ of a subformula C of the principal member A of App(ρ) of a logical rule ρ is an antecedent (consequent) occurrence of C in the conclusion X of App(ρ), and if there is a corresponding occurrence $\beta(C')$ of C' in the constituent member B of a premiss of App(ρ), then $\beta(C')$ is an antecedent (consequent) occurrence of C' in the premiss Y of App(ρ).*

For any sequence X we say that X satisfies the *HRSK-condition* or that it is an *HRSK-sequence* (HRSK for Harrop, Rasiowa, Sikorski and Kleene) iff in X there are neither antecedent occurrences of $\vee$ or $\exists$ nor consequent occurrences of & or $\forall$. If an HRSK-sequence X consists of a single formula A, we shall say that A is an *HRSK-formula.*

It should be clear that in a derivation of an HRSK-sequence the only branching we may have is induced by the $\rightarrow$-rule, in which the members of the conclusion sequent occur in only one of the premiss sequents.

As a consequence of Lemma 1, we have the following facts: $A \rightarrow B$ is an HRSK-formula iff $\neg A$ and B are; $\neg(A \rightarrow B)$ is an HRSK-formula iff A and $\neg B$ are. Similar facts are true for the remaining connectives and quantifiers.

Theorem 1. *If* (a) $X, A\&B, Y$ *is derivable and* X, Y *is an HRSK-sequence, then either* (b) X, A, Y *or* (c) X, B, Y *is derivable.*

Proof. Obviously, (a) is no axiom. Proceed by double induction on the degree of $A\&B$ and on the weight of (a) in the given derivation $\mathcal{T}$.

If $A\&B$ is not principal, the use of the induction hypothesis (on the weight of (a)) is trivial, since X, Y is an HRSK-sequence.

Suppose that the displayed occurrence of $A\&B$ is introduced in $\mathcal{T}$ via (a). If (a) is obtained by &, the theorem is true.

If (a) is obtained by &$\vee$, then B is $C \vee D$ and (a) is obtained from (a′) $X, A\&C, Y$ and (a″) $X, A\&D, Y$. By induction hypothesis (on the degree of $A\&B$), either (b′) X, A, Y is derivable or both (c′) X, C, Y and (c″) X, D, Y are derivable. If (b′) is derivable, the theorem is true; if (c′) and (c″) are derivable, then we obtain (b) by using $\vee$.

We proceed similarly if (a) is obtained by $\vee$&.

If (a) is obtained by &$\exists$ from $X, A\&B(\frac{x}{y}), Y$, then by the induction hypothesis, either (b) X, A, Y or (c′) $X, B(\frac{x}{y}), Y$ is derivable. If (c′), we obtain (c) by using $\exists$.

We proceed similarly if (a) is obtained by $\exists$&.

Theorem 2. *If* (a) $X, \neg(A \vee B), Y$ *is derivable and* X, Y *is an HRSK-sequence, then either* (b) $X, \neg A, Y$ *or* (c) $X, \neg B, Y$ *is derivable.*

Theorem 3. *If* (a) $X, \forall x A, Y$ *is derivable and* X, Y *is an HRSK-sequence, then there is a term t free for x in A such that* (b) $X, A(^x_t), Y$ *is derivable.*

Proof. If $\forall x A$ is principal in (a), the theorem is proved. Otherwise, the use of the induction hypothesis (on the weight of (a)) is trivial, since X, Y is an HRSK-sequent.

Theorem 4. *If* $X, \neg\exists x A, Y$ *is derivable and* X, Y *is an HRSK-sequence, then there is a term t free for x in A such that* $X, \neg A(^x_t), Y$ *is derivable.*

4 Closure Under a Restricted Cut Rule

We shall say that sequents X_1 and X_2 are derivable with *combined weight* w if X_i, $1 \leq i \leq 2$, is derivable with weight w_i and $w = w_1 + w_2$.

For any formula A let us define A^*: if A is of the form $\neg B$, then A^* is B; otherwise, A^* is $\neg A$.

Theorem 5. *If* (a) X, p, Y *and* (b) U, $\neg p$, V *are derivable, so is* (c) X, U, Y, V; *if* (a′) X, $\neg p$, Y *and* (b′) U, p, V *are derivable, so is* (c′) X, U, Y, V.

Proof. By induction on the weight of (b) and (b') in the given derivation.

Theorem 6. *Suppose that the following conditions are satisfied:*

the sequent (1) X, A, Y *is derivable;*

the sequent (2) U, A^*, V *is derivable;*

both A *and* A^* *are HRSK-formulas.*

Then (3) X, U, Y, V *is derivable.*

Proof. If either A or A^* is atomic, we can apply Theorem 5. Hence, we may assume that neither (1) nor (2) is an axiom.

Proceed by double induction as usual. Let the degree of A and A^* be d and d^*, respectively, and let $w = w_1 + w_2$. Our induction hypotheses are:

Hyp 1) the theorem holds for any B if $\min(d(B), d(B^*)) < \min(d, d^*)$;

Hyp 2) the theorem holds for any formula B such that $\min(d(B), d(B^*)) = \min(d, d^*)$ and any combined weight $w' < w$.

We can distinguish two cases: (I) both members A and A^* are introduced in the corresponding derivations via (1) and (2), respectively, and (II) otherwise. In II the proof is standard and therefore omitted. Let us consider I.

Since both A and A^* are HRSK-formulas, A contains no occurrences of either & or $\vee$ or quantifiers. The examination of the rules is trivial.

Theorem 7. *Suppose that the following conditions are satisfied:*

> *the sequent* (1) X, A, Y *is derivable*;
>
> *the sequent* (2) U, A^*, V *is derivable*;
>
> *either* (α) X, Y *is an HRSK-sequence and for any consequent occurrence of a subformula* $B \rightarrow C$ *of* A *either* B *or* C *contains no occurrences of either* $\vee$ *or* & *or quantifiers*;
>
> *or* (β) U, V *is an HRSK-sequence and for any consequent occurrence of a subformula* $B \rightarrow C$ *of* A^* *either* B *or* C *contains no occurrences of either* $\vee$ *or* & *or quantifiers.*

Then (3) X, U, Y, V *is derivable.*

Proof. We proceed as in the proof of Theorem 6, with the same induction hypotheses. However, in Case I there are more subcases to consider. The treatment of standard rules is standard and hence omitted; we shall examine here the new ones. Moreover, some rules are mutually dual; for example, &$\vee$ and $\neg\vee$&; we shall consider only one member of such pairs.

Suppose that A and A^* are $B\&(C \vee D)$ and $\neg(B\&(C \vee D))$, that (1) is obtained by &$\vee$ from (1′) $X, B\&C, Y$ and (1″) $X, B\&D, Y$, and that (2) is obtained by $\neg$& from (2′) $U, \neg B, V$ and (2″) $U, \neg(C \vee D), V$.

If α, we can apply Theorem 1 to (1′) and (1″). Hence, either (3′) X, B, Y is derivable or else X, C, Y and X, D, Y are derivable.

In the first case we can see that (3′) satisfies α; hence, we use Hyp 1, (3′) and (2′) to obtain (3).

In the second case we use first $\vee$ to derive (3″) $X, C \vee D, Y$; it is easy to see that (3″) satisfies α. Hence, we use Hyp 1, (3″) and (2″) to derive (3).

If A and A^* are $\neg(B\&(C \vee D))$ and $B\&(C \vee D)$, we proceed similarly.

If β, we can apply Theorem 2 to (2″). Hence, either U, $\neg C$, V or U, $\neg D$, V is derivable; by using $\neg$& we derive either U, $\neg(B\&C)$, V or $U, \neg(B\&D), V$, say (3′) $U, \neg(B\&C)$, V. It is easy to see that (3′) satisfies β; hence, we can use Hyp 1, (3′) and (1′) to derive (3).

Suppose that A and A^* are $B\&\exists xC$ and $\neg(B\&\exists xC)$, that (1) is obtained by &$\exists$ from (1′) $X, B\&C(\frac{x}{y}), Y$, where y does not occur free in $X, B\&C, Y$, and (2) is obtained by $\neg\&$ from (2′) $U, \neg B, V$ and (2″) $U, \neg\exists xC, V$.

If α, we can apply Theorem 1 to (1′). Hence, either (3′) X, B, Y or (3″) $X, C(^x_y), Y$ is derivable. Both (3′) and (3″) satisfy α. If (3′) is derivable, we use Hyp 1, (3′) and (2′) to derive (3). If (3″) is derivable, we use $\exists$ to derive $X, \exists xC, Y$ first, and then we use Hyp 1 and (2″) to derive (3).

If β, apply Theorem 4 to (2″). There is a term t free for x in C such that (3′) $U, \neg C(^x_t)$, V is derivable. Now we use $\neg\&$ to derive (3″) U, $\neg(B\&C(^x_t))$, V. On the other hand, by 2.13 of [21] and (1′), we see that (1″) X, $B\&C(^x_t)$, Y is derivable. Obviously, (3″) satisfies β; hence we use Hyp 1, (3″) and (1″) to derive (3).

Suppose that A and A^* are $B \to C$ and $\neg(B \to C)$, and that (2) is obtained by $\neg\to$ from (2′) $U, B, \neg C, V$.

Suppose that (1) is obtained by $\to$ from (1′) $X', \neg B, Y'$ and (1″) X'', C, Y'', where X and Y are X', X'' and Y', Y'', respectively.

Let α be satisfied. It is easy to see that (1′) and (1″) satisfy α. By Hyp 1, (1′), (1″) and (2′), (3) is derivable.

Let β be satisfied; then either (β.1) B contains no occurrences of $\vee$ or & or quantifiers, or (β.2) $\neg C$ contain no occurrences of $\vee$ or & or quantifiers.

If β.1, by Hyp 1, (1″) and (2′), we derive (3′) X'', U, B, Y'', V; hence we obtain (3) by using Theorem 6, (1′) and (3′).

If β.2, then (1′) and (2′) satisfy β. Hence, by Hyp 1 we obtain (3′) X', U, $\neg C$, Y', V. Now from (1″) and (3′) we obtain (3) by Theorem 6.

Theorem 8. *Suppose that the following conditions are satisfied:*

the sequent (1) X, A, Y *is derivable*;

the sequent (2) U, A^*, V *is derivable*;

both X, Y *and* U, V *are HRSK-sequences*;

(γ) *for any consequent occurrence of a subformula* $B \to C$ *of either* A *or* A^*, *either* B *or* $\neg C$ *is an HRSK-formula.*

Then (3) X, U, Y, V *is derivable.*

Proof. We proceed exactly as in the proof of Theorem 7. However, in the treatment of $\to$ we use Hyp 1 instead of Theorem 6. Note that by Lemmas 1 and 2 the eliminated member in the Cut rule satisfies γ.

5 Decidability

In this section we shall show that **GQRW** is decidable, by adapting the method of [9].

Let us say that a term t is *used* in a sequent X iff there is a member A of X such that t occurs in A. An occurrence of a term t is *free* in X iff this occurrence is in a member $A(\frac{x}{t})$ of X such that t is free for x in A. A term t is free in X iff t has at least one free occurrence in X. An occurrence of t in X is *bound in* X iff this occurrence of t is in a member A of X and in the scope of an occurrence of a quantifier Qy, where y is a variable of t. A term is *bound in* X iff at least one of its occurrences in X is bound.

By $BT(X)$ and $FT(X)$ we understand the sets of bound and free terms in X, respectively. According to Lemma 2.10 of [21] we may assume that the sets $FV(X)$ and $BV(X)$ of free and bound variables used in X, respectively, are disjoint; hence, $BT(X) \cap FT(X) = \emptyset$. The notation $BT(*)$, $FT(*)$, $BV(*)$ and $FV(*)$, where $*$ is either a term or a formula or a derivation, is defined similarly. Also, in $BV(*)$ and $FV(*)$, $*$ may be a set of terms or formulas.

In the sequel we assume that $BT(X)$ is the set of bound terms of a sequent X and that FT is any finite set of terms that are not bound in X; i.e., if $t \in FT$ is used in X, then $t \notin BT(X)$. By a *term substitution* we understand a function $\varphi : BT(X) \to FT$, defined by induction on the length of terms and on the degree of formulas.

If t is a variable, then $\varphi(t) \in FT$, if $t \in \mathrm{dom}(\varphi)$, and $\varphi(t) = t$, if $t \notin \mathrm{dom}(\varphi)$.

If $t = f(s_1, \ldots, s_n)$, where f is a function symbol and $s_1, \ldots, s_n \in BT(X)$, then $\varphi(t) \in FT$, if $t \in \mathrm{dom}(\varphi)$, and $\varphi(t) = f(\varphi(s_1), \ldots, \varphi(s_n))$ if $t \notin \mathrm{dom}(\varphi)$.

$\varphi(\neg A) = \neg\varphi(A)$ and $\varphi(A * B) = \varphi(A) * \varphi(B)$, where $*$ is a binary connective. Let Q be a quantifier; then $\varphi(QxA) = Qx(\varphi'(A))$, where $\varphi' = \varphi$ if $x \notin BV(\mathrm{dom}(\varphi))$, and $\varphi' = \varphi - (\frac{s}{\varphi(s)})$ if $x \in BV(s)$ and $s \in \mathrm{dom}(\varphi)$.

We write $(\frac{s_1,\ldots,s_n}{t_1,\ldots,t_n})$ for $\varphi : \{s_1, \ldots, s_n\} \to \{t_1, \ldots, t_n\}$; also, we shall write $u(\frac{s_1,\ldots s_n}{t_1,\ldots t_n})$ and $A(\frac{s_1,\ldots s_n}{t_1,\ldots t_n})$ for the result of a term substitution in a term u and in a formula A, respectively.

If φ_1 and φ_2 are term substitutions and $\mathrm{dom}(\varphi_1) \cap \mathrm{dom}(\varphi_2) = \emptyset$, then $\varphi_1 \cup \varphi_2$ is a term substitution, where $\varphi_1 \cup \varphi_2$: $\mathrm{dom}(\varphi_1) \cup \mathrm{dom}(\varphi_2) \to \mathrm{range}(\varphi_1) \cup \mathrm{range}(\varphi_2)$. Also, if φ_1 and φ_2 are term substitutions, then $\varphi_1 - \varphi_2$ is a term substitution, where $\varphi_1 - \varphi_2 = (\mathrm{dom}(\varphi_1) \times \mathrm{range}(\varphi_1)) \setminus (\mathrm{dom}(\varphi_2) \times \mathrm{range}(\varphi_2))$.

If φ_1 and φ_2 are term substitutions, then $\varphi_1 \cdot \varphi_2$ is a term substitution such that $\mathrm{dom}(\varphi_1 \cdot \varphi_2) = \mathrm{dom}(\varphi_1)$ and for $t \in \mathrm{dom}(\varphi_1)$, $(\varphi_1 \cdot \varphi_2)(t) = \varphi_2(\varphi_1(t))$.

A term $t \in FT$ is *minimal with respect to* X iff there is no $s \in BT(X)$ and no term substitution $\varphi : BV(s) \to FT$ such that $t = \varphi(s)$. Let $M(FT)$ be the set of terms in FT minimal with respect to X. Since FT is finite, $M(FT)$ is nonempty if FT is nonempty.

Lemma 3. *Let $s \in BT(X)$ and let φ_1 and φ_2 be term substitutions such that $BV(s) \subseteq \mathrm{dom}(\varphi_1)$ and $\mathrm{range}(\varphi_1) \subseteq FT$ and $\mathrm{dom}(\varphi_2) \subseteq M(FT)$; then $\varphi_2(\varphi_1(s)) = (\varphi_1 \cdot \varphi_2)(s)$.*

Proof. By induction on the length of s. Let s be an individual variable or an individual constant. If $s \in \mathrm{dom}(\varphi_1)$, then $\varphi_2(\varphi_1(s)) = \varphi_2(s(^{s}_{\varphi_1(s)})) = s(^{s}_{\varphi_2(\varphi_1(s))}) = (\varphi_1 \cdot \varphi_2)(s)$.

If $s \notin \mathrm{dom}(\varphi_1)$, then $\varphi_2(\varphi_1(s)) = \varphi_2(s) = s$, since neither individual constants nor individual variables used in X are minimal with respect to X (for $BT(X) \cap FT = \emptyset$) and we have $s \notin \mathrm{dom}(\varphi_2)$. Since $s \notin \mathrm{dom}(\varphi_1)$, we have $(\varphi_1 \cdot \varphi_2)(s) = s$.

Let $s = f(s_1, \ldots, s_n)$; then

$$\begin{aligned}
\varphi_2(\varphi_1(s)) &= \varphi_2(f(\varphi_1(s_1), \ldots, \varphi_1(s_n))), && \text{for } BV(s) \subseteq \mathrm{dom}(\varphi_1),\\
&= f(\varphi_2(\varphi_1(s_1)), \ldots, \varphi_2(\varphi_1(s_n))), && \text{for } \varphi_1(s) \notin M(FT),\\
&= f((\varphi_2 \cdot \varphi_1)(s_1), \ldots, (\varphi_2 \cdot \varphi_1)(s_n)), && \text{by induction hypothesis,}\\
& && s_1, \ldots, s_n \in BT(X);\\
&= (\varphi_1 \cdot \varphi_2)(s), && \text{for } s \notin \mathrm{dom}(\varphi_1 \cdot \varphi_2).
\end{aligned}$$

Corollary. *Let A be a member of X, let φ_1 and φ_2 be term substitutions defined on $BV(A)$ and $M(FT)$, respectively, such that $\mathrm{range}(\varphi_1) \subseteq FT$; then $A(^{x_1}_{\varphi_2(\varphi_1(x_1))} {}^{\ldots}_{\cdots} {}^{x_n}_{\varphi_2(\varphi_1(x_n))})$ is $A(^{x_1}_{(\varphi_1 \cdot \varphi_2)(x_1)} {}^{\ldots}_{\cdots} {}^{x_n}_{(\varphi_1 \cdot \varphi_2)(x_n)})$.*

Let $\varphi(X, FT) = \{\varphi(A) \mid$ *A is a subformula of a member of X and* $\varphi : BV(A) \to FT\}$.

Lemma 4. *Let $QxB \in \varphi_1(X, FT)$, $t \in FT$ and $\mathrm{dom}(\varphi_2) \subseteq M(FT)$; then $\varphi_2(B(^{x}_{t})) = (\varphi_2(B))(^{x}_{\varphi_2(t)})$.*

Proof. Let B be $\varphi_1(A)$. If $x \in \mathrm{dom}(\varphi_1)$, then $x \notin BV(B), x \notin BV(\varphi_2(B))$ and the lemma holds. If $x \notin \mathrm{dom}(\varphi_1)$, then

$$\begin{aligned}
\varphi_2((\varphi_1(A))(^{x}_{t})) &= \varphi_2((\varphi_1 \cup (^{x}_{t}))(A))\\
&= ((\varphi_1 \cdot \varphi_2) \cup (^{x}_{\varphi_2(t)}))(A)\\
&= ((\varphi_1 \cdot \varphi_2)(A))(^{x}_{\varphi_2(t)})\\
&= (\varphi_2(B))(^{x}_{\varphi_2(t)}).
\end{aligned}$$

Lemma 5. *Let φ_2 be a term substitution defined on $M(FT)$ for some φ_1 and let $\varphi_2(FT) = \{\varphi_2(t) \mid t \in FT\}$; then $M(\varphi_2(FT)) \subseteq \mathrm{range}(\varphi_2)$.*

Proof. If $t \in FT$ is not minimal with respect to X, then $\varphi_2(t) \in \varphi_2(FT)$ is also not minimal with respect to X. For, suppose that t is not minimal with respect to X; then $t = \varphi_1(s)$ for some $s \in BT(X)$ and $\varphi_1 : BV(s) \to FT$. By Lemma 3, $\varphi_2(t) = \varphi_2(\varphi_1(s)) = (\varphi_1 \cdot \varphi_2)(s)$. Since $\mathrm{range}(\varphi_1 \cdot \varphi_2) \subseteq \varphi_2(FT)$ and $\mathrm{dom}(\varphi_1 \cdot \varphi_2) = \mathrm{dom}(\varphi_1) = BV(s)$, the term $\varphi_2(t) \in \varphi_2(FT)$ is not minimal with respect to X. Therefore, all minimal elements of $\varphi_2(FT)$ are of the form $\varphi_2(t)$, $t \in M(FT)$; i.e., they are elements of the set $\{\varphi_2(t) \mid t \in M(FT)\} = \mathrm{range}(\varphi_2)$.

By a *critical* occurrence of a quantifier in a sequent X we understand either an antecedent occurrence of $\forall$ or a consequent occurrence of $\exists$.

In the rest of the paper we assume that X is a sequent, that $BV(X) \cap FV(X) = \emptyset$, that p is the number of noncritical occurrences of quantifiers in X, that q is the number of critical occurrences of quantifiers in X and that l is the number of occurrences γ of connectives in X such that γ is either an antecedent occurrence of $\vee$ or a consequent occurrence of &.

Let FV be any set of individual variables such that $FV \cap (BV(X) \cup FV(X)) = \emptyset$. Now we define the sets $T(i)$ of terms, $0 \leq i \leq q \cdot 2^l$:

$T(0) = FV$;

$T(i+1) = \{\varphi(t) \mid t \in BT(X) \cup FV,\ \varphi : BV(t) \to T(i)\}$.

Obviously, $T(i) \subseteq T(i+1)$ and $T(q \cdot 2^l) \cap BT(X) = \emptyset$.

Lemma 6. *Let FT be any finite nonempty set of terms, let FV be a finite set of individual variables that are not bound in X and such that $M(FT) \subseteq FV$ and let $k = \mathrm{card}(FT - FV)$; then $FT \subseteq T(k)$.*

Proof. Proceed by induction on k. If $FT \subseteq FV$, then $FT \subseteq T(0) \subseteq T(k)$. Let $FT - FV \neq \emptyset$; then there is a term $t \in FT - FV$ such that t is a subterm of no term in FT. For, otherwise, there is an infinite sequence of terms in $FT - FV$ such that every term is a subterm of the next member of the sequence, which contradicts the fact that FT is finite.

Let $t_0 \in FT - FV$ be a term that is a proper subterm of no term in FT. The set $FT' = FT - \{t_0\}$ is nonempty, for $\emptyset \neq FV \subseteq FT'$. Also, $M(FT') \subseteq FV$. For, if $t' \in FT'$ is not minimal in FT; i.e., if $t' = \varphi(s)$ and $\mathrm{range}(\varphi) \subseteq FT$, then t' is not minimal in FT', since $t_0 \notin \mathrm{range}(\varphi)$ by the choice of t_0.

Now, FT' satisfies the condition of the lemma and $\mathrm{card}(FT' - FV) = k-1$. By the induction hypothesis, $FT' \subseteq T(k-1)$. Since $t_0 \notin M(FT)$, we have $t_0 = \varphi(s)$ for some $s \in BT(X)$ and $\varphi : BV(s) \to FT$. By the choice of t_0, $t_0 \notin \text{range}(\varphi)$; hence, φ: $BV(s) \to FT' \subseteq T(k-1)$, and $t_0 \in T(k)$, by definition of $T(k)$. But $T(k-1) \subseteq T(k)$, and this concludes the proof of the lemma.

Let $\mathcal{T}$ be a derivation of X and let ter($\mathcal{T}$) be the set of all terms discharged in $\mathcal{T}$.

Lemma 7. card(ter($\mathcal{T}$))$\leq q \cdot 2^l$.

Proof. By induction on the weight of a node Y in $\mathcal{T}$. If Y is an axiom, the theorem is trivial. Suppose that Y is obtained from Y_1 (and Y_2) by derivations $\mathcal{T}_1$ (and $\mathcal{T}_2$) such that card(ter($\mathcal{T}_1$)) $\leq q_1 \cdot 2^{l_1}$ (and card(ter($\mathcal{T}_2$)) $\leq q_2 \cdot 2^{l_2}$), where q_i and l_i are the numbers of critical occurrences of quantifiers and either antecedent occurrences of $\vee$ or consequent occurrences of & in $\mathcal{T}_1$ and $\mathcal{T}_2$, respectively, $1 \leq i \leq 2$ (induction hypothesis).

If Y is obtained by a one-premiss rule, except $\forall$ and $\neg\exists$, then ter($\mathcal{T}$) = ter($\mathcal{T}_1$), $q = q_1$, and $l = l_1$; the theorem follows by the induction hypothesis.

If Y is obtained by a two-premiss rule, except $\to$, then ter($\mathcal{T}$) = ter($\mathcal{T}_1$)$\cup$ter($\mathcal{T}_2$). Hence, card(ter($\mathcal{T}$)) $\leq q_1 \cdot 2^{l_1} + q_2 \cdot 2^{l_2} \leq q_1 \cdot 2^{l-1} + q_2 \cdot 2^{l-1} \leq q \cdot 2^l$.

If Y is obtained by $\to$, then $q_1 \cdot 2^{l_1} + q_2 \cdot 2^{l_2} \leq (q_1 + q_2) \cdot 2^l = q \cdot 2^l$.

If Y is obtained either by $\forall$ or by $\neg\exists$, and if t is the term discharged by this application, then $q = q_1 + 1$, $l_1 = l$ and card(ter($\mathcal{T}_1$)$\cup \{t\}$) $\leq q_1 \cdot 2^{l_1} + 1$ $\leq q \cdot 2^l$.

Theorem 9. *For any derivation $\mathcal{T}$ of a sequent X, there is a derivation $\mathcal{T}'$ of X such that the number of eigenvariables in $\mathcal{T}'$ is not greater than p.*

Proof. By using Lemma 2.10 of [21], we can show that for any two derivations $\mathcal{T}_1$ and $\mathcal{T}_2$ of sequents X and Y there are derivations $\mathcal{T}_1'$ and $\mathcal{T}_2'$ of the same sequents, respectively, such that either all eigenvariables of $\mathcal{T}_1'$ are eigenvariables of $\mathcal{T}_2$ or all eigenvariables of $\mathcal{T}_2'$ are eigenvariables of $\mathcal{T}_1$.

Proceed by induction on the weight of X in $\mathcal{T}$. If X is an axiom, the theorem is true.

Suppose that X is obtained by a two premisses rule ρ from Y and Z, and let p_Y and p_Z be the numbers of eigenvariables in the derivations of Y and Z, respectively. By the induction hypothesis there are derivations of Y and Z such that $p_Y \leq p$ and $p_Z \leq p$. By Lemma 2.10 of [21], say, all eigenvariables of the derivation of Y are eigenvariables of the derivation of Z. Applying ρ we obtain a derivation of X; in this derivation the number of eigenvariables is $\max(p_Y, p_Z) \leq p$.

If X is obtained by a one premiss rule ρ from Y, the proof is trivial if ρ is neither $\neg\forall$ nor $\exists$. Suppose that ρ is, say, $\neg\forall$. By induction hypothesis, there is a derivation of Y such that the number of eigenvariables in this derivation is not greater than $p-1$. Obviously, the number of eigenvariables in the derivation of X is not greater than p.

Theorem 10. *For any derivation $\mathcal{T}$ of a sequent X, there is a derivation $\mathcal{T}'$ of X such that* $\text{ter}(\mathcal{T}') \subseteq T(q \cdot 2^l)$.

Proof. By Theorem 9, we may assume that the number of eigenvariables used in $\mathcal{T}$ is not greater than p. Let now FV be the set of $p+1$ variables containing the eigenvariables and a variable x_0 such that $(BV(\mathcal{T}) \cup FV(\mathcal{T})) \cap FV = \emptyset$ and let $T = \text{ter}(\mathcal{T}) \cup FV$; we have $FV \subseteq M(T)$. Let φ be a term substitution defined on $M(T)$ such that: (1) if $x \in FV$, then $\varphi(x) = x$, and (2) $\varphi(t) = x_0$ otherwise, where x_0 is the $p+1$-variable in FV. Let $\varphi(Y)$ be the result of term substitution φ performed in the node Y of $\mathcal{T}$. We want to show that $\varphi(\mathcal{T})$, obtained by such substitution, is a derivation of X. This will be done by induction on the weight of $\mathcal{T}$.

Any member of any node Y of $\mathcal{T}$ is a member of the set $\varphi(X,T) = \{\varphi(A) \mid A \textit{ is a subformula of a member of } X \textit{ and } \varphi : BV(A) \to FV \cup \{x_0\}\}$.

For, if $A * B \in \varphi(X,T)$, where $*$ is a binary connective, then $A, B \in \varphi(X,T)$; if $\neg A \in \varphi(X,T)$, then $A \in \varphi(X,T)$; if either $\forall x A \in \varphi(X,T)$ or $\exists x A \in \varphi(X,T)$, and $t \in \text{ter}(\mathcal{T})$, then $A(\binom{x}{t}) \in \varphi(X,T)$. It follows that if all members of a node Y of $\mathcal{T}$ are members of $\varphi(X,T)$, so are the members of the premiss(es) of Y. Obviously, the members of X are members of $\varphi(X,T)$; hence, all formulas used in $\mathcal{T}$ are members of $\varphi(X,T)$.

If Y is either an axiom or obtained by a propositional rule, so is $\varphi(Y)$. Let consider the case where Y is obtained by $\forall$ from $Y_1, A(\binom{x}{s}), Y_2$, where the term s is free for x in A; obviously, Y is $Y_1, \forall x A, Y_2$. By the induction hypothesis, $\varphi(Y_1), \varphi(A(\binom{x}{s})), \varphi(Y_2)$ is derivable; by Lemma 4, so is $\varphi(Y_1), (\varphi(A))(\binom{x}{\varphi(s)}), \varphi(Y_2)$; hence, we derive $\varphi(Y_1), \forall x(\varphi(A)), \varphi(Y_2)$ by using $\forall$, provided that $\varphi(s)$ is free for x in A. This can be shown by induction on the degree of s. If s is a variable, then $s \in M(T)$ and $\varphi(s)$ is free for x in A. Let s be $f(s_1, \ldots, s_n)$ and suppose that it is not minimal; then $\varphi(s) = f(\varphi(s_1), \ldots \varphi(s_n))$ by definition of term substitution. By the induction hypothesis, the terms $\varphi(s_1), \ldots \varphi(s_n)$ are free for x in A; hence, so is s.

If Y is obtained by $\neg\exists$, we proceed similarly.

If Y is obtained by $\exists$ from $Y_1, A(\binom{x}{y}), Y_2$, where y is an eigenvariable, by the induction hypothesis, $\varphi(Y_1), \varphi(A(\binom{x}{y})), \varphi(Y_2)$ is derivable. For $y \in FV$ and $\varphi(y) = y$ we have $\varphi(A(\binom{x}{y})) = \varphi(A)(\binom{x}{y})$. Hence, $\varphi(Y_1), \varphi(A)(\binom{x}{y}), \varphi(Y_2)$ is derivable, and by using $\neg\forall$ we derive $\varphi(Y)$.

If Y is obtained by $\neg\forall$, we proceed similarly.

Therefore, $\varphi(\mathcal{T})$ is a derivation of X (notice that $\varphi(X) = X$, since X contains no occurrence of a discharged term). We have $FV(\varphi(\mathcal{T})) = FV(\mathcal{T}) \subseteq FV$ as well as $\text{ter}(\varphi(\mathcal{T})) \subseteq \varphi(T)$. By Lemma 5, $M(\varphi(\mathcal{T})) \subseteq \text{range}(\varphi) = FV \cup \{x_0\}$. By Lemma 6 applied to $\varphi(T)$, we have $\varphi(T) \subseteq T(k)$, where $k = \text{card}(\varphi(T) - (FV \cup \{x_0\}))$. But $\varphi(T) - (FV \cup \{x_0\}) \subseteq \varphi(\text{ter}(\mathcal{T}))$ and hence by Lemma 7, $k \leq q \cdot 2^l$.

Now we can show that **GQRW** is decidable.

By a *search proof tree* $\mathcal{ST}(X)$ for a sequent X we understand a tree, branching upward, the nodes of which are sequents and such that:

(1) X is at the origin of $\mathcal{ST}(X)$;

(2) for any node Y which is not an end node there is a rule ρ of **GQRW** such that Y is the conclusion and the nodes immediately above Y are the premisses of ρ;

(3) if Y is an end node, then the members of Y are either atomic formulas or negations of such formulas.

It is clear that a search proof tree $\mathcal{ST}(X)$ for X is a derivation of X iff it is finite and every end node Y of $\mathcal{ST}(X)$ is an axiom.

By inspection of the rules we prove that in each premiss of a rule ρ the number of occurrences of connectives or quantifiers is smaller than the number of occurrences of connectives or quantifiers in the conclusion. Therefore, every search proof tree is finite. It remains to show that for any given sequent there is only a finite number of search proof trees. But the number of such trees depends on how many constituent members of a rule we may have. Any constituent member of a propositional rule can be obtained from the principal member either by deleting some symbols such that the remaining string of symbols is a formula or by regrouping some subformulas of the principal member. This can be done in a finite number of ways only. Hence, the propositional part **GRW** of **GQRW** is decidable.

To show that **GQRW** is decidable we may apply Theorems 9 and 10. By Theorem 9, the number of eigenvariables in a search proof tree for a sequent X is not greater than p. Take the first $p+1$ variables in alphabetic order that do not occur in X and construct the set $T(q \cdot 2^l)$. This can be done effectively, for q and l are given by X. By Theorem 10, X is derivable iff there is a derivation $\mathcal{T}$ of X such that $\text{ter}(\mathcal{T}) \subseteq T(q \cdot 2^l)$. This means that it is sufficient to examine the search proof trees with terms in the set $T(q \cdot 2^l)$ only, for the number of terms needed in the derivation of a sequent X is not greater than the number of quantifier occurrences in X where $\binom{x}{t}$ can be substituted multiplied by the maximum number of branches above X, as determined by the number of antecedent occurrence of $\vee$ and the number of consequent occurrences of &. But the set of such search proof trees is finite and effectively given. Hence, **GQRW** is decidable.

6 Herbrand's Theorem

Now we are concerned with Herbrand's theorem for **GQRW**.

Theorem 11. *If* (a) X, $\alpha(C(\forall xA))$, Y *is derivable, so is* (b) X, $\alpha(C(A(\substack{x\\y})))$, Y *provided that the displayed occurrence of* $\alpha(\forall xA)$ *is a consequent occurrence and the variable* y *is the first individual variable in alphabetical order not used in* (a).

Proof. Suppose that $\mathcal{T}$ is a regular derivation of (a) and proceed by double induction, one on the degree of C and another on the weight of (a) in $\mathcal{T}$. Obviously, (a) is not an axiom. If C is not principal, the use of the induction hypothesis (on weight) is trivial.

Suppose that C is principal. By Lemma 2, $\alpha(\forall A)$ is an occurrence in the sub-principal member of (a); if there is a corresponding occurrence of $\forall xA$ in a constituent member of a premiss of (a), the use of the induction hypothesis (on the degree of C) is trivial.

If $\alpha(\forall xA)$ has no corresponding occurrence in a premiss of (a), then it may be introduced in (a) in one of the following four ways:

1. by $\neg\forall$ from (a′) $X, \neg A(\substack{x\\z}), Y$, where z is the eigenvariable and C is $\neg\forall xA$;

2. by $\neg\vee\forall$ from (a′) $X, \neg(B \vee A(\substack{x\\z})), Y$, where z is the eigenvariable and C is $\neg(B \vee \forall xA)$;

3. by & from (a′) X, B, Y, where C is $B\&D(\forall xA)$;

4. by $\neg\vee$ from (a′) $X, \neg B, Y$, where C is $\neg(B \vee D(\forall xA))$.

If α is introduced either as in case 1 or as in case 2, we may use Lemma 2.10 of [20] and (a′) to obtain (b).

If α is introduced as in case 3 or as in case 4, we may use C', & and $\neg\vee$, respectively, to derive (b), where C' is obtained from C by substitution of $A(\substack{x\\y})$ for $\forall xA$ at the relevant place.

Theorem 12. *If* (a) X, $\alpha(C(\exists xA))$, Y *is derivable, so is* (b) X, $\alpha(C(A(\substack{x\\y})))$, Y *provided that the displayed occurrence of* $\exists xA$ *is an antecedent occurrence and the variable* y *is the first individual variable in alphabetical order not used in* (a).

Let us call a sequence X *unsaturated* iff the members of X contain neither antecedent occurrences of $\exists$ nor consequent occurrences of $\forall$.

Theorem 13. *If X is derivable, then there is an unsaturated derivable sequent X' obtained from X by using Theorems* 11 *and* 12.

Note that the converse of Theorem 13 does not hold in **GQRW**. Let y have no free occurrences in A and no occurrences in B; the formula $A \to B(^x_y) \to . \exists y(A \to B(^x_y))$ is a theorem of **GQRW** for any A and B, but $A \to \exists x B \to . \exists y(A \to B(^x_y))$ is not; the derivation of the latter in Gentzen-style in classical predicate logic involves the use of contraction.

Theorem 14. *If an unsaturated sequent* (1) $X, \forall x A, Y$ *is derivable, then there are terms* $t_1, \ldots, t_n \in T(q \cdot 2^l)$ *such that* (2) $X, A(^x_{t_1})\&\ldots\&A(^x_{t_n}), Y$ *is derivable.*

Proof. Suppose that (1) is derivable; by Theorem 9, there is a derivation $\mathcal{T}$ of (1) such that $\text{ter}(\mathcal{T}) \subseteq T(q \cdot 2^l)$. Now proceed by induction on the weight of (1) in $\mathcal{T}$. Obviously, (1) is no axiom. If (1) is obtained by $\forall$ from (1′) $X, A(^x_t), Y$, $t \in T(q \cdot 2^l)$, the theorem is proved.

If it is obtained by a rule such that $\forall x A$ is parametric in (1), the use of the induction hypothesis on weight is trivial, if (1) is obtained either by a one premiss rule or by $\to$.

Suppose that (1) is obtained from (1′) X', C, $\forall x A$, Y and (1″) X', D, $\forall x A$, Y either by $\vee$ or by $\neg\&$. By the induction hypothesis, there are terms $r_1, \ldots, r_k, s_1, \ldots, s_m, \in T(q \cdot 2^l)$ such that (2′) X', C, $A(^x_{r_1})\& \ldots\&A(^x_{r_k})$, Y and (2″) X', C, $A(^x_{s_1})\& \ldots \&A(^x_{s_m})$, Y are derivable. Let $\{t_1, \ldots, t_n\} = \{r_1, \ldots, r_k\} \cup \{s_1, \ldots s_m\}$; by using & we derive (3′) X', C, $A(^x_{t_1})\& \ldots \& A(^x_{t_n})$, Y and (3″) X', D, $A(^x_{t_1})\& \ldots\& A(^x_{t_n})$, Y. Now we derive (2) by using either $\vee$ or $\neg\&$, respectively.

If $\forall x A$ is principal in (1), the theorem is true.

Theorem 15. *If an unsaturated sequent* (1) $X, \neg\exists x A, Y$ *is derivable, then there are terms* $t_1, \ldots, t_n \in T(q \cdot 2^l)$, *such that* (2) X, $\neg(A(^x_{t_1}) \vee \ldots \vee A(^x_{t_n})), Y$ *is derivable.*

Theorems 14 and 15 can be generalized to the following:

Theorem 16. (a) *If an unsaturated sequent* (1) $X, \alpha(C(\forall x A)), Y$ *is derivable, then there are terms* $t_1, \ldots, t_n \in T(q \cdot 2^l)$ *such that*

$$(2) \quad X, \alpha(C(A(^x_{t_1})\& \ldots \&A(^x_{t_n}))), Y$$

is derivable if α is an antecedent occurrence of $\forall x A$.

(b) *If an unsaturated sequent* (1) X, $\alpha(C(\exists xA))$, Y *is derivable, then there are terms* $t_1, \ldots, t_n \in T(q \cdot 2^l)$ *such that*

$$(2) \quad X, \alpha(C(A(^x_{t_1}) \vee \ldots \vee A(^x_{t_n}))), Y$$

is derivable if α *is a consequent occurrence of* $\exists xA$.

The converse to Theorem 16 does not hold in general.

Theorem 17. *If an unsaturated sequent* X, $A(^x_{t_1})\&\ldots\&A(^x_{t_n})$, Y *is derivable,* $n > 1$, *so is* $X, \forall xA, Y$ *if one of the following conditions is satisfied:*

> X, Y *is an HRSK-sequent;*
>
> $\forall xA$ *is an HRSK-sequence and, for any consequent occurrence of a subformula* $B \rightarrow C$ *of* A, *either* B *or* C *contains no occurrence of either* $\vee$ *or* $\&$ *or quantifiers.*

Theorem 18. *If an unsaturated sequent* X, $\neg(A(^x_{t_1}) \vee \ldots \vee A(^x_{t_n}))$, Y *is derivable,* $n > 1$, *so is* X, $\neg\exists xA$, Y, *under the proviso as above.*

7 Craig-Lyndon Interpolation Theorem

In the sequel we need the following lemma:

Lemma 8. *Let* t *be a closed term used in* X *and let* X' *be obtained by substitution* (^t_x), *for all occurrences of* t *in* X, *where* x *is the first variable not used in* X; *then* X' *is derivable.*

Proof. By induction on the weight of X in the given derivation.

We prove a theorem analogous to Belnap's variable-sharing theorem. Let $\pi(A)$ ($\pi(X)$) be the set of all predicate symbols used in A (in X).

Theorem 19. *For any nonempty sequences* X *and* Y, *if* X, Y *is derivable, then* $\pi(X) \cap \pi(Y) \neq \emptyset$.

Proof. By an easy induction on the weight of X, Y in the given derivation.

Corollary. *If* $A \rightarrow B$ *is a theorem of* **GQRW**, *then there is a predicate symbol common to* A *and* B.

Proof. If $A \to B$ is a theorem of **GQRW**, then $\neg(A \to B)$ is derivable in **GQRW**; hence, so is $A, \neg B$. Therefore, by Theorem 19 there is a predicate symbol common to A and B.

Let $\sigma(X)$ denote the set of all predicate symbols and closed terms used in the members of X.

An occurrence $\alpha(F)$ of a predicate symbol F in a sequent X is an *antecedent* (*consequent*) occurrence of F iff α is an antecedent (consequent) occurrence of an atomic subformula $F(t_1, \ldots, t_n)$.

Theorem 20. *For any derivable sequence X, if S and T are nonempty sequences such that X is obtained from S, T by Perm, then there is a formula D (the interpolation formula) such that*

(1) *both S, D and D^*, T are derivable and $\sigma(D) \subseteq \sigma(S) \cap \sigma(T)$;*

(2) *if a predicate symbol F has an antecedent (consequent) occurrence in D, then F has a consequent (antecedent) occurrence in S;*

(3) *if a predicate symbol F has an antecedent (consequent) occurrence in D^*, then F has a consequent (antecedent) occurrence in T.*

Proof. We proceed by induction on the weight of X in the given derivation.

The theorem is true if X is an axiom.

If X is obtained by *Perm*, the proof is trivial.

Suppose that X is obtained by a rule ρ and that the theorem holds for the premisses of ρ.

If ρ is a one-premiss rule, we have

$$\frac{Y, U, Z}{Y, A, Z}$$

where the member(s) of U is (are) the constituent member(s) and A is the principal member of ρ. Let S' and T be sequences such that Y, A, Z can be obtained from S', A, T by *Perm*, where S is S', A. Hence, Y, U, Z can be obtained by *Perm* from S', U, T.

(1) By the induction hypothesis, there is a formula D such that (a) S', U, D and (b) D^*, T are derivable and $\sigma(D) \subseteq \sigma(S', U) \cap \sigma(T)$. If ρ is a propositional rule, we have $\sigma(D) \subseteq \sigma(S', A) \cap \sigma(T)$; it is obvious that we can use ρ to obtain S', A, D from (a).

Suppose that ρ is $\forall$; then U is $B(\frac{x}{t})$, where t is free for x in B and A is $\forall x B$. If $t \in \sigma(S)$, we use $\forall$ to derive $S', \forall x B, D$ from (a).

Suppose that D is $D(\frac{z}{t})$, $t \notin \sigma(S)$, and $t \in \sigma(S, D(\frac{z}{t})) \cap \sigma(T)$, where t is a closed term. By Lemma 8, (a′) $S', B(\frac{x}{y}), D(\frac{z}{y})$ is derivable. Obviously, by using $\forall$, we derive (a″) $S', \forall x B, D(\frac{z}{y})$ from (a′); hence, from (a″) we obtain $S', \forall x B, \exists z D$ by $\exists$. On the other hand, if D^* is $\neg D$, (b) we obtain $\neg\exists z D, T$ by $\neg\exists$. If D^* does not begin with $\neg$, then it is some D', and D is $\neg D'$; we use $\neg\neg$ to obtain (b′) $\neg\neg D', T$ first; then we derive $\neg\exists z D, T$ from (b′) by using $\neg\exists$.

If ρ is any other one-premiss quantificational rule, we proceed analogously. The proof is similar if T is A, T' for some T'.

Let ρ be a two-premiss rule except $\to$; then we have

$$\frac{Y, B, Z \qquad Y, C, Z}{Y, A, Z}$$

where B and C are the constituents and A the principal member of ρ. Let S' and T be sequences such that Y, A, Z can be obtained from S', A, T by *Perm*, where S is S', A. The premisses of ρ can be obtained by permutation from S', B, T and S', C, T, respectively. By the induction hypothesis, there are formulas D and E such that (a) S', B, D and (b) D^*, T, as well as (c) S', C, E and (d) E^*, T are derivable, where $\sigma(D) \subseteq \sigma(S, B) \cap \sigma(T)$ and $\sigma(E) \subseteq \sigma(S, C) \cap \sigma(T)$.

Let neither D nor E begin with $\neg$; from (a) and (c) we obtain

$$\text{(a}'\text{)} \quad S', B, D \& E \quad \text{and} \quad \text{(c}'\text{)} \quad S', C, D \& E$$

by &; hence, by using ρ we derive $S', A, D \& E$. On the other hand, from (b) and (d) we can derive $\neg(D \& E), T$ by $\neg$&.

If D and E are $\neg D'$ and $\neg E'$, respectively, then from (a) and (c) we derive (a′) $S', B, \neg(D' \vee E')$ and (c′) $S', C, \neg(D' \vee E')$, respectively. Hence, by using ρ we derive $S', A, \neg(D' \vee E')$. On the other hand, from (b) and (d) we derive $D' \vee E', T$ by using $\vee$.

If D and E^* are $\neg D'$ and $\neg E$, respectively, from (a) and (c) we derive (a′) $S', B, \neg D' \& E$ and (c′) $S', C, \neg D' \& E$ by using &. From (b) we derive (b′) $\neg\neg D', T$; hence, from (b′) and (d) we derive $\neg(\neg D' \& E)$, T.

We proceed similarly if D^* and E are $\neg D$ and $\neg E'$, respectively.

We also have $\sigma(D \& E) \subseteq \sigma(D) \cup \sigma(E) \subseteq (\sigma(S, B) \cap \sigma(T)) \cup (\sigma(S, C) \cap \sigma(T)) \subseteq \sigma(S, A) \cap \sigma(T)$.

We proceed similarly if T is A, T' for some T'.

Let ρ be $\to$; then we have

$$\frac{Y, \neg A, Z \qquad U, B, V}{Y, U, A \to B, Z, V}$$

and let S' and T be sequences such that $Y, U, A \to B, Z, V$ is obtained from $S', A \to B, T$ by p ermutation, where S is $S', A \to B$. Let S' and T

be S'_Y, S'_Z, S'_U, S'_V and T_Y, T_Z, T_U, T_V, respectively. The premisses of $\to$ can be obtained by *Perm* from $S'_Y, T_Y, \neg A, S'_Z, T_Z$ and S'_U, T_U, B, S'_V, T_V, respectively.

By the induction hypothesis there are formulas D and E such that

$$\text{(a) } S'_Y, \neg A, S'_Z, D \quad \text{(b) } D^*, T_Y, T_Z \quad \text{(c) } S'_U, B, S'_V, E \quad \text{(d) } E^*, T_U, T_V$$

are derivable, where

$$\sigma(D) \subseteq \sigma(S'_Y, S'_Z, A) \cap \sigma(T_Y, T_Z) \quad \text{and} \quad \sigma(E) \subseteq \sigma(S'_U, S'_V, B) \cap \sigma(T_U, T_V).$$

Suppose that D^* and E^* are $\neg D$ and $\neg E$. From (a) and (c) we derive $S'_Y, S'_U, A \to B, S'_Z, D, S'_V, E$ by $\to$, and hence we obtain $S', A \to B, \neg(D \to \neg E)$ by using $\neg\neg$, $\neg\to$ and *Perm*. On the other hand, from (b) and (d) we derive $D \to \neg E, T$, by using $\to$ and *Perm*.

Let D and E^* be $\neg D'$ and $\neg E$, respectively. From (a) and (c) we obtain $S'_Y, S'_U, A \to B, S'_Z, \neg D', S'_V, E$ by $\to$, and hence we derive $S', A \to B, \neg(E \to D')$ by $\neg\to$. On the other hand, from (d) and (b) we obtain $E \to D', T$ by $\to$.

Let D^* and E be $\neg D$ and $\neg E'$, respectively. From (a) and (c) we obtain $S', A \to B, D, \neg E'$, and hence we derive $S', A \to B, \neg(D \to E')$. On the other hand, from (b) and (d) we derive $D \to E', T$ by $\to$.

Let D and E be $\neg D'$ and $\neg E'$, respectively. From (a) and (c) we obtain $S', A \to B, \neg D', \neg E'$; hence we derive $S', A \to B, \neg(\neg D' \to E')$. On the other hand, from (b) we obtain (b′) $\neg\neg D', T_Y, T_Z$ by $\neg\neg$. Hence, from (b′) and (d) we derive $\neg D' \to E', T$.

We also have $\sigma(D \to E) = \sigma(D) \cup \sigma(E) \subseteq (\sigma(S'_Y, S'_Z, A) \cap \sigma(T_Y, T_Z)) \cup (\sigma(S'_U, S'_V, B) \cap \sigma(T_U, T_V)) \subseteq \sigma(S'_Y, S'_Z, A) \cap \sigma(T) \cup \sigma(S'_U, S'_V, B) \cap \sigma(T) \subseteq (\sigma(S', A) \cup \sigma(S', B)) \cap \sigma(T) = \sigma(S', A \to B) \cap \sigma(T)$.

We proceed similarly if T is $A \to B, T'$ for some T'.

This completes the proof of (1). Clauses (2) and (3) hold for axioms. Proceed by induction on weight and use Lemma 2.

References

[1] Anderson, A. R. and Belnap, N. D., Jr., *Entailment, The Logic of Relevance and Necessity, Vol. I*, Princeton, 1975.

[2] Belnap, N. D., Jr., Display logic, *Journal of Philosophical Logic 11* (1982), 375–417.

[3] Brady, R. T., The gentzenization and decidability of RW, *Journal of Philosophical Logic 19* (1990), 35–73.

[4] Bull, R. A., Survey of generalizations of Urquhart semantics, *Notre Dame Journal of Formal Logic 28* (1987), 220–237.

[5] Dunn, J. M., Consecution formulation of positive R with co-tenability and **t**, in [1], 381–391.

[6] Giambrone, S., TW_+ and RW_+ are decidable, *Journal of Philosophical Logic 14* (1985), 235–254.

[7] Grishin, V. N., A nonstandard logic and its application to set theory (in Russian), *Issledovaniya po formalizovannym yazykam i neklassicheskim logikam*, Nauka, Moscow, 1974, 135–171.

[8] ———, The algebraic semantics of logic without contractions (in Russian), *Issledovaniya po teorii mnozhestv i neklassicheskim logikam*, Nauka, Moscow, 1976, 247–264.

[9] ———, Herbrand's theorem for logic without contractions (in Russian), *Issledovaniya po neklassicheskim logikam i teorii mnozhestv*, Nauka, Moscow, 1979, 316–329.

[10] ———, *Referativniĭ Zhurnal' MAT* 80:8 A37.

[11] Harrop, R., On disjunctions and existential statements in intuitionistic systems of logic, *Mathematische Annalen 132* (1956), 347–361.

[12] ———, Concerning formulas of the types $A \to B \vee C$, $A \to (\exists x)B$ in intuitionistic formal systems, *The Journal of Symbolic Logic 25* (1960), 27–32.

[13] Ketonen, J. and Weyhrauch, R., A decidable fragment of predicate calculus, *Theoretical Computer Science 32* (1984), 297–307.

[14] Kleene, S. C., Disjunction and existence under implication in elementary intuitionistic formalisms, *The Journal of Symbolic Logic 27* (1962), 11–18.

[15] Kripke, S. A., The problem of entailment, abstract, *The Journal of Symbolic Logic 24* (1959), 324.

[16] Kron, A., Deduction theorems for T, E and R reconsidered, *Zeitschrift für Mathematische Logik und Grundlagen der Mathematik 22* (1976), 261–264.

[17] Mints, G. E., A cut-elimination theorem for relevant logics (in Russian), in: *Zapiski nauchnykh seminarov Leningradskogo Otdeleniya Matematicheskogo Instituta. Issledovaniya po konstruktivnoĭ matematike i matematicheskoĭ logike*, 1972, 90–97.

[18] Rasiowa, H. and Sikorski, R., On existential theorems in non-classical functional calculi, *Fundamenta Mathematicae 41* (1954), 21–28.

[19] Robinson, T. T., Disjunction under implication in the intuitionistic predicate calculus Pd, *American Mathematical Society Notices 10*, no. 2, issue 66, 197.

[20] ———, Interpretations of Kleene's metamathematical predicate $\Gamma|A$ in intuitionistic arithmetic, *The Journal of Symbolic Logic 30* (1965), 140–154.

[21] Takeuti, G., *Proof Theory*, North-Holland, Amsterdam, 1975, 1986.

[22] Urquhart, A., The undecidability of entailment and relevant implication, *The Journal of Symbolic Logic 49* (1984), 1059–1073.

[23] Zaslavskiĭ, I. D., *Symmetric Constructive Logic* (in Russian), Izdatel'stvo Akademii Nauk Armyanskoĭ SSR, Erevan, 1978.

Logic Without Structural Rules (Another Look at Cut Elimination)

Joachim Lambek

Department of Mathematics and Statistics, McGill University[1]

Dedicated to Stephen C. Kleene[2]

In [L 1958], Gentzen's sequent calculus was extended from intuitionistic logic to what I then called the 'syntactic calculus', a form of bidirectional categorial grammar, now also seen to be the intuitionistic fragment of a non-symmetric version of Girard's [1987] linear logic, and the cut-elimination theorem was proved in this generality. In [L 1969], this proof was lifted to the categorical level, where the sequents were interpreted as multilinear operations. Here we shall take another look at this proof and see that it becomes more transparent if in place of the operations we consider terms in an inductively defined language [L 1989].

1 Context-Free Grammars

We shall talk about sequents $f : A_1 \ldots A_m \to A_{m+1}$, without subjecting them to Gentzen's three structural rules: thinning, contraction and interchange; see Kleene's book [1952], where I first learned about this, as I did about many other things. Depending on what application one has in mind, one may think of such a sequent

(1) as a *context-free derivation* where the arrows point as in a categorial grammar [e.g. L 1958], in the direction opposite to that in a generative grammar [Chomsky 1957];

---

1 This research was supported by the Natural Sciences and Engineering Research Council of Canada. I would also like to thank the Bulgarian mathematicians for their hospitality.

2 This paper was presented at the 1990 conference in honour of S. C. Kleene at Varna, the proceedings of which unfortunately did not materialize.

(2) as a *deduction* in a variant of intuitionistic logic without structural rules, namely the intuitionistic fragment of Girard's [1987] linear logic, but without the interchange rule;

(3) as a *multilinear operation* in algebra, as introduced by Bourbaki [1948] to explicate the tensor product [see also L 1961];

(4) as an *arrow* (or *multiarrow*) in a multicategory, abstracting from (3) above [L 1969].

The A_i have different names in these four applications: they are called *types* in (1), *formulas* in (2), *sorts* (or *bimodules*) in (3) and just *objects* in (4).

It should be pointed out that I always think of a sequent as what Gentzen calls 'a proof of the sequent'. Thus, there can be several sequents $A_1 \ldots A_m \to A_{m+1}$ for given A_i. If we think of these as forming a set, we should also consider the notion of *equality* between two sequents sharing the same source and target. This idea is evident in situation (3) and led to an elaborate definition of equality in situation (4). It took time to mature in situations (2) [see Prawitz 1965] and (1) [see Hotz 1966].

To avoid getting lost in a terminological jungle, we shall largely concentrate on situation (1) and think of sequents as context-free derivations. Let us make a definition.

A *context-free grammar* consists first of all of two sets, the set of *sequents* S and the set of *types* T, together with two mappings:

$$\text{source} : S \to T^*, \qquad \text{target} : S \to T,$$

where T^* is the free monoid generated by T, consisting of strings of types, including the empty string. If $f \in S$ and $A_i \in T$, we write $f : A_1 \ldots A_m \to A_{m+1}$ for short in place of the two statements:

$$\text{source } f = A_1 \ldots A_m, \qquad \text{target } f = A_{m+1}.$$

Among the sequents there is given one associated with each type A, the *identity* sequent $1_A : A \to A$, and a rule, the *cut*, for manufacturing new sequents from old:

$$\frac{f : \Lambda \to A \qquad g : \Gamma A \Delta \to B}{g\langle f\rangle : \Gamma\Lambda\Delta \to B}$$

We use capital Greek letters to represent strings of types, say

$$\Gamma = A_1 \ldots A_m, \quad \Delta = B_1 \ldots B_n, \quad \Lambda = C_1 \ldots C_k.$$

The notation $g\langle f\rangle$ for the resulting sequent of a cut is of course ambiguous, as it does not show where f has been substituted into g, but we shall use it nonetheless in order to avoid making the notation too heavy.

A *multicategory* is a context-free grammar with a notion of equality between sequents satisfying four laws. To state these in our abbreviated notation, we assume f and g given as above and also $h : \Gamma' B \Delta' \to C$, $f' : \Lambda' \to A'$, $g' : \Gamma A \Delta A' \Delta' \to C$. The four laws are then stated as follows:

$$1_A\langle f\rangle = f, \qquad g\langle 1_A\rangle = g,$$
$$h\langle g\langle f\rangle\rangle = h\langle g\rangle\langle f\rangle,$$
$$g'\langle f\rangle\langle f'\rangle = g'\langle f'\rangle\langle f\rangle.$$

For further details see [L 1969]. However, as pointed out in [L 1989], there is a natural way of obtaining these equations for free. One forms the *internal language* of a context-free grammar (or multicategory) by adjoining countably many variables of each type and interpreting the sequents as (multilinear) operations. We shall do this here and save the reader from having to read up on multicategories.

We define *terms* of all types inductively: each variable is a term of its type and, if $f : A_1 \ldots A_m \to A_{m+1}$ is a sequent and a_i a term of type A_i, for $i = 1, \ldots, m$, then $fa_1 \ldots a_m$ is a term of type A_{m+1}. (Note that we have admitted variables of each type, but so far not variable types themselves. These will be touched upon in Section 7.)

Equality (or rather *provable equality*) between terms of the same type is a congruence relation which satisfies the usual law of substituting terms for free variables. Thus, from $\varphi(x) = \psi(x)$ one may infer $\varphi(a) = \psi(a)$, where a is any term of the same type as the variable x. In principle, one should *declare* variables and write $\varphi(x) =_x \psi(x)$, for example; but, in the absence of Gentzen's thinning rule, this is not important. In addition to the usual rules for equality we also stipulate the following for the identity sequent and the cut:

$$1_A x = x, \quad g\langle f\rangle uwv = gufwv.$$

Here, and elsewhere in this paper, x is a variable of type A, $u = x_1 \ldots x_m$ is a string of variables of type $\Gamma = A_1 \ldots A_m$, v a string of variables of type Δ and w a string of variables of type Λ.

The description of the internal language is now complete. Equality between sequents is now definable: two sequents $f, g : A_1 \ldots A_m \to A_{m+1}$ are said to be *equal* if provably $fx_1 \ldots x_m = gx_1 \ldots x_m$ in the internal language. It is important here that the x_i are all distinct, even if two of the A_i happen to coincide. The identities of a multicategory mentioned earlier are now easily deducible, as was in fact done in [L 1989]. For example, if $f : \Lambda \to A, g : \Gamma A \Delta \to B$ and $h : \Gamma' B \Delta' \to C$, we calculate

$$\begin{aligned} h\langle g\langle f\rangle\rangle u'uwvv' &= hu'g\langle f\rangle uwvv' \\ &= hu'gufwvv' \\ &= h\langle g\rangle u'ufwvv' \\ &= h\langle g\rangle\langle f\rangle u'uwvv'. \end{aligned}$$

2 Categorial Grammars

We now wish to enhance our context-free grammar or multicategory by certain type-forming operations, called *connectives* in logic. In particular, we shall study the following, which tend to be represented by different symbols in logic and algebra. Here are the most common notations in use:

algebra	$\otimes$	$/$	$\backslash$	$\times$	$+$	I	1	0
logic		$\Leftarrow$	$\Rightarrow$	$\wedge$	$\vee$		$\top$	$\bot$

(The absence of symbols for $\otimes$ and I in logic are explained by the fact that, in the presence of Gentzen's structural rules, these connectives will coincide with $\times$ and 1; also $/$ and $\backslash$ are then just converses of one another.) The first three connectives have been used in linguistics for a long time, turning the grammar into what has been called a *categorial grammar*. Possible linguistic applications of some of the remaining connectives have been discussed elsewhere [L 1989b]. We shall now discuss the properties of these eight connectives in turn, first from the viewpoint of multicategories.

(1) One obtains a biunique correspondence between sequents $g : \Gamma(A \otimes B)\Delta \to C$ and sequents $f : \Gamma AB\Delta \to C$ by introducing a sequent $\mathbf{m}_{AB} : AB \to A\otimes B$ and stipulating that for each f there is a unique g such that $g\langle \mathbf{m}\rangle = f$. If we put $g = f^{\S}$, this amounts to postulating $f^{\S}\langle \mathbf{m}\rangle = f$ and $(g\langle \mathbf{m}\rangle)^{\S} = g$. In particular, $\mathbf{m}^{\S}_{AB} = 1_{A\otimes B}$.

(2) One obtains a biunique correspondence between sequents $g : \Gamma \to A/B$ and sequents $f : \Gamma B \to A$ by introducing a sequent $\mathbf{e}_{AB} : (A/B)B \to A$ and stipulating that for each f there should exist a unique g such that $\mathbf{e}\langle g\rangle = f$. If we put $g = f^*$, this amounts to postulating $\mathbf{e}\langle f^*\rangle = f$ and $(\mathbf{e}\langle g\rangle)^* = g$. In particular, $\mathbf{e}^*_{AB} = 1_{A/B}$.

(3) One obtains a biunique correspondence between sequents $g : \Gamma \to B\backslash A$ and sequents $f : B\Gamma \to A$ by introducing a sequent $\mathbf{e}'_{BA} : B(B\backslash A) \to A$ and stipulating that for each f there should exist a unique g such that $\mathbf{e}'\langle g\rangle = f$. If we put $g = f^{\dagger}$, this amounts to postulating $\mathbf{e}'\langle f^{\dagger}\rangle = f$ and $(\mathbf{e}'\langle g\rangle)^{\dagger} = g$. In particular, ${\mathbf{e}'_{BA}}^{\dagger} = 1_{B\backslash A}$.

(4) One obtains a biunique correspondence between sequents $g : \Gamma \to A \times B$ and pairs of sequents $f_0 : \Gamma \to A,\ f_1 : \Gamma \to B$ by introducing two

sequents $\mathbf{p}_{AB} : A \times B \to A$ and $\mathbf{q}_{AB} : A \times B \to B$ and stipulating that for each pair f_0, f_1 there should exist a unique g such that $\mathbf{p}\langle g\rangle = f_0$ and $\mathbf{q}\langle g\rangle = f_1$. If we put $g = \langle f_0, f_1\rangle$, this amounts to postulating $\mathbf{p}\langle f_0, f_1\rangle = f_0$, $\mathbf{q}\langle f_0, f_1\rangle = f_1$ and $\langle \mathbf{p}g, \mathbf{q}g\rangle = g$. In particular, $\langle \mathbf{p}_{AB}, \mathbf{q}_{AB}\rangle = 1_{A\times B}$.

(5) One obtains a biunique correspondence between sequents $g : \Gamma(A+B)\Delta \to C$ and pairs of sequents $f_0 : \Gamma A\Delta \to C$, $f_1 : \Gamma B\Delta \to C$ by introducing two sequents $\mathbf{k}_{AB} : A \to A+B$ and $\mathbf{l}_{AB} : B \to A+B$ and stipulating that for each pair f_0, f_1 there should exist a unique g such that $g\langle \mathbf{k}\rangle = f_0$ and $g\langle \mathbf{l}\rangle = f_1$. If we put $g = [f_0, f_1]$, this amounts to postulating $[f_0, f_1]\langle \mathbf{k}\rangle = f_0$, $[f_0, f_1]\langle \mathbf{l}\rangle = f_1$ and $[g\langle \mathbf{k}\rangle, g\langle \mathbf{l}\rangle] = g$. In particular, $[\mathbf{k}_{AB}, \mathbf{l}_{AB}] = 1_{A+B}$.

(6) One obtains a biunique correspondence between sequents $g : \Gamma \mathrm{I}\Delta \to C$ and sequents $f : \Gamma\Delta \to C$ by introducing a sequent $\mathbf{i} : \quad \to \mathrm{I}$ (note the empty string before the arrow) and stipulating that for each f there should exist a unique g such that $g\langle \mathbf{i}\rangle = f$. If we put $g = f^\sharp$, this amounts to postulating $f^\sharp\langle \mathbf{i}\rangle = f$ and $(g\langle \mathbf{i}\rangle)^\sharp = g$. In particular, $\mathbf{i}^\sharp = 1_\mathrm{I}$.

(7) One stipulates a unique sequent $g : \Gamma \to 1$ and puts $g = \circ_\Gamma$. In particular $\circ_1 = 1_1$.

(8) One stipulates a unique sequent $g : \Gamma 0\Delta \to C$ and puts $g = \Box^C_{\Gamma\Delta}$. In particular, $\Box^0 = 1_0$ (where Γ and Δ are empty strings).

In the internal language, the uniqueness of a sequent has to be expressed as a rule of inference, hence the above properties of the eight connectives are written as follows:

(1) $$f^\S u\mathbf{m}xyv = fuxyv; \qquad \frac{g_1u\mathbf{m}xyv = g_2u\mathbf{m}xyv}{g_1us_1v = g_2us_1v}$$

where s_1 is a variable of type $A \otimes B$.

(2) $$\mathbf{e}f^*uy = fuy; \qquad \frac{\mathbf{e}g_1uy = \mathbf{e}g_2uy}{g_1u = g_2u}$$

(3) $$\mathbf{e}'yf^\dagger u = fyu; \qquad \frac{\mathbf{e}'yg_1u = \mathbf{e}'yg_2u}{g_1u = g_2u}$$

(4) $$\begin{matrix}\mathbf{p}\langle f_0, f_1\rangle u = f_0u, \\ \mathbf{q}\langle f_0, f_1\rangle u = f_1u;\end{matrix} \qquad \frac{\mathbf{p}g_1u = \mathbf{p}g_2u \quad \mathbf{q}g_1u = \mathbf{q}g_2u}{g_1u = g_2u}$$

(5) $$\begin{matrix}[f_0, f_1]u\mathbf{k}xv = f_0uxv, \\ [f_0, f_1]u\mathbf{l}yv = f_1uyv;\end{matrix} \qquad \frac{g_1u\mathbf{k}xv = g_2u\mathbf{k}xv \quad g_1u\mathbf{l}yv = g_2u\mathbf{l}yv}{g_1us_5v = g_2us_5v}$$

where s_5 is a variable of type $A + B$.

(6) $$f^{\sharp}u\mathbf{i}v = fuv; \qquad \frac{g_1 u\mathbf{i}v = g_2 u\mathbf{i}v}{g_1 u s_6 v = g_2 u s_6 v}$$

where s_6 is a variable of type I.

(7) $$gu = \circ_{\Gamma} u.$$

(8) $$gus_8v = \Box us_8v, \text{ where } s_8 \text{ is a variable of type } 0.$$

Perhaps this is the place to comment on an alternative notation in situation (2). If $f : \Gamma B \to A$, it is customary to write $\lambda_{y\in B} fwy$ for f^*w and something like gw 'y for $\mathbf{e}gwy$. The two multicategory equations $\mathbf{e}\langle f^*\rangle = f$ and $(\mathbf{e}\langle g\rangle)^* = g$ can then both be written as the equations (β) and (η) of the usual $\lambda\eta$-calculus, the former after substituting an arbitrary term of type B for y:

$$\lambda_{y\in B} fwy\text{`}b = fwb, \quad \lambda_{y\in B}(gw\text{`}y) = gw.$$

For situation (3) one could similarly introduce $\lambda'_{y\in B}$ with symmetric properties. But what about the other situations? In (1), for example, one would need a notation which replaces two free variables of types A and B by a single free variable of type $A\otimes B$. Undoubtedly, it is possible to extend the λ-notation to the other situations, but we shall refrain from doing so here.

3 Freely Generated Categorial Grammars

Given a multicategory (i.e. context-free grammar with equality) $\mathcal{G}$, we wish to study the categorial grammar $F(\mathcal{G})$ freely generated by $\mathcal{G}$, equipped with some or all of the connectives studied in the previous section. How is $F(\mathcal{G})$ constructed? Its objects (or sorts) are obtained inductively from those of $\mathcal{G}$ by means of the connectives $\otimes, /, \backslash$, etc. Its sequents are obtained inductively from those of $\mathcal{G}$ by adjoining the basic sequents $1_A, \mathbf{m}_{AB}, \mathbf{e}_{AB}, \mathbf{e}'_{AB}$, etc., and forming new operations from old by the cut rule and the rules $(-)^{\S}, (-)^*, (-)^{\dagger}$, etc. Actually, we don't need the basic sequents 1_A, since these are definable inductively from sequents in $\mathcal{G}$ by means of the equations $1_{A\otimes B} = \mathbf{m}^{\S}_{AB}, 1_{A/B} = \mathbf{e}^*_{AB}$, etc., established in Section 2. (As we shall see later, the cut rule can also be eliminated, provided $\mathbf{m}, \mathbf{e}, \mathbf{e}'$, etc., are replaced by appropriate rules.) Finally, the equations between comparable sequents in $F(\mathcal{G})$ are those and only those which follow from the equations in $\mathcal{G}$ and the multicategory equations

$$h\langle g\langle f\rangle\rangle = h\langle g\rangle\langle f\rangle, \qquad g'\langle f\rangle\langle f'\rangle = g\langle f'\rangle\langle f\rangle$$

as well as the equations

$$f^{\S}\langle \mathbf{m}\rangle = f, \qquad (f\langle \mathbf{m}\rangle)^{\S} = f, \qquad \text{etc.}$$

mentioned under (1) to (8) above.

It might be pointed out here that the categorial grammars which some linguists believe to be adequate for the description of natural languages are of the form $F(\mathcal{G})$, freely generated by a 'discrete' grammar $\mathcal{G}$, in which the only sequents are identity sequents $1_A : A \to A$. Of course, F will be left adjoint to the forgetful functor U from categorial grammars to context-free ones. Ultimately we hope to show that the multifunctor $\mathcal{G} \to UF(\mathcal{G})$ is full and faithful. (Multifunctors are the obvious generalization of functors to multicategories.) Given types A_i of $\mathcal{G}$, the fullness will tell us that there are no new sequents $A_1 \ldots A_m \to A_{m+1}$ in $UF(\mathcal{G})$ not already in $\mathcal{G}$ and the faithfulness will tell us that there are no new equations between two such sequents.

Assuming, more generally, that the A_i are types of $UF(\mathcal{G})$, that is, built up inductively from types of $\mathcal{G}$ by means of the connectives under consideration, we aim to show how to find all sequents $A_1 \ldots A_m \to A_{m+1}$ in $UF(\mathcal{G})$ and all equations between them.

At the moment, the only obstacle to finding all sequents $A_1 \ldots A_m \to A_{m+1}$ is the cut rule: if a sequent $g\langle f\rangle : \Gamma\Lambda\Delta \to B$ is derived from $f : \Lambda \to A$ and $g : \Gamma A \Delta \to B$, there are infinitely many possibilities for A. We cannot eliminate the cut rule completely, but we can restrict its use to such instances where A already appears in the conclusion. We do this by replacing the special sequents $\mathbf{m}, \mathbf{e}, \mathbf{e}', \mathbf{p}, \mathbf{q}, \mathbf{k}, \mathbf{l}$ by more general recipes for forming sequents.

(1) We replace $\mathbf{m}_{AB} : AB \to A \otimes B$ by the rule

$$\frac{f_0 : \Gamma \to A \qquad f_1 : \Delta \to B}{\{f_0, f_1\} : \Gamma\Delta \to A \otimes B}$$

where $\{f_0, f_1\} = \mathbf{m}\langle f\rangle\langle g\rangle$ is defined in the internal language thus:

$$\{f_0, f_1\}uv = \mathbf{m} f_0 u f_1 v.$$

Clearly $\mathbf{m}_{AB} = \{1_A, 1_B\}$.

(2) We replace $\mathbf{e}_{AB} : (A/B)B \to A$ by the rule

$$\frac{f : \Lambda \to A \qquad g : \Gamma B \Delta \to C}{g[f] : \Gamma(B/A)\Lambda\Delta \to C}$$

where $g[f] = g\langle \mathbf{e}\langle f \rangle\rangle$ is defined thus:

$$g[f]us_2wv = gu\mathbf{e}s_2fwv,$$

s_2 being a variable of type B/A. Clearly $\mathbf{e}_{AB} = 1_B[1_A]$.

(3) We replace $\mathbf{e}'_{BA} : B(B\backslash A) \to A$ by the rule

$$\frac{f : \Lambda \to A \qquad g : \Gamma B \Delta \to C}{g\{f\} : \Gamma\Lambda(A\backslash B)\Delta \to C}$$

where $g\{f\} = g\langle \mathbf{e}'\langle f \rangle\rangle$ is defined thus:

$$g\{f\}uws_3v = \mathbf{g}u\mathbf{e}'fws_3v,$$

s_3 being a variable of type $A\backslash B$. Clearly $\mathbf{e}'_{AB} = 1_B\{1_A\}$.

(4) We replace $\mathbf{p}_{AB} : A \times B \to A$ and $\mathbf{q}_{AB} : A \times B \to B$ by the rules

$$\frac{f : \Gamma A \Delta \to C}{f_p : \Gamma(A \times B)\Delta \to C} \qquad \frac{g : \Gamma B \Delta \to C}{g_q : \Gamma(A \times B)\Delta \to C}$$

where $f_p = f\langle \mathbf{p} \rangle$ and $g_q = g\langle \mathbf{q} \rangle$ are defined thus:

$$f_pus_4v = fu\mathbf{p}s_4v, \quad g_qus_4v = gu\mathbf{q}s_4v,$$

s_4 being a variable of type $A \times B$. Clearly $\mathbf{p}_{AB} = (1_A)_p$, $\mathbf{q}_{AB} = (1_B)_q$.

(5) We replace $\mathbf{k}_{AB} : A \to A + B$ and $\mathbf{l}_{AB} : B \to A + B$ by the rules

$$\frac{f : \Lambda \to A}{{}_kf : \Lambda \to A + B}, \qquad \frac{g : \Lambda \to B}{{}_\ell g : \Lambda \to A + B},$$

where ${}_kf = \mathbf{k}\langle f \rangle$ and ${}_\ell g = \mathbf{l}\langle g \rangle$ are defined thus:

$${}_kfw = \mathbf{k}fw, \quad {}_\ell gw = \mathbf{l}gw.$$

Clearly $\mathbf{k}_{AB} = {}_k(1_A)$, $\mathbf{l}_{AB} = {}_\ell(1_B)$.

(6,7,8) We do not replace the sequents $\mathbf{i} : \ \to \mathrm{I}$, $\mathrm{o}_\Gamma : \Gamma \to 1$ and $\square^C_{\Gamma\Delta} : \Gamma 0 \Delta \to C$ associated with the nullary connectives I, 1 and 0 at all.

The *cut-elimination theorem* now takes the following form:

Theorem 3.1. *If* $\mathbf{m}, \mathbf{e}, \mathbf{e}', \mathbf{p}, \mathbf{q}, \mathbf{k}$ *and* $\mathbf{l}$ *are replaced by the more general recipes as above, then any sequent in* $UF(\mathcal{G})$ *constructed with cut is* equal *to one constructed without cut.*

We shall leave the proof to the next section.

Thus sequents in $UF(\mathcal{G})$ can be constructed without using the cut rule. They can also be constructed without using the identity rule $1_A : A \to A$. For, using the equations $1_{A\otimes B} = \mathbf{m}^{\S}_{AB}$, $1_{A/B} = \mathbf{e}^{*}_{AB}$, etc., established in Section 2, we easily prove by induction that the only identity sequents required are those already in $\mathcal{G}$.

Corollary 3.2. *The multifunctor* $\mathcal{G} \to UF(\mathcal{G})$ *is full.*

Proof. A cut-free sequent $g : A_1 \ldots A_m \to A_{m+1}$ is constructed from sequents in $\mathcal{G}$ by means of the rules $(-)^{\S}$, $\{-,-\}$, etc. If the A_i are in $\mathcal{G}$, then these rules cannot be involved in the construction, as for instance the source of $f^{\S}$ and the target of $\{f_0, f_1\}$ contain the connective $\otimes$. Hence g must already live in $\mathcal{G}$.

The problem of finding all equations between sequents $f_1, f_2 : A_1 \ldots A_m \to C$ in $UF(\mathcal{G})$ will ultimately be considered in Section 5; for our present purpose, namely to prove the cut-elimination theorem, it reduces to that of finding all equations between the corresponding terms $c_i = f_i x_1 \ldots x_m$ of type C in the internal language, where it is assumed that there is no repetition among the x_j. In view of the uniqueness conditions in situations (1) to (8), if we wish to compare c_1 and c_2, we may carry out the following replacements:

(1) replace any variable of type $A \otimes B$ by $\mathbf{m}xy$, with x of type A and y of type B;

(2) if C is A/B, replace c_i by $\mathbf{e}c_i y$, with y of type B;

(3) if C is $B\backslash A$, replace c_i by $\mathbf{e}' y c_i$;

(4) if C is $A \times B$, replace c_i by the pair $(\mathbf{p}c_i, \mathbf{q}c_i)$;

(5) replace any variable of type $A + B$ first by $\mathbf{k}x$, then by $\mathbf{l}y$;

(6) replace any variable of type I by $\mathbf{i}$;

(7) if C is 1, we already know that $c_1 = c_2$;

(8) if there is a variable of type 0, we already know that $c_1 = c_2$.

The problem of comparing two terms of type C, containing the same variables in the same order and without repetition, thus reduces to the special case when C is neither A/B, $B\backslash A$, $A \times B$ nor 1 and when none of the variables in question are of type $A \otimes B$, $A + B$, I or 0.

4 Proof of the Cut-Elimination Theorem

To prove this, we may assume first of all, without loss in generality, that $f : \Lambda \to A$ and $g : \Gamma A \Delta \to B$ have already been constructed without cut. We shall then show that $g\langle f \rangle$ is equal to a sequent whose construction involves cuts of smaller 'degree', the *degree* of the cut $g\langle f \rangle$ being defined as

$$d(\Lambda) + d(\Gamma) + d(\Delta) + d(A) + d(B),$$

where $d(A)$ is the number of occurrences of all connectives in A and $d(\Gamma) = d(A_1) + \cdots + d(A_m)$. There are now five cases.

Case 0. Both f and g are in $\mathcal{G}$.

Case 1. The last step in the construction of f introduces one connective on the left of the arrow.

Case 2. The last step in the construction of g introduces one connective on the left of the arrow but not into A.

Case 3. The last step in the constructions of g introduces one connective on the right of the arrow.

Case 4. The last steps in the constructions of both f and g introduce one connective into A.

In Case 0, $g\langle f \rangle$ is already in $\mathcal{G}$; for $\mathcal{G}$, being a context-free grammar, is closed under cuts. We shall now discuss the other four cases, one connective at a time.

(1) The connective $\otimes$.

Case 1. $\Lambda = \Gamma' A' \otimes B' \Delta'$ and $f = f'^{\S}$, so our cut appears in the following context:

$$\cfrac{\cfrac{f' : \Gamma' A' B' \Delta' \to A}{f'^{\S} : \Gamma' A' \otimes B' \Delta' \to A} \qquad g : \Gamma A \Delta \to B}{g\langle f'^{\S} \rangle : \Gamma \Gamma' A' \otimes B' \Delta' \Delta \to B}\ cut$$

We replace this by the following:

$$\cfrac{\cfrac{f' : \Gamma' A' B' \Delta' \to A \qquad g : \Gamma A \Delta \to B}{g\langle f' \rangle : \Gamma \Gamma' A' B' \Delta' \Delta \to B}\ cut}{(g\langle f' \rangle)^{\S} : \Gamma \Gamma' A' \otimes B' \Delta' \Delta \to B}$$

where the new cut has smaller degree than the old one. Moreover, $g\langle f'^{\S} \rangle = (g\langle f' \rangle)^{\S}$ as is shown in the internal language as follows. In place of a variable of type $A' \otimes B'$ we may put $\mathbf{m}x'y'$, where x' and y' are variables of types A' and B' respectively. We then check that

$$g\langle f'^{\S} \rangle uu'\mathbf{m}x'y'v'v = (g\langle f' \rangle)^{\S} uu'\mathbf{m}x'y'v'v$$

by showing that both sides simplify to $guf'u'x'y'v'v$, using nothing but the definitions of $\langle\ \rangle$ and §.

Case 2. Say $\Gamma = \Gamma' A' \otimes B' \Delta'$ and $g = f'^{\S}$. (The case when $\Delta = \Gamma' A' \otimes B' \Delta'$ is similar and will be omitted.) We are given:

$$\cfrac{f : \Lambda \to A \qquad \cfrac{f' : \Gamma' A' B' \Delta' A \Delta \to B}{f'^{\S} : \Gamma' A' \otimes B' \Delta' A \Delta \to B}}{f'^{\S}\langle f \rangle : \Gamma' A' \otimes B' \Delta' \Lambda \Delta \to B}\ cut$$

We replace this by:

$$\cfrac{\cfrac{f : \Lambda \to A \qquad f' : \Gamma' A' B' \Delta' A \Delta \to B}{f'\langle f \rangle : \Gamma' A' B' \Delta' \Lambda \Delta \to B}\ cut}{(f'\langle f \rangle)^{\S} : \Gamma' A' \otimes B' \Delta' \Lambda \Delta \to B}$$

where the new cut has smaller degree. Moreover,

$$f'^{\S}\langle f \rangle u'\mathbf{m}x'y'v'wv = f'\langle f \rangle^{\S} u'\mathbf{m}x'y'v'wv,$$

since both sides reduce to $f'u'x'y'v'fwv$.

Case 3. $B = A' \otimes B'$ and $g = \{f_0, f_1\} : \Gamma'\Delta' \to A' \otimes B'$. We shall assume that A is contained in Γ'. (The case when it occurs in Δ' is similar and will be omitted.) We are given:

$$\cfrac{f : \Lambda \to A \qquad \cfrac{f_0 : \Gamma_1 A \Gamma_2 \to A' \qquad f_1 : \Delta' \to B'}{\{f_0, f_1\} : \Gamma_1 A \Gamma_2 \Delta' \to A' \otimes B'}}{\{f_0, f_1\}\langle f \rangle : \Gamma_1 \Lambda \Gamma_2 \Delta' \to A' \otimes B'}\ cut$$

We replace this by:

$$\cfrac{\cfrac{f : \Lambda \to A \qquad f_0 : \Gamma_1 A \Gamma_2 \to A'}{f_0\langle f \rangle : \Gamma_1 \Lambda \Gamma_2 \to A'}\ cut \qquad f_1 : \Delta' \to B'}{\{f_0\langle f \rangle, f_1\} : \Gamma_1 \Lambda \Gamma_2 \Delta' \to A' \otimes B'}$$

where the new cut has smaller degree. Moreover,

$$\{f_0, f_1\}\langle f\rangle u_1 w u_2 v' = \{f_0\langle f\rangle, f_1\}u_1 w u_2 v',$$

since both sides reduce to $\mathbf{m} f_0 u_1 f w u_2 f_1 v'$, in view of the definitions of $\{\ ,\ \}$ and $\langle\ \rangle$.

Case 4. $A = A' \otimes B'$, $f = \{f_0, f_1\}$ and $g = f'^{\S}$. We are given:

$$\frac{\dfrac{f_0 : \Gamma' \to A' \qquad f_1 : \Delta' \to B'}{\{f_0, f_1\} : \Gamma'\Delta' \to A' \otimes B'} \qquad \dfrac{f' : \Gamma A'B'\Delta \to B}{f'^{\S} : \Gamma A' \otimes B'\Delta \to B}}{f'^{\S}\langle\{f_0, f_1\}\rangle : \Gamma\Gamma'\Delta'\Delta \to B}\ cut$$

We replace this by:

$$\frac{f_1 : \Delta' \to B' \qquad \dfrac{f_0 : \Gamma' \to A' \qquad f' : \Gamma A'B'\Delta \to B}{f'\langle f_0\rangle : \Gamma\Gamma'B'\Delta \to B}\ cut}{f'\langle f_0\rangle\langle f_1\rangle : \Gamma\Gamma'\Delta'\Delta \to B}\ cut$$

where both new cuts have smaller degree. Moreover

$$f'^{\S}\langle\{f_0, f_1\}\rangle uu'v'v = f'\langle f_0\rangle\langle f_1\rangle uu'v'v,$$

which is seen as follows. The left-hand side reduces to $f'^{\S} u \mathbf{m} f_0 u' f_1 v' v$, in view of the definition of $\langle\ \rangle$ and $\{\ ,\}$. Now use the fact that $f'^{\S}\langle\mathbf{m}\rangle = f'$, then this reduces to $f' u f_0 u' f_1 v' v$, which is also what the right-hand side reduces to, in view of the definition of $\langle\ \rangle$ only.

(2) The connective $/$.

Case 1. $\Lambda = \Gamma' A'/B'\Lambda'\Delta'$ and $f = g'[f']$. We are given:

$$\frac{\dfrac{f' : \Lambda' \to B' \quad g' : \Gamma'A'\Delta' \to A}{g'[f'] : \Gamma'A'/B'\Lambda'\Delta' \to A} \qquad g : \Gamma A\Delta \to B}{g\langle g'[f']\rangle : \Gamma\Gamma'A'/B'\Lambda'\Delta'\Delta \to B}\ cut$$

We replace this by:

$$\frac{f' : \Lambda' \to B' \qquad \dfrac{g' : \Gamma'A'\Delta' \to A \qquad g : \Gamma A\Delta \to B}{g\langle g'\rangle : \Gamma\Gamma'A'\Delta'\Delta \to B}\ cut}{g\langle g'\rangle[f'] : \Gamma\Gamma'A'/B'\Lambda'\Delta'\Delta \to B}$$

where the new cut has smaller degree. Moreover,

$$g\langle g'[f']\rangle uu's'w'v'v = g\langle g'\rangle[f']uu's'w'v'v,$$

where s' is a variable of type A'/B', since both sides reduce to $gug'u'\mathbf{e}s'w'v'v$, in view of the definitions of $\langle\ \rangle$ and $[\,]$.

Case 2. $\Gamma A \Delta = \Gamma' A'/B' \Lambda' \Delta'$, $g = g'[f']$.
We shall first assume that A occurs in $\Delta' = \Delta_1 A \Delta_2$. We are given:

$$\dfrac{f : \Lambda \to A \qquad \dfrac{f' : \Lambda' \to B' \quad g' : \Gamma' A' \Delta_1 A \Delta_2 \to B}{g'[f'] : \Gamma' A'/B' \Lambda' \Delta_1 A \Delta_2 \to B}}{g'[f']\langle f \rangle : \Gamma' A'/B' \Lambda' \Delta_1 \Lambda \Delta_2 \to B}\, cut$$

We replace this by:

$$\dfrac{f' : \Lambda' \to B' \qquad \dfrac{f : \Lambda \to A \quad g' : \Gamma' A' \Delta_1 A \Delta_2 \to B}{g'\langle f \rangle : \Gamma' A' \Delta_1 \Lambda \Delta_2 \to B}\, cut}{g'\langle f \rangle [f'] : \Gamma' A'/B' \Lambda' \Delta_1 \Lambda \Delta_2 \to B}$$

where the new cut has smaller degree. Moreover,

$$g'[f']\langle f \rangle u' s' w' v_1 w v_2 = g'\langle f \rangle [f'] u' s' w' v_1 w v_2,$$

since both sides reduce to $g' u' \mathbf{e} s' f' w' v_1 f w v_2$.

The case when A occurs in Γ' is similar and will be omitted. However, there is still the case when A occurs in $\Lambda' = \Lambda_1 A \Lambda_2$. Then we are given:

$$\dfrac{f : \Lambda \to A \qquad \dfrac{f' : \Lambda_1 A \Lambda_2 \to B' \quad g' : \Gamma' A' \Delta' \to B}{g'[f'] : \Gamma' A'/B' \Lambda_1 A \Lambda_2 \Delta' \to B}}{g'[f']\langle f \rangle : \Gamma' A'/B' \Lambda_1 \Lambda \Lambda_2 \Delta' \to B}\, cut$$

We replace this by:

$$\dfrac{\dfrac{f : \Lambda \to A \quad f' : \Lambda_1 A \Lambda_2 \to B'}{f'\langle f \rangle : \Lambda_1 \Lambda \Lambda_2 \to B'}\, cut \qquad g' : \Gamma' A' \Delta' \to B}{g'[f'\langle f \rangle] : \Gamma' A'/B' \Lambda_1 \Lambda \Lambda_2 \Delta' \to B}$$

where the new cut is of smaller degree. Moreover,

$$g'[f']\langle f \rangle u' s' w_1 w w_2 v' = g'[f'\langle f \rangle] u' s' w_1 w w_2 v',$$

since both sides reduce to $g' u' \mathbf{e} s' f' w_1 f w w_2 v'$, in view of the definitions of $[\;]$ and $\langle\;\rangle$.

Case 3. $B = A'/B'$, $g = f'^*$. We are given

$$\frac{f : \Lambda \to A \qquad \dfrac{f' : \Gamma A \Delta B' \to A'}{f'^* : \Gamma A \Delta \to A'/B'}}{f'^*\langle f\rangle : \Gamma \Lambda \Delta \to A'/B'}\ cut.$$

We replace this by:

$$\cfrac{\cfrac{f : \Lambda \to A \qquad f' : \Gamma A \Delta B' \to A'}{f'\langle f\rangle : \Gamma \Lambda \Delta B' \to A'}\ cut}{(f'\langle f\rangle)^* : \Gamma \Lambda \Delta \to A'/B'}$$

where the new cut has smaller degree. Moreover,

$$\mathbf{e} f'^*\langle f\rangle uwvy' = \mathbf{e}(f'\langle f\rangle)^* uwvy',$$

since the left-hand side reduces to $\mathbf{e} f'^* ufwvy'$, in view of the definition of $\langle\ \rangle$. Now use the fact that $\mathbf{e}\langle f'^*\rangle = f'$, then this reduces to $f'ufwvy'$, which is also what the right-hand side reduces to in two steps. Finally, use the uniqueness part of (2).

It may be instructive to look at this argument in the λ-notation. Then

$$\begin{aligned} f'^*\langle f\rangle uwv &= \lambda_{y'\in B'} f'ufwvy' \\ &= \lambda_{y'\in B'} f'\langle f\rangle uwvy' = (f'\langle f\rangle)^* uwv. \end{aligned}$$

This does not make full use of the η-rule, but only of the substitution rule:

$$\frac{\varphi(x) =_x \psi(x)}{\lambda_{x\in A}\varphi(x) = \lambda_{x\in A}\psi(x)}.$$

Case 4. $A = A'/B'$, $f = f'^*$ and $g = g'[h]$. We are given:

$$\frac{\dfrac{f' : \Lambda B' \to A'}{f'^* : \Lambda \to A'/B'} \qquad \dfrac{h : \Lambda' \to B' \quad g' : \Gamma A' \Delta' \to B}{g'[h] : \Gamma A'/B' \Lambda' \Delta' \to B}}{g'[h]\langle f'^*\rangle : \Gamma \Lambda \Lambda' \Delta' \to B}\ cut$$

We replace this by:

$$\cfrac{\cfrac{h : \Lambda' \to B' \quad f' : \Lambda B' \to A'}{f'\langle h\rangle : \Lambda \Lambda' \to A'}\ cut \qquad g' : \Gamma A' \Delta' \to B}{g'\langle f'\langle h\rangle\rangle : \Gamma \Lambda \Lambda' \Delta \to B}\ cut$$

where both new cuts have smaller degree. Moreover, we calculate:

$$g'[h]\langle f'^*\rangle uww'v' = g'u\mathbf{e} f'^* whw'v'$$

and

$$g'\langle f'\langle h\rangle\rangle uww'v' = g'uf'whw'v'.$$

These two terms are equal in view of the identity $\mathbf{e}\langle f'^*\rangle = f'$.

(3) The connective $\backslash$ is just the mirror dual of $/$ and is treated symmetrically.

(4) The connective $\times$.

Case 1. Say $\Lambda = \Lambda_1 A' \times B' \Lambda_2$ and $f = f'_p$. We are given:

$$\dfrac{\dfrac{f' : \Lambda_1 A' \Lambda_2 \to A}{f'_p : \Lambda_1 A' \times B' \Lambda_2 \to A} \qquad g : \Gamma A \Delta \to B}{g\langle f'_p \rangle : \Gamma \Lambda_1 A' \times B' \Lambda_2 \Delta \to B}\ cut$$

We replace this by:

$$\dfrac{\dfrac{f' : \Lambda_1 A' \Lambda_2 \to A \quad g : \Gamma A \Delta \to B}{g\langle f' \rangle : \Gamma \Lambda_1 A' \Lambda_2 \Delta \to B}}{(g\langle f' \rangle)_p : \Gamma \Lambda_1 A' \times B' \Lambda_2 \Delta \to B}\ cut$$

where the new cut has smaller degree. Moreover, if t' is a variable of type $A' \times B'$,

$$g\langle f'_p \rangle u w_1 t' w_2 v = (g\langle f' \rangle)_p u w_1 t' w_2 v,$$

as both sides reduce to $guf'w_1\mathbf{p}t'w_2v$, by means of the definitions of $\langle\ \rangle$ and f'_p.

Case 2. Say $\Gamma = \Gamma_1 A' \times B' \Gamma_2$ and $g = g'_p$. We are given:

$$\dfrac{f : \Lambda \to A \qquad \dfrac{g' : \Gamma_1 A' \Gamma_2 A \Delta \to B}{g'_p : \Gamma_1 A' \times B' \Gamma_2 A \Delta \to B}}{g'_p\langle f \rangle : \Gamma_1 A' \times B' \Gamma_2 \Lambda \Delta \to B}\ cut$$

We replace this by:

$$\dfrac{\dfrac{f : \Lambda \to A \qquad g' : \Gamma_1 A' \Gamma_2 A \Delta \to B}{g'\langle f \rangle : \Gamma_1 A' \Gamma_2 \Lambda \Delta \to B}\ cut}{(g'\langle f \rangle)_p : \Gamma_1 A' \times B' \Gamma_2 \Lambda \Delta \to B}$$

where the new cut has smaller degree. Moreover,

$$g'_p\langle f \rangle u_1 t' u_2 w v = (g'\langle f \rangle)_p u_1 t' u_2 w v,$$

since both sides reduce to $g'u_1\mathbf{p}t'u_2fwv$ as in Case 1.

Case 3. $B = A' \times B'$ and $g = \langle g_0, g_1 \rangle$. We are given:

$$\dfrac{f : \Lambda \to A \qquad \dfrac{g_0 : \Gamma A \Delta \to A' \quad g_1 : \Gamma A \Delta \to B'}{\langle g_0, g_1 \rangle : \Gamma A \Delta \to A' \times B'}}{\langle g_0, g_1 \rangle \langle f \rangle : \Gamma \Lambda \Delta \to A' \times B'}\ \mathit{cut}$$

We replace this by:

$$\dfrac{\dfrac{f : \Lambda \to A \quad g_0 : \Gamma A \Delta \to A'}{g_0 \langle f \rangle : \Gamma \Lambda \Delta \to A'}\ \mathit{cut} \qquad \dfrac{f : \Lambda \to A \quad g_1 : \Gamma A \Delta \to B'}{g_1 \langle f \rangle : \Gamma \Lambda \Delta \to B'}\ \mathit{cut}}{\langle g_0 \langle f \rangle, g_1 \langle f \rangle \rangle : \Gamma \Lambda \Delta \to A' \times B'}$$

where both cuts have smaller degrees. Moreover,

$$\mathbf{p} \langle g_0, g_1 \rangle \langle f \rangle uwv = \mathbf{p} \langle g_0 \langle f \rangle, g_1 \langle f \rangle \rangle uwv$$

(and similarly for $\mathbf{q}$), as the left-hand side reduces to

$$\mathbf{p} \langle g_0, g_1 \rangle ufwv = g_0 ufwv$$

and the right-hand side likewise.

Case 4. $A = A' \times B'$, $f = \langle f_0, f_1 \rangle$ and $g = g'_p$ say. We are given:

$$\dfrac{\dfrac{f_0 : \Lambda \to A' \quad g_1 : \Lambda \to B'}{\langle f_0, f_1 \rangle : \Lambda \to A' \times B'} \qquad \dfrac{g' : \Gamma A' \Delta \to B}{g'_p : \Gamma A' \times B' \Delta \to B}}{g'_p \langle \langle f_0, f_1 \rangle \rangle : \Gamma \Lambda \Delta \to B}\ \mathit{cut}$$

We replace this by:

$$\dfrac{f_0 : \Lambda \to A' \qquad g' : \Gamma A' \Delta \to B}{g' \langle f_0 \rangle : \Gamma \Lambda \Delta \to B}\ \mathit{cut}$$

where the new cut has smaller degree. Moreover,

$$\begin{aligned} g'_p \langle \langle f_0, f_1 \rangle \rangle uwv &= g'_p u \langle f_0, f_1 \rangle wv \\ &= g' u \mathbf{p} \langle f_0, f_1 \rangle wv \\ &= g' u f_0 wv = g' \langle f_0 \rangle uwv. \end{aligned}$$

(5) The connective $+$.

Case 1. $\Lambda = \Gamma' A' + B' \Delta'$ and $f = [f_0, f_1]$. We are given:

$$\dfrac{\dfrac{f_0 : \Gamma' A' \Delta' \to A \quad f_1 : \Gamma' B' \Delta' \to A}{[f_0, f_1] : \Gamma' A' + B' \Delta' \to A} \qquad g : \Gamma A \Delta \to B}{g\langle [f_0, f_1] \rangle : \Gamma \Gamma' A' + B' \Delta' \Delta \to B}\ cut$$

We replace this by:

$$\dfrac{\dfrac{f_0 : \Gamma' A' \Delta' \to A \quad g : \Gamma A \Delta \to B}{g\langle f_0 \rangle : \Gamma \Gamma' A' \Delta' \Delta \to B}\ cut \qquad \dfrac{f_1 : \Gamma' B' \Delta' \to A \quad g : \Gamma A \Delta \to B}{g\langle f_1 \rangle : \Gamma \Gamma' B' \Delta' \Delta \to B}\ cut}{[g\langle f_0 \rangle, g\langle f_1 \rangle] : \Gamma \Gamma' A' + B' \Delta' \Delta \to B}$$

where both new cuts have smaller degrees. Moreover, we have

$$g\langle [f_0, f_1] \rangle u u' \mathbf{k} x' v' v = [g\langle f_0 \rangle, g\langle f_1 \rangle] u u' \mathbf{k} x' v' v$$

(and similarly if $\mathbf{k}x'$ is replaced by $\mathbf{l}y'$), since both sides reduce to $guf_0u'x'v'v$.

Case 2. $\Gamma = \Gamma' A' + B' \Delta'$, say, and $g = [g_0, g_1]$ We are given:

$$\dfrac{f : \Lambda \to A \qquad \dfrac{g_0 : \Gamma' A' \Delta' A \Delta \to B \quad g_1 : \Gamma' B' \Delta' A \Delta \to B}{[g_0, g_1] : \Gamma' A' + B' \Delta' A \Delta \to B}}{[g_0, g_1]\langle f \rangle : \Gamma' A' + B' \Delta' \Lambda \Delta \to B}\ cut$$

We replace this by:

$$\dfrac{\dfrac{f : \Lambda \to A \quad g_0 : \Gamma' A' \Delta' A \Delta \to B}{g_0\langle f \rangle : \Gamma' A' \Delta' \Lambda \Delta \to B}\ cut \qquad \dfrac{f : \Lambda \to A \quad g_1 : \Gamma' B' \Delta' A \Delta \to B}{g_1\langle f \rangle : \Gamma' B' \Delta' \Lambda \Delta \to B}\ cut}{[g_0\langle f \rangle, g_1\langle f \rangle] : \Gamma' A' + B' \Delta' \Lambda \Delta \to B}$$

where each new cut has smaller degree. Moreover,

$$[g_0, g_1]\langle f \rangle u' \mathbf{k} x' v' w v = [g_0\langle f \rangle, g_1\langle f \rangle] u' \mathbf{k} x' v' w v$$

(and similarly for $\mathbf{l}y'$ in place of $\mathbf{k}x'$), since both sides reduce to $g_o u' x' v' f w v$.

Case 3. $B = A' + B'$ and $g = {}_k g'$, say. We are given:

$$\dfrac{f : \Lambda \to A \qquad \dfrac{g' : \Gamma A \Delta \to A'}{{}_k g' : \Gamma A \Delta \to A' + B'}}{{}_k g'\langle f \rangle : \Gamma \Lambda \Delta \to A' + B'}\ cut$$

We replace this by:

$$\cfrac{\cfrac{f : \Lambda \to A \qquad g' : \Gamma A \Delta \to A'}{g'\langle f\rangle : \Gamma\Lambda\Delta \to A'}}{{}_k(g'\langle f\rangle) : \Gamma\Lambda\Delta \to A' + B'}\ cut$$

where the new cut has smaller degree. Moreover,

$${}_k g'\langle f\rangle uwv = {}_k(g'\langle f\rangle)uwv,$$

since both sides reduce to $\mathbf{k}g'ufwv$.

Case 4. $A = A' + B'$, $f = {}_k f'$ say and $g = [g_0, g_1]$. We are given

$$\cfrac{\cfrac{f' : \Lambda \to A'}{{}_k f' : \Lambda \to A' + B'} \qquad \cfrac{g_0 : \Gamma A'\Delta \to B \quad g_1 : \Gamma B'\Delta \to B}{[g_0, g_1] : \Gamma A' + B'\Delta \to B}}{[g_0, g_1]\langle {}_k f'\rangle : \Gamma\Lambda\Delta \to B}$$

We replace this by:

$$\cfrac{f' : \Lambda \to A' \qquad g_0 : \Gamma A'\Delta \to B}{g_0\langle f'\rangle : \Gamma\Lambda\Delta \to B}\ cut$$

where the new cut has smaller degree. Moreover,

$$[g_0, g_1]\langle {}_k f'\rangle uwv = [g_0, g_1]u\mathbf{k}f'wv = g_0uf'wv = g_0\langle f'\rangle uwv.$$

(6) The nullary connective I. Note that Case 3 is not possible.

Case 1. $\Lambda = \Gamma'\mathrm{I}\Delta'$ and $f = f'^{\sharp}$. We are given:

$$\cfrac{\cfrac{f' : \Gamma'\Delta' \to A}{f'^{\sharp} : \Gamma'\mathrm{I}\Delta' \to A} \qquad g : \Gamma A\Delta \to B}{g\langle f'^{\sharp}\rangle\Gamma\Gamma'\mathrm{I}\Delta'\Delta \to B}\ cut$$

We replace this by:

$$\cfrac{\cfrac{f' : \Gamma'\Delta' \to A \quad g : \Gamma A\Delta \to B}{g\langle f'\rangle : \Gamma\Gamma'\Delta'\Delta \to B}\ cut}{(g\langle f'\rangle)^{\sharp} : \Gamma\Gamma'\mathrm{I}\Delta'\Delta \to B}$$

where the new cut has smaller degree. Moreover,

$$g\langle f'^{\sharp}\rangle uu'\mathbf{i}v'v = guf'u'v'v$$

and

$$(g\langle f'\rangle)^{\sharp}uu'\mathbf{i}v'v = g\langle f'\rangle uu'v'v$$

reduces to the same thing.

Case 2. $\Gamma = \Gamma' \mathrm{I} \Delta'$, say, and $g = g'^{\sharp}$. We are given:

$$\cfrac{f : \Lambda \to A \qquad \cfrac{g' : \Gamma' \Delta' A \Delta \to B}{g'^{\sharp} : \Gamma' \mathrm{I} \Delta' A \Delta \to B}}{g'^{\sharp}\langle f \rangle : \Gamma' \mathrm{I} \Delta' \Lambda \Delta \to B}\ cut.$$

We replace this by:

$$\cfrac{\cfrac{f : \Lambda \to A \qquad g' : \Gamma' \Delta' A \Delta \to B}{g'\langle f \rangle : \Gamma' \Delta' \Lambda \Delta \to B}\ cut}{(g'\langle f \rangle)^{\sharp} : \Gamma' \mathrm{I} \Delta' \Lambda \Delta \to B}$$

where the new cut has smaller degree. Moreover,

$$g'^{\sharp}\langle f \rangle u' \mathbf{i} v' w v = g' u' v' f w v = (g'\langle f \rangle)^{\sharp} u' \mathbf{i} v' w v.$$

Case 4. $A = \mathrm{I}$, $f = \mathbf{i}$ and $g = g'^{\sharp}$. We are given:

$$\cfrac{\mathbf{i} : \ \to \mathrm{I} \qquad \cfrac{g' : \Gamma \Delta \to B}{g'^{\sharp} : \Gamma \mathrm{I} \Delta \to B}}{g'^{\sharp}\langle \mathbf{i} \rangle : \Gamma \Delta \to B}\ cut$$

We see that $g'^{\sharp}\langle \mathbf{i} \rangle = g'$ is constructible without cut.

(7) The nullary connective 1. Only Case 2 survives. We are given:

$$\cfrac{f : \Lambda \to A \qquad \circ : \Gamma A \Delta \to 1}{\circ\langle f \rangle : \Gamma \Lambda \Delta \to 1}\ cut.$$

We replace this by $\circ : \Gamma \Lambda \Delta \to 1$, constructed without cut, and note that $\circ\langle f \rangle = \circ$.

(8) The nullary connective 0. Only two cases survive.

Case 1. We are given:

$$\cfrac{\Box : \Gamma' 0 \Delta' \to A \qquad g : \Gamma A \Delta \to B}{g\langle \Box \rangle : \Gamma \Gamma' 0 \Delta' \Delta \to B}\ cut.$$

We replace this by $\Box : \Gamma \Gamma' 0 \Delta' \Delta \to B$, constructed without cut, and note that $g\langle \Box \rangle = \Box$.

Case 2. Say $\Gamma = \Gamma' 0 \Delta'$. We are given

$$\cfrac{f : \Lambda \to A \qquad \Box : \Gamma' 0 \Delta' A \Delta \to C}{\Box\langle f \rangle : \Gamma' 0 \Delta' \Lambda \Delta \to C}\ cut.$$

We replace this by $\Box : \Gamma' 0 \Delta' A \Delta \to C$, constructed without cut, and note that $\Box\langle f \rangle = \Box$.

5 Equality Between Cut-Free Sequents[3]

Let us summarize what we have done so far. We have constructed the objects of $UF(\mathcal{G})$ from those of $\mathcal{G}$ inductively by means of the following *connectives* (operations on objects):

$$A \otimes B,\ A/B,\ A\backslash B,\ A \times B,\ A + B,\ \mathrm{I},\ 1,\ 0.$$

For argument's sake, we shall assume that all this structure is present; otherwise we just leave out what is not needed.

We have also constructed the sequents of $UF(\mathcal{G})$ from those of $\mathcal{G}$ inductively by means of the following operations on sequents:

$$1_A, g\langle f\rangle, \mathbf{m}_{AB}, f^{\S}, \mathbf{e}_{AB}, f^*, \mathbf{e}'_{AB}, f^{\dagger},$$

$$\mathbf{p}_{AB}, \mathbf{q}_{AB}, \langle f, g\rangle, \mathbf{k}_{AB}, \mathbf{l}_{AB}, [f,g], f^{\sharp}, \mathbf{i}, \circ_{\Gamma}, \Box^{A}_{\Gamma\Delta}.$$

It should be borne in mind that this is just an abbreviated notation for the usual proof-trees. While most of it is unambiguous, it fails to distinguish between

$$f^{\S} : \Gamma A \otimes B\Delta CD\Lambda \to E \text{ and } f^{\S'} : \Gamma AB\Delta C \otimes D\Lambda \to E,$$

for example. Fortunately we need not distinguish between $f^{\S\S'}$ and $f^{\S'\S}$, as those two constructions denote the 'same' sequent. In this section it will be convenient to define equality between sequents without using the internal language, namely as the smallest congruence relation satisfying the *multicategory identities*:

$$(0.1)\quad f\langle 1_A\rangle = f, \quad 1_B\langle f\rangle = f, \quad \text{where } f : A \to B;$$

$$(0.2)\quad h\langle g\rangle\langle f\rangle = \begin{cases} h\langle f\rangle\langle g\rangle & \text{if } A' \text{ is in } \Gamma \text{ or } \Delta, \\ h\langle g\langle f\rangle\rangle & \text{if } A' \text{ is in } \Lambda, \end{cases}$$

where $f : \Lambda' \to A'$, $g : \Lambda \to A$, $h : \Gamma A\Delta \to B$; as well as the following *reductions*:

$$\begin{array}{llllll}
(1) & f^{\S}\langle \mathbf{m}\rangle = f, & (1') & (f\langle \mathbf{m}\rangle)^{\S} = f, & & \\
(2) & \mathbf{e}\langle f^*\rangle = f, & (2') & (\mathbf{e}\langle f\rangle^* = f, & & \\
(3) & \mathbf{e}'\langle f^{\dagger}\rangle = f, & (3') & (\mathbf{e}'\langle f\rangle)^{\dagger} = f, & & \\
(4a) & \mathbf{p}\langle f, g\rangle = f, & (4b) & \mathbf{q}\langle g, f\rangle = f, & (4') & \langle \mathbf{p}\langle f\rangle, \mathbf{q}\langle f\rangle\rangle = f, \\
(5a) & [f,g]\langle \mathbf{k}\rangle = f, & (5b) & [g,f]\langle \mathbf{l}\rangle = f, & (5') & [f\langle \mathbf{k}\rangle, f\langle \mathbf{l}\rangle] = f, \\
(6) & f^{\sharp}\langle \mathbf{i}\rangle = f, & (6') & (f\langle \mathbf{i}\rangle)^{\sharp} = f, & & \\
(7) & f = \circ_{\Gamma}, & (8) & f = \Box^{C}_{\Gamma\Delta}. & &
\end{array}$$

3 This section, as well as the following one, owes its present form to criticism of an earlier version by Grigori Mints and Djordje Čubrić and leans on the idea of Zucker [1974].

We recall from Sections 3 and 4 how every *sequent construction* can be translated into cut-free form. (When we say 'sequent construction', we want to emphasize that the equality relation between sequents is to be ignored.) More precisely, a *cut-free sequent construction* is defined inductively from the sequents of $\mathcal{G}$ by means of the following operations:

$$\{f,g\},\ f^{\S},\ g[f],\ f^{*},\ g\{f\},\ f^{\dagger},\ f_p,\ f_q,$$
$$\langle f,g\rangle,\ {}_kf,\ {}_\ell f,\ [f,g],\ \mathbf{i},\ f^{\sharp}, \circ_\Gamma,\ \Box^A_{\Gamma\Delta}.$$

In what follows, we shall write

$$\begin{aligned}
\varphi(h) &= h^{\S}, h^{*}, h^{\dagger}, h^{\sharp}, h_p, h_q, {}_kh, {}_\ell h;\\
\psi(h,h') &= \langle h,h'\rangle, [h,h'];\\
\chi(h,h') &= \{h,h'\}, h[h'], h\{h'\};\\
\kappa &= \circ, \Box.
\end{aligned}$$

The proof of the cut-elimination theorem in Section 4 may be summarized by the following 'reductions' between sequent constructions, where the degree of the cut on the left-hand side is greater than the degrees of the cuts on the right-hand side. The equal sign between sequents is here replaced by an arrow between sequent constructions.

Case 1:

$$\begin{aligned}
&g\langle\varphi(h)\rangle \ \rightarrow\ \varphi(g\langle h\rangle),\\
&g\langle\psi(h,h')\rangle \ \rightarrow\ \psi(g\langle h\rangle, g\langle h'\rangle),\\
&g\langle\chi(h,h')\rangle \ \rightarrow\ \chi(g\langle h\rangle, h') \text{ or } \chi(h, g\langle h'\rangle),\\
&g\langle\kappa\rangle \ \rightarrow\ \kappa.
\end{aligned}$$

Case 2 or 3:

$$\begin{aligned}
&\varphi(g)\langle h\rangle \ \rightarrow\ \varphi(g\langle h\rangle),\\
&\psi(g,g')\langle h\rangle \ \rightarrow\ \psi(g\langle h\rangle, g'\langle h\rangle),\\
&\chi(g,g')\langle h\rangle \ \rightarrow\ \chi(g\langle h\rangle, g') \text{ or } \chi(g, g'\langle h\rangle),\\
&\kappa\langle h\rangle \ \rightarrow\ \kappa.
\end{aligned}$$

Case 4:

$$g^{\S}\langle\{h,h'\}\rangle \ \rightarrow\ g\langle h\rangle\langle h'\rangle,$$

$$
\begin{aligned}
&g[g']\langle h^*\rangle \;\to\; g\langle h\langle g'\rangle\rangle,\\
&g\{g'\}\langle h^\dagger\rangle \;\to\; g\langle h\langle g'\rangle\rangle,\\
&g_p\langle h,h'\rangle \;\to\; g\langle h\rangle,\; g_q\langle h,h'\rangle \;\to\; g\langle h'\rangle,\\
&[g,g']\langle {}_k h\rangle \;\to\; g\langle h\rangle,\; [g,g']\langle {}_\ell h\rangle \;\to\; g'\langle h\rangle,\\
&g^\sharp\langle \mathbf{i}\rangle \;\to\; g.
\end{aligned}
$$

As was pointed out in Section 1, the identity symbols 1_A for composite A, can also be eliminated, thus $1_{A\otimes B} = \mathbf{m}^\sharp_{AB}$. But $\mathbf{m}_{AB} = \{1_A, 1_B\}$, so $1_{A\otimes B} = \{1_A, 1_B\}^\S$. In this way we obtain:

$$
\begin{aligned}
1_{A\otimes B} &= \{1_A, 1_B\}^\S,\\
1_{A/B} &= (1_B[1_A])^*,\\
1_{A\backslash B} &= (1_A\{1_B\})^\dagger,\\
1_{A\times B} &= \langle (1_A)_p, (1_B)_q\rangle,\\
1_{A+B} &= [{}_k(1_A), {}_\ell(1_B)],\\
1_{\mathrm{I}} = \mathbf{i}^\sharp,\quad & 1_1 = \circ_1,\quad 1_0 = \square^0.
\end{aligned}
$$

Since every sequent is *equal* to a cut-free one, the cut-free form of a sequent is of course uniquely determined. It is however of interest to describe the equality relation between cut-free sequent constructions intrinsically, without referring to sequents that may contain cuts and without referring to the internal language.

Theorem 5.1. *Equality between cut-free sequent constructions is the smallest congruence relation $\simeq$ such that*

$$
\begin{aligned}
&\varphi(\varphi'(h)) \simeq \varphi'(\varphi(h)),\\
&\varphi(\psi(h,h')) \simeq \psi(\varphi(h),\varphi(h')),\\
&\varphi(\chi(h,h')) \simeq \chi(\varphi(h),h') \text{ or } \chi(h,\varphi(h')),\\
&\psi(\psi'(h_0,h_1)\psi'(h_0',h_1')) \simeq \psi'(\psi(h_0,h_0'),\psi(h_1,h_1')),\\
&\chi(\psi(h_0,h_1),h) \simeq \psi(\chi(h_0,h),\chi(h_1,h)),\\
&\chi(h,\psi(h_0,h_1)) \simeq \psi(\chi(h,h_0),\chi(h,h_1)),\\
&\chi(\chi'(h_0,h_1),h) \simeq \chi'(\chi(h_0,h),h_1) \text{ or } \chi'(h_0,\chi(h_1,h)),\\
&\chi(h,\chi'(h_0,h_1)) \simeq \chi'(\chi(h,h_0),h_1) \text{ or } \chi'(h_0,\chi(h,h_1)),\\
&\varphi(\kappa) \simeq \kappa, \psi(\kappa,\kappa) \simeq \kappa,\\
&\chi(\kappa,h) \simeq \kappa, \chi(h,\kappa) \simeq \kappa, \kappa \simeq \kappa'.
\end{aligned}
$$

Before proving this, we obtain a corollary.

Corollary 5.2. *The multifunctor $\mathcal{G} \to UF(\mathcal{G})$ is faithful.*

Proof. Suppose $f, g : \Gamma \to A$ are sequents in $\mathcal{G}$ and $f = g$ holds in $UF(\mathcal{G})$. Since none of φ, ψ, χ or κ occur, the equation must already hold in $\mathcal{G}$.

In what follows we shall extend the reduction symbol $\to$ to contexts, so that $f \to g$ implies $\ldots f \ldots \to \ldots g \ldots$, and make it transitive as well.

6 Proof of the Equality Characterization Theorem

We shall prove Theorem 5.1 in two steps. First we show that the proof of the cut-elimination theorem translates each sequent construction into a unique cut-free form, unique modulo $\simeq$, the congruence relation defined in Theorem 5.1. Secondly we show that any equation between sequents induces their cut-free forms to be in the relation $\simeq$.

(I) Any possible ambiguity in the translation to cut-free form arises from an overlap between Case 1 and Case 2 or 3 in Section 5.

For example, according to Case 1, we have

$$\varphi(g)\langle\varphi'(h)\rangle \;\to\; \varphi'(\varphi(g)\langle h\rangle),$$

while, according to Case 2, we have

$$\varphi(g)\langle\varphi'(h)\rangle \;\to\; \varphi(g\langle\varphi'(h)\rangle).$$

We claim that the right-hand sides of both inequalities reduce to the same cut-free form module $\simeq$. Indeed,

$$\varphi'(\varphi(g)\langle h\rangle) \;\to\; \varphi'(\varphi(g\langle h\rangle)),$$

by Case 2 or 3, and

$$\varphi(g\langle\varphi'(h)\rangle) \;\to\; \varphi(\varphi'(g\langle h\rangle)),$$

by Case 1. Suppose $g\langle h\rangle$ has cut-free form f, then the former reduces to $\varphi'(\varphi(f))$ and the latter to $\varphi(\varphi'(f))$. But $\varphi(\varphi'(f)) \simeq \varphi'(\varphi(f))$, as was to be shown.

A similar argument applies when the cut whose degree is being reduced involves ψ, χ or κ.

(II) We claim that the two sides of equations (0.1), (0.2), (1) to (6), (1′) to (6′), (7) and (8) reduce to the same cut-free form modulo $\simeq$.

(0.1) Consider $f : \Lambda \to A$ and $g : \Gamma A \Delta \to B$, where f and g are cut-free. We shall prove that $1_A\langle f\rangle$ has cut-free form f and leave the proof that $g\langle 1_A\rangle$ has cut-free form g to the reader. The proof will be by induction on the number of operation symbols occurring in f or g.

If f contains no operation symbols, then f is in $\mathcal{G}$ and there is nothing to prove. If the last step in the construction of f does not introduce the main connective of A, we have $f = \varphi(h)$, for example, hence

$$1_A\langle f\rangle = 1_A\langle \varphi(h)\rangle \;\to\; \varphi(1_A\langle h\rangle),$$

by Case 1, and this reduces to $\varphi(h) = f$ by inductional assumption. The possibilities $f = \psi(h, h'), \chi(h, h')$ or κ are treated similarly. If the last step in the construction of f does introduce the main connective of A, take $A = A_0 \otimes A_1$, for example. Then

$$1_A\langle f\rangle = \{1_{A_0}, 1_{A_1}\}^\S\langle\{f_0, f_1\}\rangle \;\to\; \{1_{A_0}, 1_{A_1}\}\langle f_0\rangle\langle f_1\rangle,$$

by Case 4, and this reduces to $\{1_{A_0}\langle f_0\rangle, f_1\}$ and $\{f_0, 1_{A_1}\langle f_1\rangle\}$, by Case 3, which reduces to $\{f_0, f_1\} = f$, by inductional assumption. The possibilities $A = A_0/A_1$ etc. are handled similarly.

(0.2) Consider cut-free sequents $f : \Gamma A \Delta \to B$, $g : \Lambda \to A$ and $h : \Lambda' \to A'$, where A' is either in Γ or Δ or in Λ. Then

$$f\langle g\rangle\langle h\rangle = f\langle h\rangle\langle g\rangle \text{ or } f\langle g\langle h\rangle\rangle$$

respectively. We claim that, in either case, the two sides of the equation have the same cut-free form modulo $\simeq$. We shall carry out the proof in the second case, leaving the first to the reader.

Write $\Lambda = \Lambda_0\ A'\Lambda_1$. Our proof will be by induction on the degree of the cut $f\langle g\rangle$ plus $d(\Lambda')$ (see Section 4). We may assume that f, g and h are not all in $\mathcal{G}$, then we have the following possibilities, not necessarily mutually exclusive:

(a) the construction of f introduces a connective into Γ, Δ or B;

(b) the construction of g introduces a connective into Λ_0 or Λ_1;

(c) the construction of h introduces a connective into Λ';

(d) the constructions of f and g introduce the main connective into A;

(e) the constructions of g and h introduce the main connective into A'.

(a) Say $f = \varphi(f')$, for example. By Case 2 or 3 of the cut-elimination theorem, we have

$$\begin{aligned} f\langle g\rangle\langle h\rangle &\rightarrow \varphi(f'\langle g\rangle)\langle h\rangle \rightarrow \varphi(f'\langle g\rangle\langle h\rangle), \\ f\langle g\langle h\rangle\rangle &\rightarrow \varphi(f'\langle g\langle h\rangle\rangle). \end{aligned}$$

By inductional assumption, $f'\langle g\rangle\langle h\rangle$ and $f'\langle g\langle h\rangle\rangle$ reduce to the same cut-free form modulo $\simeq$. The possibilities $f = \psi(f', f''), \chi(f', f'')$ or κ are handled similarly.

(b) and (c) are similar to (a), except that Case 1 is used partly in (b) and entirely in (c).

(d) For example, assume $A = A_0 \otimes A_1$, $\Lambda = \Lambda_0\Lambda_1$, $f = f'^{\S}, f' : \Gamma A_0 A_1 \Delta \rightarrow B$, $g = \{g_0, g_1\}, g_i : \Lambda_i \rightarrow A_i$. Then, by Case 2 or 3,

$$g\langle h\rangle \rightarrow \{g_0\langle h\rangle, g_1\} \text{ or } \{g_0, g_1\langle h\rangle\},$$

say the former. Moreover, by Case 4,

$$\begin{aligned} f\langle g\rangle\langle h\rangle &\rightarrow f'\langle g_0\rangle\langle g_1\rangle\langle h\rangle, \\ f\langle g\langle h\rangle\rangle &\rightarrow f'\langle g_0\langle h\rangle\rangle\langle g_1\rangle. \end{aligned}$$

Now, by (b), the former has the same cut-free form as $f'\langle g_0\rangle\langle h\rangle\langle g_1\rangle$, hence, by inductional assumption, as the latter. The possibilities $A = A_0/A_1$ etc. are treated similarly.

(e) For example, assume $A' = A'_0 \otimes A'_1$, $\Lambda' = \Lambda'_0\Lambda'_1$, $h = \{h_0, h_1\}$, $h_i : \Lambda'_i \rightarrow A'_i$, $g = g'^{\S}$, $g' : \Lambda_0 A'_0 A'_1 \Lambda_1 \rightarrow A$. Then, by Case 1, $f\langle g\rangle \rightarrow (f\langle g'\rangle)^{\S}$, hence

$$f\langle g\rangle\langle h\rangle \rightarrow (f\langle g'\rangle)^{\S}\langle h_0, h_1\rangle \rightarrow f\langle g'\rangle\langle h_0\rangle\langle h_1\rangle,$$

by Case 4. Also, by Case 4,

$$f\langle g\langle h\rangle\rangle \rightarrow f\langle g'\langle h_0\rangle\langle h_1\rangle\rangle.$$

Now, by inductional assumption, the former has the same cut-free form modulo $\simeq$ as $f\langle g'\langle h_0\rangle\rangle\langle h_1\rangle$ and, again by inductional assumption, this has the same cut-free form modulo $\simeq$ as the latter. (Note that the degree of $f\langle g'\langle h_0\rangle\rangle$ plus $d(\Lambda'_1)$ is less than the degree of $f\langle g\langle h\rangle\rangle$ plus $d(\Lambda')$.)

(1) to (6). For example, in (1), the left-hand side is

$$f^{\S}\langle \mathbf{m}\rangle = f^{\S}\langle\{1_A, 1_B\}\rangle \rightarrow f\langle 1_A\rangle\langle 1_B\rangle,$$

by Case 4. If f is assumed to be cut-free, this reduces to f, by two applications of (0.1). (2) to (6) are handled similarly.

(1′) to (6′). Consider (1′) : $(f\langle \mathbf{m}\rangle)^{\S} = f$, for example. We proceed by induction on the number of operation symbols in the construction of f, which is again assumed to be cut-free. If $f = g^{\S}$, we already know, from the discussion of (1) above, that $g^{\S}\langle \mathbf{m}\rangle$ has the same cut-free form modulo $\simeq$ as g, hence $(f\langle \mathbf{m}\rangle)^{\S} = (g^{\S}\langle \mathbf{m}\rangle)^{\S}$ as $g^{\S} = f$. Otherwise, $f\langle \mathbf{m}\rangle$ is reduced by Case 2 or 3, say $f = \varphi(g)$, for example. Then $(f\langle \mathbf{m}\rangle)^{\S} \to (\varphi(g\langle \mathbf{m}\rangle))^{\S}$, and this reduces to $(\varphi(h))^{\S} \simeq \varphi(h^{\S})$, where h is the cut-free form of $g\langle \mathbf{m}\rangle$. Now $(g\langle \mathbf{m}\rangle)^{\S}$ has cut-free form $h^{\S}$; so, by inductional assumption, $h^{\S} \simeq g$; hence $\varphi(h^{\S}) \simeq \varphi(g) = f$. The possibilities $f = \psi(g, g'), \chi(g, g')$ or κ are handled similarly. (2′) to (6′) are treated in the same fashion as (1′).

(7) and (8). We shall consider (7), leaving (8) to the reader. We are given $f : \Gamma \to 1$ and claim that $f \simeq \circ_{\Gamma}$. We assume f cut-free and proceed by induction on the number of operation symbols occurring in f. Suppose, for example, that $f = \varphi(g)$, where $g : \Delta \to 1$. Then, by inductional assumption, $g \simeq \circ_{\Delta}$. Hence $f \simeq \varphi(\circ_{\Delta}) \simeq \circ_{\Gamma}$. The possibilities $f = \psi(g, g'), \chi(g, g')$ or κ are handled similarly.

I would have liked ending Section 6 with a theorem saying that two cut-free (proofs of) sequents are equal if and only if they have the same *generality*. This had been asserted in [L 1968], although the proof there had some serious gaps, and was taken up again by Szabo [1978]. All I can say now is that I hope to return to this question in the future.

7 Some Final Remarks

As pointed out in the introduction, this paper merely starts putting the finishing touches on a program started by the author more than thirty years ago. In particular, it provides details for an approach using the internal language of a multicategory announced in [L 1989]. Not surprisingly, a number of other people have also contributed to this program, sometimes allowing one or even two of Gentzen's structural rules. As it is not feasible to do justice to all these contributions here, for a survey of recent work the reader is referred to the comprehensive article by Ono [1990]. I would just like to single out for attention the monograph by Szabo [1978], the work by Mints [1976, 1977], by Komori [1986] and by Ono himself, as well as the use of an internal language by Jay [1989].

The linear logic of Girard [1987, 1989], aside from permitting Gentzen's interchange rule, is related to our categorial grammar as classical logic is to intuitionistic logic, its sequents being of the form $f : \Gamma \to \Delta$, where Δ is not necessarily of length 1. To discuss the internal language of such a 'polycategory' [see Szabo 1975], one has to consider not only variables but also 'covariables'. I hope to return to this topic on another occasion.

Unfortunately, limitations of time (and space) do not allow me to give here a proper treatment of *variable types*. If type variables $X, Y, \ldots$ are admitted, one must then also consider *polynomial types* $X \otimes A$, X/Y, etc., and *polynomial sequents* such as $\mathbf{m}_{XA}$ and $\mathbf{e}_{XY}$. One may then introduce *quantifiers* $\forall_X$ and $\exists_X$ (or $\prod_X$ and $\coprod_X$ in algebraic notation), as in [L 1989b]. The cut-elimination theorem will still hold; but I have not yet looked at the reduction of polynomial sequents to normal form.

References

N. Bourbaki, *Algèbre Multilinéaire*, Hermann, Paris, 1948.

N. Chomsky, *Syntactic Structures*, Mouton, 's-Gravenhage, 1957.

J. Y. Girard, Linear logic, *Theoretical Computer Science 50* (1987), 1–102.

———, Towards a geometry of interaction, *Contemporary Math. 92* (1989), 69–108.

G. Hotz, Eindeutigkeit und Mehrdeutigkeit formaler Sprachen, *Elektronische Informationsbearbeitung und Kybernetik 2* (1966), 235–247.

C. B. Jay, Languages for monoidal categories, *J. of Pure and Applied Algebra 59* (1989), 61–85.

S. C. Kleene, *Introduction to Metamathematics*, Van Nostrand, New York, 1952.

Y. Komori, Predicate logics without structural rules, *Studia Logica 45* (1986), 393–404.

J. Lambek, The mathematics of sentence structure, *Amer. Math. Monthly 65* (1958), 154–169.

———, On the calculus of syntactic types, *Amer. Math. Soc. Proc. Symposia Appl. Math. 12* (1961), 166–178.

———, Deductive systems and categories, I, *J. Math. Systems Theory 2* (1968), 278–318.

———, Deductive systems and categories, II, *Springer LNM 86* (1969), 76–122.

———, Multicategories revisited, *Contemporary Math. 92* (1989), 217–239.

———, On a connection between algebra, logic and linguistics, in: *Les Actes des Journées d'Étude 'Esquisses, Logique, Informatique Théorique'*, Université Paris 7, Juin 1988, *Diagrammes 22* (1989), 59–75.

G. E. Mints, Cut-elimination theorem for relevant logics, translated from *Zapiski Nauchnykh Seminarov Leningradskogo Otdeleniya Mat. Instituta im. V. A. Steklova AN SSSR 32* (1972), 90–97, in *J. Soviet Math. 6* (1976), 422–428.

———, Closed categories and the theory of proofs, translated from *Zapiski Nauchnykh Seminarov Leningradskogo Otdeleniya Mat. Instituta im. V. A. Steklova AN SSSR 68* (1977), 83–114, in *J. Soviet Math. 15* (1981), 45–62.

M. Moortgat, *Categorial Investigations*, Foris Publ., Dordrecht, 1988.

H. Ono, Structural rules and a logical hierarchy, in: *Proc. Summer School and Conference on Mathematical Logic 'Heyting '88'*, ed. by P. P. Petkov, Plenum Press, New York, 1990, 95–104.

D. Prawitz, *Natural Deduction*, Almquist and Wiksell, Stockholm, 1965.

M. E. Szabo (ed.), *The Collected Papers of Gerhard Gentzen*, Studies in Logic and the Foundations of Mathematics, North-Holland, Amsterdam, 1969.

M. E. Szabo, Polycategories, *Communications in Algebra 3* (1975), 663–689.

———, *Algebra of Proofs*, Studies in Logic and the Foundations of Mathematics, North-Holland, Amsterdam, 1978.

J. Zucker, The correspondence between cut-elimination and normalization, II, *Ann. Math. Logic 7* (1974), 113–155.

From Categorial Grammar to Bilinear Logic

Joachim Lambek

Department of Mathematics and Statistics, McGill University[1]

1 Introduction

The *syntactic calculus*, also known as 'bidirectional categorial grammar', is a kind of logic without any structural rules, other than the obligatory reflexive law and cut-rule. It had been inspired by multilinear algebra and non-commutative ring theory and was developed with applications to linguistics in mind. Here we shall confine attention to the associative version, although a non-associative version has also been studied [L 1961, Kandulski 1988, Došen 1988, 1989]. It differs from a very rudimentary form of Girard's linear logic [Girard 1987, 1989] by the absence of the interchange rule which licenses commutativity. Because of its roots in non-commutative algebra and syntax, all appearance of commutativity is forbidden. The notation, which goes back to a pre-historic collaboration with George Findlay, was carefully chosen to reflect this absence of commutativity: the order of two letters was never to be wantonly interchanged, yet all rules were to be preserved under left-right symmetry, the guiding slogan being 'symmetry without commutativity'. It must be admitted, however, that the notation, proposed for linguistics [L 1958] and ring theory [L 1966], caught on in neither field (with some recent exceptions in linguistics).

Girard's linear logic, while admitting the commutative law, contains some additional operations which were missing from the syntactic calculus, as neither linguistic nor ring-theoretical applications seemed to require them. Surprisingly, some of these additional operations do occur in the standard models for the syntactic calculus, although it is still not apparent that they have useful applications. *Bilinear logic* arose from a desire to

1 This research has been supported by the Engineering and Natural Sciences Research Council of Canada and, in part, by the Quebec Department of Education.

unify the syntactic calculus and linear logic. Earlier attempts to produce such a unification were discussed at a McGill seminar in 1988, as was the paper by Yetter [1990], who confined himself to a classical form of bilinear logic, calling it 'cyclic logic', in which the additional operations can be defined by De Morgan's rule. Here we shall discuss several versions of bilinear logic, only one of which admits Yetter's cyclic law. Similar ideas are being pursued by Cockett and Seely, by de Paiva (and Hyland) and by Pelletier (and Mulvey), as became apparent at a recent conference in Montreal in June 1991.

After distributing a preliminary version of this paper, I learned from Kosta Došen of the pioneering article (in Russian) by Grishin [1983], which not only anticipated the proposed axioms for bilinear logic, but seems to antedate the invention of linear logic! There remains the hope that this revised exposition will provide some motivation for Grishin's axioms, by showing that they are dictated by a Gentzen style presentation and by illustrating them with some natural models, one of which extends the original linguistic application.

While preparing this article for the typist, I discovered some impressive manuscripts by Abrusci [1990] and Ono [1989, 1990] hiding on my desk under a pile of other papers. These two authors also deal with bilinear logic in their own way and, in some respects, go much beyond the present exposition. In particular, they prove completeness of bilinear logic with respect to certain classes of models, a question which is not addressed here.

2 Recalling the Syntactic Calculus

The *syntactic calculus* deals with types, also called 'formulas', and arrows between them, which logicians think of as entailments or deductions and linguists as derivations. The types or formulas are constructed from basic ones by certain operations or connectives, of which we shall here consider three nullary and five binary ones, although not all of these operations occur in every exposition of the syntactic calculus. From a number of notational variants, we have here chosen the following:

$$\mathrm{I},\ \otimes,\ /,\ \backslash,\ \top,\ \wedge,\ \bot,\ \vee.$$

These operations are subject to the following axioms and rules of inference. (For purposes of this elementary exposition, we have not labeled the arrows, although a more advanced exposition would require this.)

$$A \to A, \qquad \frac{A \to B \quad B \to C}{A \to C},$$

$$A \otimes \mathrm{I} \leftrightarrow A \leftrightarrow \mathrm{I} \otimes A, \quad (A \otimes B) \otimes C \leftrightarrow A \otimes (B \otimes C),$$

$$A \otimes B \to C \quad \text{iff} \quad A \to C/B \quad \text{iff} \quad B \to A\backslash C,$$

$$A \to \top, \quad A \wedge B \to A, \quad A \wedge B \to B, \quad \frac{C \to A \qquad C \to B}{C \to A \wedge B},$$

$$\bot \to A, \quad A \to A \vee B, \quad B \to A \vee B, \quad \frac{A \to C \qquad B \to C}{A \vee B \to C}.$$

Certain derived rules of inference are quite useful. For example, the rule

$$\frac{A \to B \qquad C \to D}{A \otimes C \to B \otimes D}$$

expresses what categorists call the 'bifunctoriality' of the operation $\otimes$. In the present context, it is connected with the fact, also stated in the language of category theory, that the functors $A \otimes -$ and $- \otimes B$ have right adjoints. Here is a proof of the derived rule under consideration:

$$\dfrac{\dfrac{\dfrac{A \to B \qquad \dfrac{B \otimes C \to B \otimes C}{B \to (B \otimes C)/C}}{A \to (B \otimes C)/C}}{A \otimes C \to B \otimes C} \qquad \dfrac{\dfrac{C \to D \qquad \dfrac{B \otimes D \to B \otimes D}{D \to B\backslash(B \otimes D)}}{C \to B\backslash(B \otimes D)}}{B \otimes C \to B \otimes D}}{A \otimes C \to B \otimes D}$$

In a similar way, one can prove the following derived rules:

$$\frac{A \to B \quad C \to D}{A/D \to B/C}, \quad \frac{A \to B \quad C \to D}{B\backslash C \to A\backslash D}, \quad \frac{A \to B \quad C \to D}{A \wedge C \to B \wedge D}, \quad \frac{A \to B \quad C \to D}{A \vee C \to B \vee D}.$$

For some purposes, one may also impose the following *structural axioms*:

$$A \to \mathrm{I}, \quad A \to A \otimes A, \quad A \otimes B \to B \otimes A,$$

corresponding to Gentzen's three structural rules. The axiom $A \to \mathrm{I}$ is required in non-commutative ideal theory, at least as long as one deals with integral ideals; it may be dropped when one wishes to consider fractional ideals. Of course, in commutative algebra, one postulates $A \otimes B \to B \otimes A$, which is also the only structural axiom admitted in linear logic. All three structural axioms are required for intuitionistic logic, although in relevance logic the axiom $A \to \mathrm{I}$ is dropped. It follows that, in intuitionistic logic, some of the connectives collapse:

$$\mathrm{I} \leftrightarrow \top, \quad A \otimes B \leftrightarrow A \wedge B, \quad A/B \leftrightarrow B\backslash A.$$

To show the second equivalence, for example, one may proceed as follows:

$$\frac{\dfrac{A \to \mathrm{I} \qquad B \to B}{A \otimes B \to \mathrm{I} \otimes B} \qquad \mathrm{I} \otimes B \to B}{A \otimes B \to B}$$

and similarly to show $A \otimes B \to A$. On the other hand:

$$\frac{C \to C \otimes C \qquad \dfrac{C \to A \qquad C \to B}{C \otimes C \to A \otimes B}}{C \to A \otimes B}$$

Thus, $\otimes$ satisfies the same rules as $\wedge$ and, if the latter is present, one easily infers $A \otimes B \leftrightarrow A \wedge B$.

Surprisingly, the three structural rules are also implicit in Curry's [1958, 1961] semantical calculus and provide the clue to Montague semantics. In fact, the passage from syntax to semantics may be viewed as introducing the three structural rules [L 1988].

For later reference, we point out the following easily proved equivalences:

$$A/\mathrm{I} \leftrightarrow A \leftrightarrow \mathrm{I}\backslash A,$$

$$(A\backslash B)/C \leftrightarrow A\backslash(B/C), \quad A/(B \otimes C) \leftrightarrow (A/C)/B,$$

together with the mirror image of the last.

3 Four Different View-Points

When studying systems such as the syntactic calculus, one may take one of several points of view, regarding the system as

(a) a deductive system (no equality relation between arrows is considered);

(b) a poset (all arrows from A to B are taken to be the same and one usually writes $A \leq B$ for $A \to B$ and $A = B$ for $A \leftrightarrow B$);

(c) a pre-ordered set (as in (b), except that A and B are not necessarily regarded as equal when $A \leftrightarrow B$);

(d) a category (arrows are labeled and an equality relation between arrows is carefully chosen to ensure, for example, that $\otimes$ is a bifunctor and that $- \otimes B$ is left adjoint to $-/B$).

Strictly speaking, the name 'syntactic calculus' is reserved for the deductive system. For the poset, and perhaps the pre-ordered set, the terminology will be discussed below. The category has been called 'biclosed monoidal' or 'residuated' [L 1969]. The reader is warned that, while $\leftrightarrow$ may sometimes express isomorphism in a category, it does not always do so. For example,

$$(A \otimes B) \otimes C \cong A \otimes (B \otimes C);$$

but C/B is merely a retract of $C/((C/B)\backslash C)$ and not, in general, isomorphic to it.

I believe the categorical view-point (d) to be the most important. In fact, the original motivation for looking at biclosed monoidal categories came from the study of bimodules: given a ring R (associative with unity, but not necessarily commutative), one considers bimodules ${}_RA_R$ satisfying:

$$(rr')a = r(r'a), \quad (ra)r' = r(ar'), \quad a(rr') = (ar)r',$$

for all a in A and r and r' in R, in addition to distributivity and the equations $1a = a = a1$. If ${}_RB_R$ is another bimodule, one can form bimodules ${}_R(A \otimes B)_R$, ${}_R(A/B)_R$ and ${}_R(B\backslash A)_R$, where

$$A \otimes B = A_R \otimes_R B, \quad A/B = \mathrm{Hom}(B_R, A_R), \quad B\backslash A = \mathrm{Hom}({}_RB, {}_RA).$$

However, in this survey article, we shall confine attention to view-points (a) and (b), which can be understood without any knowledge of category theory. It is hoped that occasional side-remarks about categorical concepts will not put off the reader not acquainted with categories.

When taking the view-point (b), we call the system involving $\mathrm{I}, \otimes, /$ and $\backslash$ a *residuated monoid*. When the lattice operations $\top, \wedge, \bot$ and $\vee$ are present, the usual name is 'residuated lattice'. If the residuated lattice happens to be complete, it is nowadays called a '(unital) quantale'. (I feel that, just as for rings, the unity element should be understood implicitly.) It may be worthwhile pointing out an insight gained from category theory: since $- \otimes B$ is left adjoint to $-/B$, the former preserves joins and the latter preserves meets. Thus we have the following infinite distributive laws, which the non-categorically inclined reader is invited to prove directly:

$$\Big(\bigvee_{j \in \mathcal{I}} A_j\Big) \otimes B = \bigvee_{j \in \mathcal{I}} (A_j \otimes B),$$

$$\Big(\bigwedge_{j \in \mathcal{I}} A_j\Big)/B = \bigwedge_{j \in \mathcal{I}} (A_j/B),$$

$$C/\Big(\bigvee_{j \in \mathcal{I}} B_j\Big) = \bigwedge_{j \in \mathcal{I}} (C/B_j).$$

Exploiting the left-right symmetry, we obtain three other distributive laws, in which / is replaced by \.

This may be the place to point out that, historically, adjoints precede category theory. In lattice theory, an adjunction used to be called a 'Galois correspondence', but I prefer the categorical terminology.

4 Subsets of a Monoid

A typical example of a residuated monoid is offered by the subsets of a monoid $(M, 1, \cdot)$, where $\leq$ stands for inclusion and, omitting the dot, one defines

$$\begin{aligned} A \otimes B &= \{ab \mid a \in A \wedge b \in B\}, \\ &= \{c \mid \exists_a \exists_b (c = ab \wedge a \in A \wedge b \in B\}, \\ A/B &= \{c \mid \forall_b (b \in B \Rightarrow cb \in A)\}, \\ A \backslash B &= \{c \mid \forall_a (a \in A \Rightarrow ac \in B)\}, \\ \mathrm{I} &= \{1\}. \end{aligned}$$

In fact, the syntactic calculus was so called because of its application to the free monoid generated by the vocabulary of English.

Looking at the above example with the hindsight gained from Girard's linear logic, it becomes apparent that there are at least two other operations lurking under the surface:

$$\begin{aligned} A \oplus B &= \{c \mid \forall_a \forall_b (c = ab \Rightarrow (a \in A \vee b \in B))\}, \\ \mathrm{O} &= \{c \mid \neg\, c \in \mathrm{I}\}, \end{aligned}$$

although the linguistic significance of the operation $\oplus$ is not too clear.

In a world in which classical logic prevails, we may utilize the complement $\neg A = \{b \mid \neg\, b \in A\}$, noting that $\neg\neg A = A$, and define:

$$A \oplus B = \neg(\neg A \otimes \neg B), \ \mathrm{O} = \neg \mathrm{I},$$

in the spirit of De Morgan. From these definitions one easily deduces that

$$(A \oplus B) \oplus C = A \oplus (B \oplus C), \ A \oplus \mathrm{O} = A = \mathrm{O} \oplus A.$$

Is it really necessary to adopt classical logic to prove these equations? First let us look at

$$A \oplus \mathrm{O} = \{c \mid \forall_a \forall_b (c = ab \Rightarrow (a \in A \vee b \neq 1)\}.$$

Assuming that equality in M is *decidable*, we know that either $b = 1$ or $b \neq 1$, and so the above can be rewritten:

$$\begin{aligned} A \oplus \mathrm{O} &= \{c \mid \forall_a \forall_b (c = ab \wedge b = 1) \Rightarrow a \in A\} \\ &= \{c \mid \forall_a (c = a \Rightarrow a \in A)\} = A. \end{aligned}$$

Next, look at

$$\begin{aligned} &(A \oplus B) \oplus C \\ &= \{e \mid \forall_d \forall_c (e = dc \Rightarrow (\forall_a \forall_b (d = ab \Rightarrow (a \in A \vee b \in B)) \vee c \in C)\}. \end{aligned}$$

In order to prove the associative law, we hope to transform this expression into

$$\begin{aligned} &\{e \mid \forall_d \forall_c \forall_a \forall_b ((e = dc \wedge d = ab) \Rightarrow (a \in A \vee b \in B \vee c \in C))\} \\ &= \{e \mid \forall_a \forall_b \forall_c \, (e = abc \Rightarrow (a \in A \vee b \in B \vee c \in C))\}. \end{aligned}$$

Decidability of equality will help, inasmuch as it allows us to rewrite $d = ab \Rightarrow \cdots$ as $d \neq ab \vee \cdots$; but we also need the following 'co-intuitionistic distributive law':

$$\forall_a (\Phi(a) \vee P) \leftrightarrow \forall_a \Phi(a) \vee P.$$

Unfortunately, adopting this new distributive law as an additional postulate would render the underlying propositional logic Boolean, as is seen by an argument involving the internal language of a topos (see e.g. [LS 1986]), which some readers may wish to skip.

Indeed, assume in particular that

$$\forall_{x \in [R]} (Q \vee P) \leftrightarrow \forall_{x \in [R]} Q \vee P,$$

where x is a variable of type $[R]$, the subobject of 1 corresponding to the proposition $R : 1 \to \Omega$ in a topos, and where Q does not depend on x. This may be written:

$$R \Rightarrow (Q \vee P) \leftrightarrow (R \Rightarrow Q) \vee P.$$

Taking $Q = \bot$, we see that $\mathrm{Hom}(1, \Omega)$ will be Boolean.

We can also introduce two new operations $\div$ (*less*) and $\dot{-}$ (*from*), analogous to / (*over*) and \ (*under*), to satisfy

$$C \leq A \oplus B \iff C \div B \leq A \iff A \dot{-} C \leq B.$$

Indeed, we need only define

$$\begin{aligned} C \div B &= \neg(\neg C / \neg B) \\ &= \{x \mid \exists_b (xb \in C \wedge \neg\, b \in B)\}, \end{aligned}$$

and $A \dot{-} C$ similarly.

We end this section by mentioning an amusing property of the operation $\oplus$ between subsets of the free monoid generated by a set Σ. If A is any set of strings of elements of Σ, let $\ell(A)$ denote the set of all substrings of elements of A. Then $C \leq A \otimes B$ implies $\ell(C) \leq \ell(A) \oplus \ell(B)$.

5 A Weak Version of Bilinear Logic

A weak form of bilinear logic, let us call it BL1, will be obtained from the syntactic calculus described earlier by adding the following new operations:

$$\mathrm{O},\ \oplus,\ \div,\ \dot{-},$$

subject to the following new axioms and rules of inference:

$$(A \oplus B) \oplus C \leftrightarrow A \oplus (B \oplus C),\ A \oplus \mathrm{O} \leftrightarrow A \leftrightarrow \mathrm{O} \oplus A,$$

$$C \to A \oplus B \ \text{ iff } \ C \div B \to A \ \text{ iff } \ A \dot{-} C \to B.$$

The following derived rules are easily proved or just inferred from symmetry:

$$\frac{A \to B \quad C \to D}{A \oplus C \to B \oplus D},\quad \frac{A \to B \quad C \to D}{A \div D \to B \div C},\quad \frac{A \to B \quad C \to D}{B \dot{-} C \to A \dot{-} D}.$$

Models of BL1 might be called 'bi-residuated monoids' or 'bi-residuated lattices', the latter if the lattice operations are present. If, moreover, the lattice is complete, we may speak of a 'biquantale'. In a biquantale, we have three new distributive laws and their mirror duals:

$$\Big(\bigwedge_{j\in\mathcal{I}} A_j\Big) \oplus B = \bigwedge_{j\in\mathcal{I}} (A_j \oplus B),$$

$$\Big(\bigvee_{j\in\mathcal{I}} A_j\Big) \div B = \bigvee_{j\in\mathcal{I}} (A_j \div B),$$

$$C \div \Big(\bigwedge_{j\in\mathcal{I}} B_j\Big) = \bigvee_{j\in\mathcal{I}} (C \div B_j).$$

We need not have stated these laws, had we just reminded the reader that everything in BL1 is invariant under arrow reversal, replacing

$$\to,\ \mathrm{I},\ \otimes,\ /,\ \backslash,\ \top,\ \wedge$$

by

$$\leftarrow,\ \mathrm{O},\ \oplus,\ \div,\ \dot{-},\ \bot,\ \vee$$

and vice versa. In biquantales also $\bigwedge$ and $\bigvee$ are interchanged. If it so happens that we have an operation $\neg$ satisfying

$$\neg\neg A \leftrightarrow A,\quad \frac{A \to B}{\neg B \to \neg A},$$

as we do for subsets of a monoid, arrow reversal may be justified by De Morgan duality, if one defines $A \oplus B = \neg(\neg A \otimes \neg B)$ etc.

In some models of BL1 there is an operation $^{\cup}$, called 'converse', which similarly justifies the left-right symmetry. It should satisfy:

$$A^{\cup\cup} = A, \qquad \mathrm{I}^{\cup} = \mathrm{I}, \qquad (A \otimes B)^{\cup} = B^{\cup} \otimes A^{\cup},$$

$$\mathrm{O}^{\cup} = \mathrm{O}, \qquad (A \oplus B)^{\cup} = B^{\cup} \oplus A^{\cup},$$

$$\text{if} \quad A \leq B, \quad \text{then} \quad A^{\cup} \leq B^{\cup}.$$

One easily deduces

$$\top^{\cup} = \top, \qquad (A \cap B)^{\cup} = A^{\cup} \cap B^{\cup}, \qquad (A/B)^{\cup} = B^{\cup}\backslash A^{\cup},$$

etc. Such an operation exists, for example, in the free monoid generated by a set Σ. If σ_i is an element of Σ, $(\sigma_1 \cdots \sigma_n)^{\cup} = \sigma_n \cdots \sigma_1$ is the mirror image of a string. This operation plays a role in certain formal languages, though not in natural languages such as English.

We had pointed out that the syntactic calculus is turned into positive intuitionistic logic when the following structural axioms are postulated:

$$A \to \mathrm{I}, \qquad A \to A \otimes A, \qquad A \otimes B \to B \otimes A.$$

In the presence of a converse operation, these structural axioms may be replaced by the following, which imply them:

$$A \to \mathrm{I}, \; A \to A \otimes A, \; A \to A^{\cup}.$$

Indeed, we then have

$$A \otimes B \to A^{\cup} \otimes B^{\cup} \to (A^{\cup} \otimes B^{\cup})^{\cup} \to B^{\cup\cup} \otimes A^{\cup\cup} \to B \otimes A.$$

Curiously, if $\to$ is reversed, but I and $\otimes$ remain, the latter three conditions describe an equivalence relation or a subgroup (see Sections 12 and 8 respectively).

The subsets of a monoid do not constitute the most general model of BL1. More general is a model constructed from sets of arrows of a category. In the former model, though not in the latter, one can prove the principle:

$$\text{if} \quad A \otimes B \leq C \oplus D, \quad \text{then} \quad A \leq C \quad \text{or} \quad B \leq D.$$

But even the latter model is still very special, inasmuch as it admits, for example, an operation $\Rightarrow$ satisfying

$$A \wedge B \leq C \; \text{ if and only if } \; B \leq A \Rightarrow C.$$

No such operation is postulated in BL1.

6 Grishin's Laws

The subsets of a monoid satisfy more than BL1; they also satisfy

(a) $(A \oplus B)/C \to A \oplus (B/C), \quad C\backslash(B \oplus A) \to (C\backslash B) \oplus A.$

These laws were considered by Grishin, though only as being of secondary importance.

To see that the first of (a) holds for subsets of a monoid, assume $x \in (A \oplus B)/C$ and $c \in C$; then $xc \in A \oplus B$, that is, if $xc = ab$, then either $a \in A$ or $b \in B$. We claim that $x \in A \oplus (B/C)$, so suppose $x = ay$ and $a \notin A$, it suffices to show that $y \in B/C$, that is, for any $c \in C$, $yc \in B$. This follows since $xc = a(yc) \in A \oplus B$ and $a \notin A$.

Grishin obtained a number of equivalent formulations of (a), of which we shall only present the following:

(a′) $(A \div D) \otimes C \to A \div (D \otimes C), \quad C \otimes (D \div A) \to (C \otimes D) \div A,$

Indeed, the first part of (a) asserts that, for all D,

$$\text{if} \quad D \to (A \oplus B)/C, \quad \text{then} \quad D \to A \oplus (B/C).$$

This may be transformed thus:

$$\text{if} \quad A \div (D \otimes C) \to B, \quad \text{then} \quad (A \div D) \otimes C \to B.$$

This being so for all B, we obtain the first part of (a′).

By BL1(a) we shall mean BL1 together with (a). A simple model of BL1 which does not satisfy (a) is any non-Boolean *bi-Heyting algebra*, by which we mean a Heyting algebra in which not only

$$A \wedge B \to C \quad \Longleftrightarrow \quad A \to C \Leftarrow B,$$

but also

$$C \to A \vee B \quad \Longleftrightarrow \quad C \sim B \to A,$$

where we read $\Leftarrow$ as *if* and $\sim$ as *but not*.

Bi-Heyting algebras were studied long ago by Rauszer [1974] and more recently by Lawvere [1991], who points out that, in any presheaf topos, $\mathrm{Hom}(1, \Omega)$ is a bi-Heyting algebra, and by Reyes and Zolfaghari [1991]. In a bi-Heyting algebra, the first part of (a) becomes

$$C \Rightarrow (B \vee A) \to (C \Rightarrow B) \vee A.$$

Taking $B = \bot$ and $C = A$ we obtain $\top \to \neg A \vee A$, Aristotle's law of the excluded middle.

7 Sequent Calculus for Weak Bilinear Logic

BL1 may be presented by means of a sequent calculus à la Gentzen, involving sequents of the form

$$\Gamma = A_1 \cdots A_m \to B_1 \cdots B_n = \Delta,$$

without any restrictions on m and n. Of course, the intended meaning is

$$A_1 \otimes \cdots \otimes A_m \to B_1 \oplus \cdots \oplus B_n,$$

with the understanding that, when $m = 0$, the left-hand side is I and that, when $n = 0$, the right-hand side is O. The axioms and rules of inference may be separated, as usual, into structural rules and introduction rules.

Structural rules:

$$A \to A,$$

$$\frac{\Lambda \to A \quad \Gamma A \Delta \to \Theta}{\Gamma \Lambda \Delta \to \Theta}, \qquad \frac{\Lambda \to \Phi A \Psi \quad A \to \Theta}{\Lambda \to \Phi \Theta \Psi}.$$

Introduction rules:

$$\frac{\Gamma\Delta \to \Theta}{\Gamma \mathrm{I} \Delta \to \Theta}, \qquad \to \mathrm{I},$$

$$\frac{\Gamma A B \Delta \to \Theta}{\Gamma (A \otimes B) \Delta \to \Theta}, \qquad AB \to A \otimes B,$$

$$(B/A)A \to B, \qquad \frac{\Gamma A \to B}{\Gamma \to B/A},$$

$$\Lambda \to \Phi \top \Psi,$$

$$A \wedge B \to A, \qquad \frac{\Lambda \to \Phi A \Psi \qquad \Lambda \to \Phi B \Psi}{\Lambda \to \Phi (A \wedge B) \Psi}.$$

Other introduction rules may be obtained from those stated by exploiting left-right symmetry, e.g. $A \wedge B \to B$, or by arrow reversal, e.g. $A \oplus B \to AB$. Also one may absorb the two cut-rules into the introduction rules, for example, by recasting $AB \to A \otimes B$ as

$$\frac{\Gamma \to A \qquad \Delta \to B}{\Gamma \Delta \to A \otimes B},$$

for the purpose of proving a cut-elimination theorem.

In BL1 (a) we may replace the introduction rule for / on the right by the more general rule:

$$\frac{\Gamma A \to \Phi B}{\Gamma \to \Phi(B/A)}.$$

Szabo [1978] developed a sequent calculus for logics with restricted structural rules, making use of the following more general cut-rule:

$$\frac{\Lambda \to \Phi A \Psi \qquad \Gamma A \Delta \to \Theta}{\Gamma \Lambda \Delta \to \Phi \Theta \Psi}.$$

While this is acceptable for linear logic, it involves an implicit interchange between Φ and Γ and between Ψ and Δ. As we shall illustrate with an example later, it is not valid in our model. From the point of view of bilinear logic, the cut-rules must be restricted to those cases where either Φ or Γ is empty and either Ψ or Δ is empty.

8 Mixed Associative Laws

The restriction mentioned at the end of the preceding section permits, in addition to the two cut-rules already used in the sequent presentation of BL1, also the following:

$$\frac{\Lambda \to \Phi A \qquad A\Delta \to \Theta}{\Lambda\Delta \to \Phi\Theta}, \qquad \frac{\Lambda \to A\Psi \qquad \Gamma A \to \Theta}{\Gamma\Lambda \to \Theta\Psi}.$$

Under the standard interpretation of $\Gamma \to \Delta$ as $A_1 \otimes \cdots \otimes A_m \to B_1 \oplus \cdots \oplus B_n$, these new cut-rules correspond to the *mixed associative laws*:

(b) $\quad (A \oplus B) \otimes C \to A \oplus (B \otimes C), \quad A \otimes (B \oplus C) \to (A \otimes B) \oplus C.$

These laws have also been noticed independently by de Paiva and by Cockett and Seely, who refer to them as 'weak distributive laws'.

However, (b) had already been studied by Grishin, who gave it primary significance over (a). He mentions bi-Heyting algebras as models which satisfy (b) (and not (a) unless they are Boolean). Moreover, Grishin presents a number of equivalent formulations of (b), of which we single out the following:

(b′) $\quad A \oplus (D/C) \to (A \oplus D)/C, \qquad C\backslash(D \oplus A) \to (C\backslash D) \oplus A,$

which happens to be the converse of (a).

Indeed, the first of (b) gives rise to

$$A \oplus B \to (A \oplus (B \otimes C))/C,$$

hence, if we replace a B by D/C and note that $(D/C) \otimes C \to D$, to

$$A \oplus (D/C) \to (A \oplus D)/C.$$

Conversely, from this we obtain

$$(A \oplus (D/C)) \otimes C \to A \oplus D,$$

hence, upon replacing D by $B \otimes C$ and noting that $B \to (B \otimes C)/C$, the first of (b).

We shall write BL1(b) for BL1 together with (b) and BL2 for BL1 together with (a) and (b).

In BL2, when all four cut-rules are admitted, we may recast the introduction rules in a form which absorbs the cuts as follows:

$$\frac{\Gamma\Delta \to \Theta}{\Gamma \mathrm{I} \Delta \to \Theta}, \qquad \to \mathrm{I},$$

$$\frac{\Gamma A B \Delta \to \Theta}{\Gamma(A \otimes B)\Delta \to \Theta}, \qquad \frac{\Gamma \to \Phi A \qquad \Delta \to B\Psi}{\Gamma\Delta \to \Phi(A \otimes B)\Psi},$$

$$\frac{\Lambda \to A\Psi \qquad \Gamma B \Delta \to \Theta}{\Gamma(B/A)\Lambda\Delta \to \Theta\Psi}, \text{ with } \Delta \text{ or } \Psi \text{ empty}, \qquad \frac{\Gamma A \to \Phi B}{\Gamma \to \Phi(B/A)},$$

$$\Lambda \to \Phi \top \Psi,$$

$$\frac{\Gamma A \Delta \to \Theta}{\Gamma(A \wedge B)\Delta \to \Theta}, \qquad \frac{\Lambda \to \Phi A \Psi \qquad \Lambda \to \Phi B \Psi}{\Lambda \to \Phi(A \wedge B)\Psi},$$

with additional introduction rules obtained from left-right symmetry and arrow reversal. In Section 9 we shall sketch the proof of a cut-elimination theorem for BL2. Here, for example, is a cut-free proof of one of the mixed associative laws:

$$\dfrac{\dfrac{\dfrac{A \to A \qquad \dfrac{B \to B \qquad C \to C}{BC \to B \otimes C}}{(A \oplus B)C \to A(B \otimes C)}}{(A \oplus B)C \to A \oplus (B \otimes C)}}{(A \oplus B) \otimes C \to A \oplus (B \otimes C)}$$

Most of these introduction rules are already valid in BL1, with two exceptions. In the rule for introducing $\otimes$ on the right, Φ and Ψ have to be empty for BL1 and BL1(a); it is here that the new cut rules are used, in fact twice:

$$\dfrac{\Delta \to B\Psi \qquad \dfrac{\Gamma \to \Phi A \qquad AB \to A \otimes B}{\Gamma B \to \Phi(A \otimes B)}}{\Gamma\Delta \to \Phi(A \otimes B)\Psi}$$

This rule holds in BL1(b) and in BL2. Moreover, as was already pointed out, the rule for introducing / on the right depends on (a); it holds in BL1(a), hence in BL2, but not in BL1, hence neither in BL1(b).

The subsets of a group constitute a model of BL2. Indeed, suppose $uc \in (A \oplus B) \otimes C$, where $u \in A \oplus B$ and $c \in C$. We claim that $uc \in A \oplus (B \otimes C)$. For suppose that $uc = ax$ and $a \notin A$; then $u = axc^{-1}$ and therefore $xc^{-1} \in B$, whence $x \in B \otimes C$.

More generally, this is so for the subsets of a (Brandt) *groupoid*, that is, a category in which all arrows are invertible (isomorphisms). Another special case of this more general model is the algebra of binary relations on a set. We shall look at this in some detail, later, in Section 12.

9 Cut Elimination Revisited

We recall the cut-rule for BL2 and BL1(b):

$$\frac{\Lambda \to \Gamma A \Delta \qquad \Phi A \Psi \to \Theta}{\Phi \Lambda \Psi \to \Gamma \Theta \Delta}$$

with Γ or Φ empty and Δ or Ψ empty. In BL1 and BL1(a) we even have Γ and Δ empty or Φ and Ψ empty. We claim that in BL1, BL1(a), BL1(b) and BL2 every proof with cut can be replaced by one without cut. As usual, the proof will be by induction on the degree of the cut, namely

$$d(A) + d(\Lambda) + d(\Gamma) + d(\Delta) + d(\Phi) + d(\Psi) + d(\Theta),$$

where $d(A)$ is the number of occurences of connectives in A and, if $\Lambda = A_1 \cdots A_k$, $d(\Lambda) = d(A_1) + \cdots + d(A_k)$. The connectives in question are I, $\otimes$, /, \, $\top$ and $\wedge$, as well as their duals O, $\oplus$, $\div$, $\div$ and $\vee$.

We may assume that $\Lambda \to \Gamma A \Delta$ and $\Phi A \Psi \to \Theta$ have already been proved without cut and we will show that the cut in question can be replaced by one or more cuts of smaller degree, if it cannot be discarded altogether. We may look at each connective separately; but, for reasons of symmetry, we need only look at I, $\otimes$, /, $\top$ and $\wedge$. We distinguish five cases.

(0) Both $\Lambda \to \Gamma A \Delta$ and $\Phi A \Psi \to \Theta$ are axioms not containing any connectives. (As in [L 1990], it is assumed that the set of these axioms is closed under the structural rules; for example, they could all be of the form $A \to A$.)

(1) The last step in the proof of $\Lambda \to \Gamma A \Delta$ introduces the connective into Λ.

(2) The last step in the proof of $\Phi A \Psi \to \Theta$ introduces the connective into Φ or Ψ.

(3) The last step in the proof of $\Phi A \Psi \to \Theta$ introduces the connective into Θ.

(4) The last step in the proof of $\Lambda \to \Gamma A \Delta$ introduces the connective into Γ or Δ.

(5) The last steps in the proofs of both $\Lambda \to \Gamma A \Delta$ and $\Phi A \Psi \to \Theta$ introduce the main connective of A.

Now, in earlier publications [L 1958, 1969, 1990] it was assumed that Γ and Δ are empty (and also that Θ consists of a single formula, which does not affect the argument). It remains to look at case (4) and some additional subcases of other cases. Let us concentrate on (4), and let us say that the connective is being introduced into Δ. Thus Δ is not empty and therefore Ψ is empty; moreover either Γ or Φ will be empty. We distinguish several subcases depending on which connective is being discussed, and present the following three as examples:

$$(/) \qquad \frac{\dfrac{\Lambda C \to \Gamma A \Delta_1 B}{\Lambda \to \Gamma A \Delta_1 (B/C)} \qquad \Phi A \to \Theta}{\Phi \Lambda \to \Gamma \Theta \Delta_1 (B/C)}\ \text{cut}$$

may be replaced by

$$\dfrac{\dfrac{\Lambda C \to \Gamma A \Delta_1 B \qquad \Phi A \to \Theta}{\Phi \Lambda C \to \Gamma \Theta \Delta_1 B}\ \text{cut}}{\Phi \Lambda \to \Gamma \Theta \Delta_1 (B/C)}$$

$$(\top) \qquad \frac{\Lambda \to \Gamma A \Delta_1 \top \Delta_2 \qquad \Phi A \to \Theta}{\Phi \Lambda \to \Gamma \Theta \Delta_1 \top \Delta_2}\ \text{cut}$$

may be replaced by

$$\Phi \Lambda \to \Gamma \Theta \Delta_1 \top \Delta_2.$$

$$(\wedge) \qquad \frac{\dfrac{\Lambda \to \Gamma A \Delta_1 B \Delta_2 \qquad \Lambda \to \Gamma A \Delta_1 C \Delta_2}{\Lambda \to \Gamma A \Delta_1 (B \wedge C) \Delta_2} \qquad \Phi A \to \Theta}{\Phi \Lambda \to \Gamma \Theta \Delta_1 (B \wedge C) \Delta_2}\ \text{cut}$$

may be replaced by

$$\dfrac{\dfrac{\Lambda \to \Gamma A \Delta_1 B \Delta_2 \quad \Phi A \to \Theta}{\Phi\Lambda \to \Gamma\Theta\Delta_1 B\Delta_2}\text{cut} \quad \dfrac{\Lambda \to \Gamma A \Delta_1 C \Delta_2 \quad \Phi A \to \Theta}{\Phi\Lambda \to \Gamma\Theta\Delta_1 C\Delta_2}\text{cut}}{\Phi\Lambda \to \Gamma\Theta\Delta_1(B \wedge C)\Delta_2}$$

In each case, the new cuts, if any, have smaller degrees. Of course, the new cuts share the restrictions on Φ and Γ with the old cuts. In BL1 and BL1(a), both Φ and Γ are empty; in BL1(b), the cut involving / does not arise.

10 Another Axiomatization of BL2

In BL1 we have four negations

$$^{\perp}C = \mathrm{O}/C,\ C^{\perp} = C\backslash\mathrm{O}, {}^{\top}C = \mathrm{I} \div C,\ C^{\top} = C \div \mathrm{I},$$

which may be called *left annihilator, right annihilator, left creator* and *right creator* respectively, as they have the following properties:

$$(\#) \qquad {}^{\perp}C \otimes C \to \mathrm{O}, C \otimes C^{\perp} \to \mathrm{O}, \mathrm{I} \to {}^{\top}C \otimes C, \mathrm{I} \to C \otimes C^{\top}.$$

However, in BL1(a), ${}^{\top}C \to C^{\perp}$ and, in BL1(b), $C^{\perp} \to {}^{\top}C$. Thus, in BL2, $C^{\perp} \leftrightarrow {}^{\top}C$ and similarly $C^{\top} \leftrightarrow {}^{\perp}C$. Therefore, we may identify ${}^{\perp}C$ with $C^{\top}$ and ${}^{\top}C$ with $C^{\perp}$, so that there are only two negations in BL2.

Indeed, it follows from Grishin's law (a) that

$$\mathrm{I} \to C\backslash C \leftrightarrow C\backslash(\mathrm{O} \oplus C) \to (C\backslash\mathrm{O}) \oplus C,$$

so that $\mathrm{I} \div C \to C\backslash\mathrm{O}$ in BL1(a). On the other hand, assuming that $\mathrm{I} \div C \to D$, we infer from the mixed associative law (b) that

$$C\backslash\mathrm{O} \leftrightarrow \mathrm{I}\otimes(C\backslash\mathrm{O}) \to (D\oplus C)\otimes(C\backslash\mathrm{O}) \to D\oplus(C\otimes(C\backslash\mathrm{O})) \to D\oplus\mathrm{O} \leftrightarrow D,$$

for any D, so that $C\backslash\mathrm{O} \to \mathrm{I} \div C$ in BL1(b).

Recalling that in BL2

$$(A \oplus B)/C \leftrightarrow A \oplus (B/C)$$

and taking $B = \mathrm{O}$, we obtain

$$A/C \leftrightarrow A \oplus C^{\top},\ C\backslash A \leftrightarrow C^{\perp} \oplus A,$$

the second equivalence following by left-right symmetry. Moreover, by arrow reversal, we obtain

$$A \div C \leftrightarrow A \otimes C^{\top},\ C \div A \leftrightarrow C^{\perp} \otimes A.$$

If we regard these four equivalences as definitions of $/, \backslash, \div$ and $\div$ respectively, we may obtain an alternative axiomatization of BL2 in terms of $\otimes, \mathrm{I}, \oplus, \mathrm{O}$ together with $^{\perp}$ and $^{\top}$. In this section, we shall ignore the connectives $\wedge, \top, \vee$ and $\perp$.

Clearly, the following axioms are necessary, in addition to the bifunctoriality of $\otimes$ and $\oplus$:

$$\begin{array}{cc}
(A \otimes B) \otimes C \leftrightarrow A \otimes (B \otimes C), & (A \oplus B) \oplus C \leftrightarrow A \oplus (B \oplus C), \\
A \otimes \mathrm{I} \leftrightarrow A \leftrightarrow \mathrm{I} \otimes A, & A \oplus \mathrm{O} \leftrightarrow A \leftrightarrow \mathrm{O} \oplus A, \\
A \otimes (B \oplus C) \to (A \otimes B) \oplus C, & (A \oplus B) \otimes C \to A \oplus (B \otimes C), \\
C^{\top} \otimes C \to \mathrm{O}, & \mathrm{I} \to C \oplus C^{\top}, \\
C \otimes C^{\perp} \to \mathrm{O}, & \mathrm{I} \to C^{\perp} \oplus C.
\end{array}$$

To see that these axioms are also sufficient we shall check, for example, the combination of (a′) and (b):

$$(A \oplus B)/C \leftrightarrow (A \oplus B) \oplus C^{\top} \leftrightarrow A \oplus (B \oplus C^{\top}) \leftrightarrow A \oplus (B/C),$$

and the fact that $A \otimes B \to C$ if and only if $A \to C/B$. Indeed, it follows from the former (and bifunctoriality) that

$$A \leftrightarrow A \otimes \mathrm{I} \to A \otimes (B \oplus B^{\top}) \to (A \otimes B) \oplus B^{\top} \to C \oplus B^{\top} \leftrightarrow C/B,$$

and from the latter that

$$A \otimes B \to (C/B) \otimes B \leftrightarrow (C \oplus B^{\top}) \otimes B \to C \oplus (B^{\top} \otimes B) \to C \oplus \mathrm{O} \leftrightarrow C.$$

It is worth pointing out that $C^{\perp}$ and $C^{\top}$ are uniquely determined by the above properties (#). For example, suppose that $C \otimes C^{*} \to \mathrm{O}$ and $\mathrm{I} \to C^{*} \oplus C$. Then we have, in view of (b),

$$C^{\perp} \leftrightarrow \mathrm{I} \otimes C^{\perp} \to (C^{*} \oplus C) \otimes C^{\perp} \to C^{*} \oplus (C \otimes C^{\perp}) \to C^{*} \oplus \mathrm{O} \leftrightarrow C^{*},$$

and similarly $C^{*} \to C^{\perp}$, so that $C^{\perp} \leftrightarrow C^{*}$.

Replacing C by $A^{\top}$ and C^{*} by A in this argument, we also infer the following equivalences:

(§) $$A^{\top\perp} \leftrightarrow A \leftrightarrow A^{\perp\top}.$$

We also note that

(§§) $$A \to B \quad \text{iff} \quad A \otimes B^{\perp} \to \mathrm{O} \quad \text{iff} \quad B^{\top} \otimes A \to \mathrm{O}.$$

Indeed, from $A \to B$ (and functoriality) we infer that

$$A \otimes B^{\perp} \to B \otimes B^{\perp} \to \mathrm{O},$$

while from this we infer conversely that

$$A \leftrightarrow A \otimes \mathrm{I} \to A \otimes (B^{\perp} \oplus B) \to (A \otimes B^{\perp}) \oplus B \to \mathrm{O} \oplus B \leftrightarrow B.$$

It is now easy to prove that

$$\mathrm{I}^{\perp} \leftrightarrow \mathrm{O}, \qquad (A \otimes B)^{\perp} \leftrightarrow B^{\perp} \oplus A^{\perp}$$

and similarly if $^{\perp}$ is replaced by $^{\top}$. For example,

$$\begin{aligned} C \to (A \otimes B)^{\perp} \quad &\text{iff} \quad (A \otimes B)^{\perp\top} \otimes C \to \mathrm{O} \\ &\text{iff} \quad A^{\perp\top} \otimes (B \otimes C) \to \mathrm{O} \\ &\text{iff} \quad B \otimes C \to A^{\perp} \\ &\text{iff} \quad C \to B \backslash A^{\perp} \\ &\text{iff} \quad C \to B^{\perp} \oplus A^{\perp}. \end{aligned}$$

Thus we may define $A \oplus B$ as $(B^{\perp} \otimes A^{\perp})^{\top}$ and give an alternative axiomatization of BL2 in terms of $\mathrm{I}, \otimes, ^{\perp}$ and $^{\top}$ alone, using the above properties of I and $\otimes$ together with (§) and (§§). We leave it as an exercise to the reader to show that these properties imply the mixed associative laws and the fact that $(B^{\perp} \otimes A^{\perp})^{\top} \leftrightarrow (B^{\top} \otimes A^{\top})^{\perp}$.

In the sequent formulation of BL2, the annihilation and creation rules for $^{\perp}$ and $^{\top}$ are expressed most concisely thus:

$$\begin{aligned} C^{\top}C \to \quad &, \qquad\qquad \to CC^{\top}, \\ CC^{\perp} \to \quad &, \qquad\qquad \to C^{\perp}C. \end{aligned}$$

If the cut-rules are absorbed, these rules become:

$$\frac{\Delta \to C\Psi}{C^{\top}\Delta \to \Psi}, \qquad \frac{\Gamma C \to \Phi}{\Gamma \to \Phi C^{\top}},$$

$$\frac{\Gamma \to \Phi C}{\Gamma C^{\perp} \to \Phi}, \qquad \frac{C\Delta \to \Psi}{\Delta \to C^{\perp}\Psi}.$$

Although I have not made a careful check, I should point out that bilinear logic, more or less in the form presented here, has also been studied by other authors: Abrusci [1990] calls it 'pure noncommutative propositional logic', Ono [1990], who also considers quantifiers, discusses 'Gentzen linear logic without interchange' (GL^{-}), and Blute and Seely [1991] speak of 'non-commutative planar $*$-autonomous categories'. Both Abrusci and the last mentioned also look at so-called proof nets, which had been introduced by Girard [1987] in the commutative case.

11 Models of BL2

While the subsets of a group satisfy the axioms of BL2, they satisfy too much, as $C^\perp$ and $C^\top$ will there be the same. The reason for this is that, if an element of a monoid has both a left and a right inverse, then these must coincide. To get interesting models of BL2, we will have to weaken the notion of inverse.

An element a in a partially ordered monoid is said to have a *right adjoint* a^r if

$$aa^r \leq 1, \qquad 1 \leq a^r a$$

and a *left adjoint* a^ℓ if

$$a^\ell a \leq 1, \qquad 1 \leq aa^\ell.$$

Each of these, if it exists, is uniquely determined. For suppose both a^r and a^* are right adjoints, then

$$a^r = 1a^r \leq a^* aa^r \leq a^* 1 = a^*,$$

and similarly $a^* \leq a^r$.

We shall consider a lattice ordered monoid $(M, 1, \cdot, \leq, \top, \wedge, \bot, \vee, {}^r, {}^\ell)$ in which each element a has both a right adjoint a^r and a left adjoint a^ℓ. (Evidently, a lattice ordered group is a special case of this if $\leq$ is taken to mean $=$, and this case has been considered by Grishin [1983].) Let us call such a structure a *lattice ordered monoid with adjoints*. It is a model of BL2 in which

$$\mathrm{I} = \mathrm{O} = 1, \quad a \otimes b = a \oplus b = a \cdot b, \quad a^\perp = a^r, \quad a^\top = a^\ell.$$

If we don't like the collapse of connectives in a lattice ordered monoid with adjoints, we can pass to another model by taking upward closed subsets of a *partially ordered monoid with adjoints*, meaning that the lattice operations are absent from the monoid. (Downward closed subsets of partially ordered semigroups have already been studied by Došen [1985] and Buszkowski [1986] in their completeness results for the syntactic calculus.) We define the connectives occurring in the second presentation of BL2 as follows:

$$\begin{aligned} A \otimes B &= \{x \mid \exists_a \exists_b (a \cdot b \leq x \wedge a \in A \wedge b \in B)\}, \\ \mathrm{I} &= \{x \mid 1 \leq x\}, \\ A \oplus B &= \{x \mid \forall_a \forall_b (x \leq a \cdot b \Rightarrow (a \in A \vee b \in B)\}, \\ \mathrm{O} &= \{x \mid x \not\leq 1\}, \end{aligned}$$

$$A^{\perp} = \{x \mid \forall_{a\in A} a\cdot x \not\leq 1\},$$
$$A^{\top} = \{x \mid \forall_{a\in A} x\cdot a \not\leq 1\}.$$

Since upward closed subsets form a complete lattice, we also have the connectives $\top, \wedge, \perp$ and $\vee$. We shall prove that (i) $A \otimes A^{\perp} \leq \mathrm{O}$ and that (ii) $\mathrm{I} \leq A^{\perp} \oplus A$.

(i) Suppose $x \in A \otimes A^{\perp}$; then there exist $a \in A$ and $b \in A^{\perp}$ such that $a \cdot b \leq x$. But from $b \in A^{\perp}$ we infer that $a \cdot b \not\leq 1$, hence $x \not\leq 1$, so $x \in \mathrm{O}$.

(ii) Suppose $1 \leq x \leq b \cdot a$ and $b \notin A^{\perp}$; then there exists an $a' \in A$ such that $a' \cdot b \leq 1$, hence $a' \leq a' \cdot b \cdot b^{\ell} \leq b^{\ell} \leq b^{\ell} \cdot b \cdot a \leq a$, and so $a \in A$.

We end this section by presenting an example of a lattice ordered monoid with adjoints in which the left and right adjoints are not the same.

Consider all unbounded order preserving mappings $f : \mathbb{Z} \to \mathbb{Z}$, where $\mathbb{Z}$ is the ordered set of integers; they form a monoid under composition. Moreover, every such f has both a right and a left adjoint in the usual sense, also unbounded:

$$f(x) \leq y \iff x \leq f^{r}(y),\ x \leq f(y) \iff f^{\ell}(x) \leq y,$$

defined as follows:

$$f^{r}(x) = \max\{x \mid f(x) \leq y\},\ f^{\ell}(x) = \min\{y \mid x \leq f(y)\}.$$

These are adjoints also in the present sense. For

$$\begin{aligned} f \circ g \leq h &\iff \forall_x (f(g(x)) \leq h(x)) \\ &\iff \forall_x (g(x) \leq f^{\ell}(h(x))) \\ &\iff g \leq f^{\ell} \circ h. \end{aligned}$$

Moreover,

$$\begin{aligned} f \leq h \circ g^{r} &\iff g \circ h^{\ell} \leq f^{\ell}, \text{ by taking left adjoints,} \\ &\iff h^{\ell} \leq g^{\ell} \circ f^{\ell}, \text{ by the result just proved,} \\ &\iff f \circ g \leq h, \text{ by taking right adjoints.} \end{aligned}$$

Finally, we take a special case to show that $f^{r} \neq f^{\ell}$. Let $f(x) = 2x$; then

$$f^{r}(y) = \left[\frac{y}{2}\right], \qquad f^{\ell}(y) = \left[\frac{y+1}{2}\right].$$

I cannot refrain from mentioning an amusing property of this example: the sets of integers $\{f(x) + x \mid x \in \mathbb{Z}\}$ and $\{f^{r}(y) + y + 1 \mid y \in \mathbb{Z}\}$ are complementary subsets of $\mathbb{Z}$, for any unbounded order preserving mapping $f : \mathbb{Z} \to \mathbb{Z}$.

Other models of what is probably BL2 have been studied by Abrusci, by an ingenious modification of Girard's so-called 'phase semantics' (see Section 14 below), and by Ono, who looks at ideals in certain algebras, and by Sambin (see [Ono 1990]).

12 The Algebra of Binary Relations on a Set

We study binary relations on a set X, assuming that the underlying logic is classical. If $R \subseteq X \times X$, we write xRy for $(x, y) \in R$. Aside from the obvious lattice operations, we define the bilinear operations as follows:

$$
\begin{array}{ccc}
aIb & \text{iff} & a = b \\
a(R \otimes S)b & \text{iff} & \exists_z (aRz \wedge zSb) \\
a(R/S)b & \text{iff} & \forall_z (bSz \Rightarrow aRz) \\
a(S\backslash R)b & \text{iff} & \forall_z (zSa \Rightarrow zRb) \\
a\mathrm{O}b & \text{iff} & a \neq b \\
a(R \oplus S)b & \text{iff} & \forall_z (aRz \vee zSb) \\
a(R \mathbin{\dot{-}} S)b & \text{iff} & \exists_z (aRz \sim bSz) \\
a(S \mathbin{\dot{-}} R)b & \text{iff} & \exists_z (zRb \sim zSa)
\end{array}
$$

where $\sim$ means 'but not'.

We claim that all the rules of BL2 are satisfied, with $\to$ interpreted as inclusion, here written $\leq$, and give four sample proofs.

(i) $R \oplus \mathrm{O} = R$:

$$
\begin{aligned}
a(R \oplus \mathrm{O})b &\iff \forall_z (aRz \vee z \neq b) \\
&\iff \forall_z (z = b \Rightarrow aRz) \\
&\iff aRb,
\end{aligned}
$$

where the decidability of equality is used in passing to the second line.

(ii) $T \leq R \oplus S$ iff $T \mathbin{\dot{-}} S \leq R$:

Here the left-hand side asserts that

$$
\begin{array}{ll}
 & \forall_x \forall_y (xTy \Rightarrow \forall_z (xRz \vee zSy)) \\
\text{i.e.} & \forall_x \forall_y \forall_z (xTy \Rightarrow (xRz \vee zSy))
\end{array}
$$

$$\text{i.e.} \qquad \forall_x \forall_z \forall_y ((xTy \sim zSy) \Rightarrow xRz)$$
$$\text{i.e.} \qquad \forall_x \forall_z (\exists_y (xTy \sim zSy) \Rightarrow xRz),$$

which is the right-hand side. To pass to the third line of the proof we used the fact that the algebra of propositions is co-Heyting.

(iii) $(R \oplus S) \oplus T = R \oplus (S \oplus T)$:

Here a and b are in the relation on the left-hand side provided

$$\forall_z (\forall_t (aRt \vee tSz) \vee zTb)$$
$$\text{i.e.} \qquad \forall_z \forall_t (aRt \vee tSz \vee zTb),$$

using the co-intuitionistic rule that $\vee$ distributes over $\forall$. The same is true when a and b are in the relation on the right-hand side.

(iv) $R \otimes (S \oplus T) \leq (R \otimes S) \oplus T$:

Here a and b are in the relation on the left-hand side provided

$$\exists_x (aRx \wedge \forall_y (xSy \vee yTb)).$$

By ordinary intuitionistic predicate logic, this entails

$$\forall_y (\exists_x (aRx \wedge xSy) \vee yTb),$$

which means that a and b are in the relation on the right-hand side.

The binary relations on a set X satisfy more than the rules of BL2; in fact, they also satisfy

$$(*) \qquad A \otimes B \leq \mathrm{O} \quad \text{if and only if} \quad A \wedge B^{\cup} \leq \bot,$$
$$(**) \qquad \mathrm{I} \leq A \oplus B \quad \text{if and only if} \quad \top \leq A \vee B^{\cup},$$

where $B^{\cup}$ is the converse relation defined by

$$xB^{\cup}y \quad \text{if and only if} \quad yBx.$$

As an easy consequence, relations satisfy the following weak interchange rules:

$$(\dagger) \qquad A \otimes B \to \mathrm{O} \quad \text{if and only if} \quad B \otimes A \to \mathrm{O},$$
$$(\dagger\dagger) \qquad \mathrm{I} \to A \oplus B \quad \text{if and only if} \quad \mathrm{I} \to B \oplus A,$$

hence

$$A \backslash \mathrm{O} \leftrightarrow \mathrm{O}/A, \qquad A \div \mathrm{I} \leftrightarrow \mathrm{I} \div A.$$

Tarski and Givant [1987] give an axiomatic description of relation algebras. Since they insist on equations, the partial order $\leq$ does not occur explicitly in their axioms. However, the easiest way to summarize their axioms is to say that a *classical relation algebra* is a residuated lattice with a converse operation satisfying (*) and

$$\mathrm{O}/(A\backslash \mathrm{O}) = A.$$

However, it should be pointed out that the subsets of a group also satisfy these axioms, provided

$$A^{\cup} = \{x \mid x^{-1} \in A\}.$$

Cockett and Seely have proposed the following additional 'weak distributive law':

$$A \otimes (B \oplus C) \to B \oplus (A \otimes C),$$

which is also implicit in Szabo's cut-rule. In particular, taking $C = \mathrm{O}$ and $B = \mathrm{I}$, we could infer:

$$A \to \mathrm{I} \oplus (A \otimes \mathrm{O}).$$

In a relation algebra, this would mean:

$$\forall_x \forall_y (xAy \Rightarrow \forall_{t \neq x} \exists_{u \neq y} tAu),$$

for example:

If Ann likes Bill, then all others like someone else.

Obviously, this is not so, as long as Jane likes only Bill. Thus, this proposed law does not hold in BL1 or BL2 or even in BL3, to be considered in the next section.

13 Cyclic Bilinear Logic

By BL3 we shall understand BL2 together with the following rules: if $A \otimes B \to \mathrm{O}$ then $B \otimes A \to \mathrm{O}$, if $\mathrm{I} \to A \oplus B$ then $\mathrm{I} \to B \oplus A$, or, equivalently,

$$\mathrm{O}/A \leftrightarrow A\backslash \mathrm{O}, \qquad \mathrm{I} \div A \leftrightarrow A \div \mathrm{I}.$$

These rules are not independent; according to the second formulation of BL2 they amount to $A^{\perp} \leftrightarrow A^{\top}$, so that we can discard $^{\top}$ altogether. In a sequent calculus presentation, they should be expressed as additional structural rules:

$$\frac{\Gamma\Delta \to}{\Delta\Gamma \to}, \qquad \frac{\to \Phi\Psi}{\to \Psi\Phi}.$$

These rules characterize Yetter's 'cyclic (non-commutative) linear logic'.

Models of BL3 are obtained by considering any residuated lattice $(\mathcal{M}, \leq, \mathrm{I}, \otimes, /, \backslash, \top, \wedge, \bot, \vee)$ with a *symmetric dualizing element* O satisfying

$$\mathrm{O}/A = A\backslash\mathrm{O}, \qquad (\mathrm{O}/A)\backslash\mathrm{O} = A$$

for all A (we again use capital letters to denote elements) and defining

$$A \oplus B = ((\mathrm{O}/B) \otimes (\mathrm{O}/A))\backslash\mathrm{O}.$$

The laws of BL3 are easily verified.

The subsets of a Brandt groupoid form such a model, in which O is the set of all non-identity arrows; in particular, so do the binary relations on a set, with O being the complement of the identity relation.

Such models may also be constructed by considering the 'closed' elements in any residuated semilattice $(\mathcal{M}, \leq, 1, \cdot, /, \backslash, \top, \wedge)$, where we have deliberately written 1 and $\cdot$ rather than I and $\otimes$, with a *symmetric element* O, meaning that $\mathrm{O}/A = A\backslash\mathrm{O}$ for all elements A of $\mathcal{M}$ or, equivalently,

$$A \cdot B \leq \mathrm{O} \quad \text{iff} \quad B \cdot A \leq \mathrm{O}.$$

Here A is called *closed* if $(\mathrm{O}/A)\backslash\mathrm{O} = A$; Girard [1987] calls such elements 'facts'. This construction goes back to Prüfer in algebra (see e.g. Krull [1948]) and Gödel in logic (see e.g. Kleene [1952]). Note that O is closed, since $1 \leq \mathrm{O}/\mathrm{O}$, hence $(\mathrm{O}/\mathrm{O})\backslash\mathrm{O} \leq 1\backslash\mathrm{O} = \mathrm{O}$, and since always $A \leq (\mathrm{O}/A)\backslash\mathrm{O}$.

We claim that the closed elements of $\mathcal{M}$ form a model of BL3, with I and $\otimes$ defined as follows:

$$\mathrm{I} = (\mathrm{O}/1)\backslash\mathrm{O} = \mathrm{O}/\mathrm{O} = \mathrm{O}\backslash\mathrm{O},$$
$$A \otimes B = (\mathrm{O}/(A \cdot B))\backslash\mathrm{O} = ((\mathrm{O}/B)/A)\backslash\mathrm{O}.$$

These are closed because $\mathrm{O}/A = A\backslash\mathrm{O}$ is always closed, as is easily checked. Moreover, A/B and $B\backslash A$ are closed if A is, since e.g.

$$A/B = (\mathrm{O}/(A\backslash\mathrm{O}))/B = \mathrm{O}/(B \cdot (A\backslash\mathrm{O})).$$

It remains to verify that the closed elements of $\mathcal{M}$ form a residuated lattice (in which the closed element O is clearly a symmetric dualizing element).

First note that

$$A \otimes \mathrm{I} = ((\mathrm{O}/\mathrm{I})/A)\backslash\mathrm{O} = ((\mathrm{O}/1)/A)\backslash\mathrm{O} = (\mathrm{O}/A)\backslash\mathrm{O} = A,$$

and similarly that $\mathrm{I} \otimes A = A$. Next,

$$\begin{aligned} A \otimes B \leq C \quad &\text{iff} \quad \mathrm{O}/C \leq \mathrm{O}/(A \otimes B) = \mathrm{O}/(A \cdot B) \\ &\text{iff} \quad (\mathrm{O}/C) \cdot (A \cdot B) \leq \mathrm{O} \end{aligned}$$

$$\begin{aligned} &\text{iff} \quad A \cdot B \leq (0/C)\backslash 0 = C \\ &\text{iff} \quad A \leq C/B, \end{aligned}$$

and similarly iff $B \leq A\backslash C$. Finally, we deduce the associativity of $\otimes$ from that of $\cdot$ as follows:

$$\begin{aligned} (A \otimes B) \otimes C \leq D \quad &\text{iff} \quad A \otimes B \leq D/C \\ &\text{iff} \quad A \cdot B \leq D/C \\ &\text{iff} \quad (A \cdot B) \cdot C \leq D, \end{aligned}$$

and similarly

$$A \otimes (B \otimes C) \leq D \quad \text{iff} \quad A \cdot (B \cdot C) \leq D.$$

Any closure operation in a lattice preserves meets, hence $\top$ and $A \wedge B$ are closed if A and B are. We define

$$\bot = 0/\top = \top\backslash 0, \quad A \vee B = ((0/A) \wedge (0/B))\backslash 0.$$

If the original semilattice happens to be a lattice already, the original operations $\bot$ and $\vee$ should be ignored.

To find particular models of BL3 we shall look for examples of symmetric elements in residuated lattices.

(A) A Heyting algebra is a residuated lattice, with $1 = \top$, $A \cdot B = A \wedge B$, $A/B = B\backslash A = B \Rightarrow A$, and $0 = \bot$ is a symmetric element. Here $0/A = \neg A$ and the closed elements have the form $\neg\neg A$. This is the case originally studied by Gödel.

(B) The subsets of a monoid (or category) form a residuated lattice. Let $1 = \{1\}$ (or the set of all identity arrows) and 0 its complement. Then 0 will be symmetric if $ab \neq 1 \Rightarrow ba \neq 1$, which is so if $ba = 1 \Rightarrow ab = 1$, for all elements a and b. Here are some special cases:

(B1) The monoid is commutative.

(B2) For all elements $a, a^2 = 1$ implies $a = 1$; this is so in any free monoid, but then there are only two closed subsets.

(B3) The subsets of a group (or Brandt groupoid, in particular, the binary relations on a given set). If classical logic is used, all subsets are closed. But intuitionistically $(0/A)\backslash 0 = \neg\neg A$, where $\neg A$ is the complement of A, so the complemented sets (or relations) form a model of BL3.

(C) The (integral) ideals in a ring R form a residuated lattice with $1 = R = \top = \mathrm{I}$ and the usual multiplication of ideals. Note that $A + B = A \vee B$ and $A \cdot B \leq A \wedge B$. The zero ideal $\mathrm{O} = \{0\} = \bot$ is symmetric in the following cases:

(C1) R is commutative.

(C2) R is *semiprime*, that is, $C^2 \leq \mathrm{O}$ implies $C \leq \mathrm{O}$ for every ideal C. For then

$$A \cdot B \leq \mathrm{O} \Rightarrow (B \cdot A)^2 \leq \mathrm{O} \Rightarrow B \cdot A \leq \mathrm{O}.$$

(C3) R is a *$*$-ring with proper involution*. By this one understands a ring with an involution * satisfying not only

$$1^* = 1,\ 0^* = 0,\ (a \cdot b)^* = b^* \cdot a^*,\ (a + b)^* = a^* + b^*,\ a^{**} = a,$$

but also

$$a \cdot a^* = 0 \Rightarrow a = 0.$$

For then

$$A \cdot B \leq \mathrm{O} \text{ iff } A \cap B^* \leq \mathrm{O} \text{ iff } B \cap A^* \leq \mathrm{O} \text{ iff } B \cdot A \leq \mathrm{O}.$$

To see the first equivalence, suppose $A \cdot B \leq \mathrm{O}$ and $a \in A \cap B^*$; then $aa^* \in A \cdot B \leq \mathrm{O}$, hence $a = 0$, and so $A \cap B^* \leq \mathrm{O}$. Conversely, suppose $A \cap B^* \leq \mathrm{O}$ and let $a \in A,\ b \in B$; then

$$(ab)(ab)^* = abb^*a^* \in A \cdot B^* \leq A \cap B^* \leq \mathrm{O},$$

hence $ab = 0$, and so $A \cdot B \leq \mathrm{O}$.

(D) While Prüfer had originally applied his idea to fractional ideals in a commutative ring, this may be generalized to the following noncommutative situation: let R be a subring of Q and consider the lattice of all *fractional* ideals, that is, all addive subgroups A of Q such that $R \cdot A \cdot R \leq A$. Assume that, for all such A, $A \cdot B \leq R$ implies $B \cdot A \leq R$, that is, that the residual quotients R/A and $A \backslash R$ coincide (the usual notation being $R : A$). Then we have a residuated lattice in which $\top = Q, \bot = \{0\}$ and $\mathrm{O} = \mathrm{I} = R$ is a symmetric element.

14 Some Half-Baked Remarks about Girard's Modal Operators

Girard has proposed also two 'modal operators' ! and ? for his linear logic, satisfying:

$$!A \to A \qquad\qquad A \to ?A$$

$$!A \to !!A \qquad\qquad ??A \to ?A$$

$$!A \otimes !B \leftrightarrow !(A \wedge B) \qquad\qquad ?A \oplus ?B \leftrightarrow ?(A \vee B)$$

In many of the models of bilinear logic considered here, there are such operators which moreover satisfy:

$$!A \to B \quad \text{if and only if} \quad A \to ?B,$$

thus making ! left adjoint to ?. Indeed, if A is any set of arrows in a category, we may put

$$!A = A \wedge \mathrm{I}, \qquad ?A = A \vee \mathrm{O}$$

and check all the above conditions. The only ones that do not hold for trivial reasons are:

$$A \wedge B \wedge \mathrm{I} \leq (A \wedge \mathrm{I}) \otimes (B \wedge \mathrm{I}), \quad A \vee B \vee \mathrm{O} \geq (A \vee \mathrm{O}) \oplus (B \vee \mathrm{O}).$$

These hold because, for any set E of arrows in a category,

$$E \leq \mathrm{I} \Rightarrow E \leq E \otimes E, \qquad E \geq \mathrm{O} \Rightarrow E \geq E \oplus E,$$

as is easily verified.

Girard's purpose was to use such modal operators to embed intuitionistic logic into classical linear logic. His argument works for bilinear logic as well, if one introduces an intuitionistic entailment $A \overset{\text{int}}{\to} B$ to mean $!A \to B$ and defines intuitionistic implication $A \Rightarrow B$ as $!A \backslash B$. It is then easily verified that

$$A \wedge B \overset{\text{int}}{\to} C \quad \text{if and only if} \quad B \overset{\text{int}}{\to} A \Rightarrow C.$$

It seems to me that a more symmetric procedure would be to insist that ! is left adjoint to ?, as it is in the above example, so that $A \overset{\text{int}}{\to} B$ could be rendered equivalently as $A \to ?B$, and to change the definition of $A \Rightarrow B$ to $!A \backslash ?B$. One could then weaken the third rules for ! and ? as follows:

$$!(!A \otimes !B) \leftrightarrow !(A \wedge B), \qquad ?(?A \oplus ?B) \leftrightarrow ?(A \vee B).$$

Perhaps a more general construction for $!A$ in a quantale with a converse operation would be as the join of *co-equivalences* E, satisfying

$$E \leq \mathrm{I}, \qquad E \leq E^{\cup}, \qquad E \leq E \otimes E.$$

Note that co-equivalences are *projections*:

$$E \otimes E = E = E^{\cup}.$$

The construction of $!A$ as $A \wedge \mathrm{I}$ does not work in our ring theoretic examples. It may be of interest here to recall how one constructs a locale from the quantale of ideals in ring theory. This locale is of course the 'spectrum' of the ring, a topological space whose open sets have the form:

$$V(A) = \text{ set of all prime ideals containing } A,$$

A being any ideal whatsoever. This locale is isomorphic to the lattice of all 'semiprime' ideals under the correspondence which associates with $V(A)$ the 'prime radical':

$$\surd A = \bigwedge \{P \geq A \mid P \text{ prime}\}.$$

The prime radical satisfies the equations:

$$\surd(A \otimes B) = \surd A \wedge \surd B = \surd(A \wedge B).$$

15 Summary and Conclusion

Bilinear logic arose from an attempt to combine the author's syntactic calculus with Girard's linear logic. We have discussed three successive versions of bilinear logic, which differ by the number of structural rules they admit.

BL1 admits two cut-rules. BL1(a) also satisfies Grishin's Law and is illustrated by the lattice of all sets of arrows of a category; BL1(b) satisfies the mixed associative law and is illustrated by any bi-Heyting algebra.

BL2 admits four cut-rules and is best illustrated by the same examples as BL3 below, but also by the lattice of upward closed subsets of a partially ordered monoid with adjoints.

BL3 admits four cut-rules and two cyclic rules; it is illustrated by the lattice of subsets of a Brandt groupoid, in particular by the algebra of relations on a set, but also by the lattice of closed elements of a residuated monoid with a symmetric element O, for example, of the lattice of ideals in a semiprime ring or in a Baer $*$-ring.

All systems of bilinear logic studied here display two kinds of symmetry, which had already been exploited by Grishin; one may interchange left and right, while replacing / by \ etc; one may also reverse all arrows, while replacing $\otimes$ by $\oplus$ etc. (I didn't get around to incorporate universal and existential quantifiers.) There is an analogy here to the two basic symmetries allowed, with some reservations, in Physics: interchanging right- and left-hand coordinate systems and time reversal. The possibility of arrow reversal makes our bilinear logic classical; an earlier attempt to consider an intuitionistic version of $\oplus$ failed. I guess intuitionistic bilinear logic is just the syntactic calculus.

As promised, we have confined attention to bilinear logic as a deductive system, with partially ordered sets as models. In our entire discussion, we have completely ignored the fact that there may be more than one arrow from $A_1 \otimes \cdots \otimes A_m$ to $B_1 \oplus \cdots \oplus B_n$ and, consequently, the notion of equality on the set of such arrows. In case $n = 1$, such considerations gave rise to the concept of 'multicategory' [L 1969, 1989]. The most recent exposition of this [L 1990] makes use of an 'internal language', replacing the sequent $f : A_1 \cdots A_m \to B$ by the multilinear polynomial $fx_1 \cdots x_m$ of type B, where the x_i are variables of type A_i. This is essentially the same thing as replacing Gentzen's sequents by the natural deductions advocated by Prawitz. In [L 1969] the name 'polycategory' was proposed for the situation when $n \neq 1$, and this notion was elaborated by Szabo [1975]. However, his polycategories allow forms of the cut-rule which are not permitted in bilinear logic here. See also Velinov [1988], who treats the special case in which $\otimes$ and $\oplus$ coincide.

Three years ago, I began an investigation of the internal language of a polycategory, using not only variables but also 'covariables'. It was clear that the sequent $f : A_1 \cdots A_m \to B_1 \cdots B_n$ should be replaced by the term $y_n \cdots y_1 f x_1 \cdots x_m$ of a language, where the x_i are variables of type A_i and the y_i are *covariables* of type B_j. However, I soon got side-tracked by the question of how to analyze terms of this language unambiguously, there being a conflict between concatenation and composition of terms, and I did not pursue the matter further.

If we confine attention to arrows $f : A \to B$, with $m = n = 1$, we are talking about a noncommutative generalization of the $*$-autonomous categories of Barr [1979], as are Cockett and Seely [1991].

References

V. M. Abrusci, Phase semantics and sequent calculus for pure noncommutative linear propositional logic, *Rapporto Scientifico 11* (1990), Università degli Studi di Bari.

M. Barr, *∗-autonomous categories*, Springer LNM 752, Berlin, 1979.

N. Belnap, Linear logic displayed, *Notre Dame J. Formal Logic 31* (1990), 14–25.

R. Blute and R.A.G. Seely, Natural deduction and coherence for weakly distributive categories, Draft, 1991.

W. Buszkowski, Completeness results for Lambek syntactic calculus, *Zeitschr. f. math. Logik und Grundlagen d. Math. 32* (1986), 13–28.

J.R.B. Cockett and R.A.G. Seely, Weakly distributive categories, Working draft, McGill University, 1991.

H. B. Curry, Some logical aspects of grammatical structure, *AMS Proceedings Symposia Applied Math. 12* (1961), 56–68.

H. B. Curry and R. Feys, *Combinatory Logic*, North-Holland Publ. Co., Amsterdam, 1958.

V. de Paiva, A Dialectica model of the Lambek calculus, Draft, Cambridge University, 1991.

K. Došen, A completeness theorem for the Lambek calculus of syntactic categories, *Zeitschr. f. math. Logik und Grundlagen d. Math. 31* (1985), 235–241.

———, Sequent-systems and groupoid models, I and II, *Studia Logica 47* (1988), 353–385, *48* (1989), 41–65.

J.-Y. Girard, Linear logic, *Theoretical Computer Science 50* (1987), 1–102.

———, Towards a geometry of interactions, *Contemporary Math. 92* (1989), 69–108.

V. N. Grishin, On a generalization of the Ajdukiewicz-Lambek system, in: *Studies in Nonclassical Logics and Formal Systems*, Nauka, Moscow, 1983, 315–343.

M. Kandulski, The non-associative Lambek calculus, in: Buszkowski et al. (eds), *Categorial Grammar*, Benjamins, Amsterdam, 1988, 141–151.

S. C. Kleene, *Introduction to Metamathematics*, Van Nostrund, New York, 1952.

W. Krull, *Idealtheorie*, Chelsea Publ. Co., New York, 1948.

J. Lambek, The mathematics of sentence structure, *Amer. Math. Monthly 65* (1958), 154–169.

———, On the calculus of syntactic types, *AMS Proceedings Symposia Applied Math. 12* (1961), 166–178.

———, *Lectures on Rings and Modules*, Ginn, Waltham, Mass., 1966, Chelsea, New York, 1976, 1986.

———, Deductive Systems and Categories II, *Springer LNM 86* (1969), 76–122.

———, Multicategories revisited, *Contemporary Math. 92* (1989), 217–339.

———, Categorial and categorical grammars, in: R. Oehrle et al. (eds), *Categorial Grammars and Natural Language Structures*, Reidel, Dordrecht and Boston, 1988, 297–317.

———, On a connection between algebra, logic and linguistics, *Diagrammes 22* (1989), 59–75.

———, Logic without structural rules: Another look at cut elimination, (in this book) 1990.

F. W. Lawvere, Intrinsic co-Heyting boundaries and the Leibniz rule in certain toposes, Preprint, 1991.

H. Ono, Algebraic aspect of logics without structural rules, Preprint, 1989.

———, Semantics for substructural logics, (in this book) 1990.

C. Rauszer, Semi-Boolean algebras and their applications to intuitionistic logic with dual operations, *Fundamenta Mathematicae 83* (1974), 219–249.

G. E. Reyes and H. Zolfaghari, Bi-Heyting algebras, topos and modalities, Rapports de récherche 91-9, Université de Montréal, 1991.

K. I. Rosenthal, *Quantales and their Applications*, Longman Scientific and Technical, Harlow, Essex, 1990.

R.A.G. Seely, $*$-Autonomous categories and cofree co-algebras, *Contemporary Math. 92* (1989), 371–382.

K.L.N. Swamy, Dually residuated lattice ordered semigroups, *Math. Annalen 159* (1965), 105–144.

M. E. Szabo, Polycategories, *Communications in Algebra 3* (1975), 663–689.

———, *Algebra of Proofs*, North-Holland, Amsterdam, 1978.

A. Tarski and S. Givant, A formalization of set theory without variables, *AMS Colloquium Publications 41*, Providence, R.I., 1987.

Y. Velinov, An algebraic structure for derivations in rewriting systems, *Theoretical Computer Science 57* (1988), 205–224.

D. N. Yetter, Quantales and (non-commutative) linear logic, *J. Symbolic Logic 55* (1990), 41–64.

Postscript. Special thanks are due to Kosta Došen for his careful reading and constructive criticism of the original version of this article.

The Semantics of Entailment 0

Robert K. Meyer and Edwin D. Mares

Automated Reasoning Project
Australian National University

The system **NR** of relevant implication with modality was proposed in Meyer (1968) to combine the system **R** of relevant implication with the modal logic **S4**. This motivated a definition of entailment, in something like the sense of Anderson and Belnap (1975), as strict relevant implication. While **NR** exactly captures the entailment of certain fragments of the system **E** of entailment — in particular, the pure calculus $\mathbf{E}_{\rightarrow}$ of entailment — it was shown in Maksimova (1973) that there is a non-theorem of **E** which is provable on **NR** translation.

This fact, distressing for both systems, seems to have stopped research into **NR** cold for the last decade. That the '90's may improve on the '80's in this regard, we wish to suggest here two extensions of **NR** to accommodate a genuine Boolean negation. As a generic name for these systems, we use **CNR**. The two systems we propose vary in strength. For the weaker we also use **CNR** (when no ambiguity will result); for the stronger, **CR4**.[1]

Previous 'classical' relevant logics have been conservative extensions of the corresponding non-Boolean systems. This will *not* be true here. Or, to put it another way, the previous system **NR** strikes us now as significantly *incomplete* relative to its natural motivation. For the *appealing* way to formulate systems of entailment is the way chosen by Anderson and Belnap in (1975). That is, a vertebrate $\Rightarrow$ is grounded in a theory of *relevant implication* (provided, for example, by the system **R**) and a theory of *modality* (similarly provided, say, by **S4**). While previous systems of entailment (including our own) still seem to us more or less satisfactory on

1 **CR4** is the putative Boolean counterpart of the system **R4** for which the rule γ is proved admissible in Mares and Meyer (1992).

the **R** side, they now seem to us somewhat less satisfactory on the modal side.[2]

The picture that emerges from the present study is most clear in our 'Galactic' and 'Intergalactic' theorems of section IV. Briefly, this picture is as follows. A **CNR** model structure is composed of a number of *component* **CR** model structures ('galaxies'). In each of these galaxies relevant implication may be characterized in the usual semantical way, without any reference to modal considerations whatsoever. On the other hand, the galaxies are *connected* by a Kripke-style **S4** relation of relative possibility. We like to think of this picture two dimensionally. If we slice the model horizontally, we get a **CR** model structure. If we slice it vertically, we get a Kripke **S4** model structure. Aesthetically, it's neat!

Aside from the *fun* that we have had in proving some of these theorems (almost all of which resulted from attempts to prove things that are in fact *false*), we also think them important for the theory of *entailment*. Entailment, on an old tradition, is *necessary* implication. But material implication, as a ground for this theory, has been tried and found wanting. Also, as it seems to us, taking entailment as a *primitive* connective and attempting to formalize its logic has not quite worked out. The remaining alternative is to *improve* the theory of the grounding implication, while grafting it smoothly to a theory of necessity. This has been our aim here. And, since for the first time we are supplying a semantics for entailment which really is a semantics for *entailment*, we are calling this paper 'The Semantics of Entailment 0'.

Accordingly, we shall formulate **CNR** and **CR4** and provide them with a semantics. We shall show that they are conservative extensions of **R**, **CR**, and **S4**.

I

The primitive connectives of **CNR** shall be $\rightarrow$ (relevant implication), $\wedge$ (conjunction), $\sim$ (De Morgan negation), $\neg$ (Boolean negation), $\Box$ (**S4** ne-

2 In a word, we believe that an improved understanding of the *semantical* situation yields an improved **NR**. Note moreover that there is nothing privileged about the choice of **R** as our relevant implication or **S4** as our modal logic (except perhaps the observation of A. R. Anderson that, although there are many modal logics, **S4** is the true one). For example, Bacon long ago suggested providing **R** with alternative accompanying Lewis systems such as **S3**, **S5**, etc. But it is an old project to get **R** and **S4** together; and it is the one we are pursuing here. It is to be granted that the *original* project, of Ackermann, Anderson, Belnap, and others, was simply to formalize entailment directly. But, aside from the Anderson-Belnap line of separating its motivation into a modal and a relevant component, we know of no persuasive motivating line that supports this original project.

cessity), and t (Ackermann t). We introduce the following additional connectives by definition:

D∨. $A \vee B =_{\text{df}} \neg(\neg A \wedge \neg B)$
D⊃. $A \supset B =_{\text{df}} \neg A \vee B$
D⇒. $A \Rightarrow B =_{\text{df}} \neg\sim A$
D*. $^{*}A =_{\text{df}} \neg\sim A$
D≺. $A \prec B =_{\text{df}} \Box(S \supset B)$
D◊. $\Diamond A =_{\text{df}} \neg\Box\neg A$
D◊. $\Diamond A =_{\text{df}} \sim\Box\sim A$
D∘. $A \circ B =_{\text{df}} \sim(A \rightarrow \sim B)$
D↔. $A \leftrightarrow B =_{\text{df}} (A \rightarrow B) \circ (B \rightarrow A)$
Df. $\mathrm{f} =_{\text{df}} \sim \mathrm{t}$
DF. $\mathrm{F} =_{\text{df}} p \wedge \neg p$[3]
DT. $\mathrm{T} =_{\text{df}} \neg \mathrm{F}$.

Formulas A, B, C are built up as usual from a countably infinite set of atoms (for which we use p, q, r, etc.). For ease in reading formulas, we rank binary connectives (including defined ones) $\wedge$, $\circ$, $\vee$, $\supset$, $\prec$, $\rightarrow$, $\Rightarrow$, $\leftrightarrow$, in order of increasing scope. Otherwise, we associate to the left and use dots as parentheses in accordance with the conventions of Curry (1963). The defined connectives include $\vee$ (disjunction), $\supset$ (Boolean implication), $\Rightarrow$ (entailment), $\prec$ (**S4** strict implication), $\diamond$ (**S4** possibility), $\circ$ (fusion), and $\leftrightarrow$ (relevant equivalence). Note that $\vee$ may be equivalently defined using De Morgan negation $\sim$ in place of $\neg$ but that $\supset$ and $\Diamond$ may not be so defined. (We have added a distinct De Morgan possibility $\Diamond$.)

A formula is *Boolean* if no primitive logical particles occur in it but $\wedge$ and $\neg$ (at most); it is an S4 formula if none occur but $\wedge$, $\neg$ and $\Box$; it is an S4˜ formula if none occur but $\wedge$, $\neg$, $\sim$, and $\Box$; it is an R formula if none occur but $\wedge$, $\sim$, $\rightarrow$; and it is a CR formula if none occur but $\wedge$, $\sim$, $\rightarrow$, $\neg$. (Evidently the defined connectives $\supset$ and $\prec$ may occur in Boolean formulas; as well $\Diamond$ and $\prec$ may occur in S4 formulas; while $\Diamond$ can occur in S4˜ formulas.)

We shall present **CNR** as a common extension of **S4** and of the Boolean extension **CR** of **R**.[4] A formula is a Boolean tautology if it is a substitution instance of a Boolean formula which is a truth-table tautology. It is an S4 tautology if it is a substitution instance of an S4 formula that is a theorem

3 Where p is the first sentential variable. In fact, since F is the Church false constant (to be distinguished from the more interesting Ackermann false constant f in relevant logics – see Meyer (1979)), it may be defined indifferently by $A \wedge \neg A$ for any formula A. Its chief characteristic is that it entails absolutely everything.

4 The system called **CR** in Meyer and Routley (1973) was in fact a Boolean extension of the positive fragment $\mathbf{R}_{+}$ of **R**. We now apply this name to the Boolean extension of all of **R** in Meyer (1976), which was called **CR*** in Meyer and Routley (1974).

of **S4**. Here is a sufficient set of axioms for **CNR**:

A1. $A \to B$ whenever $A \supset B$ is an S4 tautology[5]
A2. $A \to B \to . B \to C \to . A \to C$
A3. $A \to . A \to B \to B$
A4. $A \to (A \to B) \to . A \to B$
A5. $(A \to B) \wedge (A \to C) \to . A \to B \wedge C$
A6. $A \to {\sim} B \to . B \to {\sim} A$
A7. ${\sim}{\sim} A \to A$
A8. $\Box(A \to B) \to . \Box A \to \Box B$
A9. $A \to (\mathrm{t} \to A)$
A10. $(\mathrm{t} \to A) \to A$
A11. If A is an axiom, $\Box A$ is also an axiom.

The rules are:

R1. From $A \to B$ and A, infer B
R2. From A and B, infer $A \wedge B$.

A11 may be generalized to the derivable rule

$\Box$I. From A, infer $\Box A$,

in view of A8 and the instance $\Box A \wedge \Box B \to \Box(A \wedge B)$ of A1.

We get **CR4** by adding to the above formulation of **CNR** the following additional axiom scheme:

A12. $\Box({\sim} A \vee B) \to {\sim} \Box A \vee \Box B$.

II

A **CNR** model structure (henceforth CNRms) is a quintuple $\mathbf{K} = \langle P, K, R, S, {}^* \rangle$, where K is a set, P is a subset of K (of 'possible' worlds), R is a ternary relation and S is a binary relation on K, and * is a unary operation on K subject to definitions and postulates to come. For ease in presenting these, we introduce the following generalized notation for relational products: Let M be an m-ary and N an n-ary relation; then when there is an x such that $Mm_1 \ldots m_{m-1}x$ and $Nn_1 \ldots n_{i-1}xn_{i+1} \ldots n_n$, we may write

(1) $Nn_1 \ldots n_{i-1}(Mm_1 \ldots m_{m-1})n_{i+1} \ldots n_n$.

And where $M(Mm_1 \ldots m_{m-1})x_2 \ldots x_m$ we may write in particular

(2) $M^2m_1 \ldots m_{m-1}x_2 \ldots x_m$.

This produces the particular definitions

DR^2. $R^2xyzw =_{\text{df}} \exists a(Rxya \;\&\; Razw)$
DS^2. $S^2xy =_{\text{df}} \exists a(Sxa \;\&\; Say)$,

---

5 Caution: We set down A1 only for S4 tautologies, *not* for S4˜ tautologies. In particular, $p \wedge {\sim} p \to q$ is excluded here, as in all relevant logics, on grounds of relevance.

in conformity with our past usage. We also define

DE. $Exyz =_{\text{df}} R(Sx)yz$
D$\leq$. $\leq xy =_{\text{df}} R(P)xy$.[6]

Our defined relations may be understood as follows. E is the ternary relation which stands to entailment $\Rightarrow$ as R stands to relevant implication $\rightarrow$. Where 'worlds' are understood as theories, $\leq$ will be just the subset relation on theories.[7]

The postulates on a CNRms **K** are the following:

P=. $a \leq b \iff a = b$

P1.	Saa	S-reflexivity
P2.	$Raaa$	R-reflexivity
P3.	$S^2ab \Rightarrow Sab$	S-transitivity
P4.	$S(P)b \Rightarrow Pb$	S-P-closure
P5.	$R^2abcd \Rightarrow R^2acbd$	Pasch's Law
P6.	$S(Rab)c \Rightarrow R(Sa)(Sb)c$	Hume's Dictum[8]
P7.	$Rabc \Rightarrow Rac^*b^*$	*-contraposition
P8.	$a^{**} = a$	*-involution

A CNRms is a CR4ms if it satisfies in addition the postulate

P++. $Sab \Rightarrow Sa^*b^*$ possibility law

The original plan of Routley and Meyer (1972) (*The Semantics of Entailment II*, henceforth [SEII]) contemplated a single 'real world' 0 and *no* distinguished subset P of K. We can get P anyway, on this plan. For we characterize a CNRms $\langle P, K, R, S, ^* \rangle$ as *simple* if there exists a $0 \in P$ such that $P = \{p : S0p\}$. We shall show below that all our semantical work can in fact be done with simple CNRms and CR4ms.

We turn now to interpretive machinery. A valuation v of **CNR** in an CNRms **K** is a function from $V \times K$, where V is the set of sentential variables of **CNR**, into $\{\text{T}, \text{F}\}$. Each valuation v in a CNRms uniquely determines an associated interpretation I from $\mathbb{F} \times K$ into $\{\text{T}, \text{F}\}$, where $\mathbb{F}$ is the set of all formulas, according to familiar homomorphic conditions about to be set down. Before doing so, we adopt the convention that $I(A, a) = \text{T}$ may be written $A_I a$ and $I(A, a) = \text{F}$ may be written $\neg A_I a$; when I is clear in context we omit the subscript. I being so fixed, we may now state the truth conditions for compound formulas at worlds a as follows, where $a \in K$ and intuitive quantifiers range over K:

Tt. (t)a iff Pa
T$\wedge$. $(A \wedge B)a$ iff Aa and Ba

---

6 Note that according to our convention $x \leq y$ is by definition $\exists a\,(Pa \mathbin{\&} Raxy)$.

7 Note that $\leq$ is somewhat differently defined in the semantics for **R** in Routley and Meyer (1973) (henceforth [SEI]), in [SEII] for **NR**, and here.

8 Hume did not exactly put it this way, but this is clearly what he had in mind.

T$\sim$. $(\sim A)a$ iff $\neg(Aa^*)$
T$\neg$. $(\neg A)a$ iff $\neg(Aa)$[9]
T$\Box$. $(\Box A)a$ iff $\forall x\,(Sax \Rightarrow Ax)$
T$\rightarrow$. $(A \rightarrow B)a$ iff $\forall x \forall y\,(Raxy \Rightarrow .\ Ax \Rightarrow By)$

A formula A is true at a on an interpretation I, if $I(A,a) = \mathrm{T}$ and is otherwise false at a; A is *verified* on I iff $A_I p$ for all $p \in P$; A is **CNR**-*valid* iff A is verified on all interpretations in all CNRms; A is **CR4**-*valid* iff A is verified on all interpretations in all CR4ms.

We now set down a list of truth conditions for defined connectives. The following hold in every CNRms:

T$\vee$. $(A \vee B)a$ iff Aa or Ba
T$\Rightarrow$. $(A \Rightarrow B)a$ iff $\forall x \forall y\,(Exya \Rightarrow .\ Ax \Rightarrow By)$
T$\circ$. $(A \circ B)a$ iff $\exists x \exists y\,(Rxya \mathbin{\&} Ax \mathbin{\&} By)$
T$\prec$. $(A \prec B)a$ iff $\forall x\,(Sax \Rightarrow .\ Ax \Rightarrow Bx)$
T$\Diamond$. $(\Diamond A)a$ iff $\exists x\,(Sax \mathbin{\&} Ax)$
T$\dot{\Diamond}$. $(\dot{\Diamond} A)a$ iff $\exists x\,(Sa^*x^* \mathbin{\&} Ax)$

Comparison of the last two truth conditions highlights the difference between **CNR** and **CR4**. We had already, in [SEII], contemplated the addition of the special axiom A12 to **NR**, to create the system that we call here **R4**.[10] And we noted in that paper that the possibility postulate P++ sufficed to verify it when added to the **NR** semantics. We were unable to show (and still do not know) whether the addition of P++ to the **NR** semantics suffices for the completeness of the extension of **NR** got by adding A12. The situation in the present paper is more satisfying. Both versions of **CNR** are sound and complete in the obvious senses. Moreover, the semantic sense of P++ is immediate. It just means that T$\dot{\Diamond}$ reduces to T$\Diamond$ in CR4ms; that is, $\dot{\Diamond}$ and $\Diamond$ are semantically indistinguishable in the **CR4** context.[11]

The present semantics has some interesting connections to the semantics of [SEII]. Why, the reader may wonder, did we not simply add the truth condition T$\neg$ to the [SEII] conditions on an interpretation in an arbitrary

9 We have heard that there are those who find that the present truth condition T$\neg$ a more appealing one for negation than our previous T$\sim$ for De Morgan negation. We marvel at such obtuseness!

10 Belnap had previously contemplated adding its equivalent to **E**.

11 This also explains our choosing 'possibility' as a tag for our postulate P. It stands more fully for 'Boolean possibility' $\neg\Box\neg A$ equals 'De Morgan possibility' $\sim\Box\sim A$. We are inclined to think that this is a good postulate, in view of analogues with other dual particles. E.g. Boolean disjunction $\neg(\neg A \wedge \neg B)$ equals De Morgan disjunction $\sim(\sim A \wedge \sim B)$ and Boolean existence $\neg\forall x \neg Ax$ equals De Morgan existence $\sim\forall x \sim Ax$, etc. But we allow the possibility, however unlikely, that we may be wrong; so we carry **CNR** as a weaker alternative to **CR4**.

NRms. The reason lies in the hereditary condition imposed on a valuation in [SEII]. In **NR**, this makes true the following associated hereditary condition on all **NR** formulas A on all interpretations I:

H. $a \leq b \Rightarrow . Aa \Rightarrow b$.

However, the inductive argument that establishes H breaks down in the presence of Boolean negation. For suppose that H holds for a particular formula A and consider the formula $\neg A$. Suppose $(\neg A)a$ and $a \leq b$. To show $(\neg A)b$ the obvious course it to assume $(A)b$ and to derive an absurdity on inductive hypothesis. But, short of asserting the consequent, none is forthcoming.

Our solution to this problem, here as in our previous papers on 'classical relevant logics', is to trivialize the $\leq$ relation. This too is in accordance with semantic motivation. For, on the intuition that 'worlds' are (particularly nice) theories, and that $\leq$ is the subset relation on such theories, $a \leq b$ holds non-trivially if the nice theory a is a proper subtheory of the nice theory b. 'Nice', in the present context, means prime. That is, a nice theory has one of A, B as a member whenever it has $A \vee B$ as a member. But, on a Boolean understanding, $A \vee \neg A$ is true at all worlds. By primeness, so is one of A, $\neg A$. Accordingly, if two prime Boolean theories are distinct, they must differ at least in that one contains A and the other contains $\neg A$ for some formula A. So neither can be a subset of the other, justifying the postulate P= in the Boolean case.

Passage from the general notion of an NRms to the more specific notion of a CNRms induces certain changes in the facts noted in [SEII]. For example, it makes the facts T2, T3, T5, T6, and T7 on page 55 trivially true. Note also that the postulates P6, P7, and P8 of [SEII] are also trivial in a CNRms. But the following is still interesting:

T1. $Rabc \Rightarrow Rbac$.

And we may generalize T4 of [SEII] as follows:

T4$'$. $\forall x\,(Px \Rightarrow x^* \leq x)$.

In fact, we may apply P= to show

PP. $\forall x\,(Px \Rightarrow x^* = x)$.

This ensures that the possible worlds, in our terminology, correspond on any interpretation to theories which are consistent and complete with respect to De Morgan negation. Moreover, as we shall see, all logical truths (in the **CNR** sense) are also true at all possible worlds. On the other hand, if a is not a possible world (i.e. if Pa fails), we can falsify $p \Rightarrow p$ at a (and also make $\sim(p \Rightarrow p)$ true) on the valuation that makes p true at exactly the worlds x such that Px. (Exercise: Verify.) So the possible worlds are those of Aristotle's choice; one may say there of what is that it is and of what is not that it is not, without denying logical truths and without truth value gluts or gaps. All other worlds we characterize as impossible.[12]

The following is an important consequence of P=, of which we make considerable use:

$\forall x \in K\ \exists p \in P\ Rpxx$ reflexivity of $\leq$

We conclude this section with the important semantic entailment fact. On any interpretation I, we have, for all formulas A and B,

SemEnt $A \Rightarrow B$ is verified on I iff $\forall x\,(A_I x \Rightarrow B_I x)$.

While SemEnt holds in all NRms, note that it is trivially verified in the CNRms context.

III

CNR and **CR4**, as presented here, rest on an amalgam of modal ideas from **S4** and relevant ideas from **R**. While it will follow anyway from more general results, we show now that **CNR** and **CR4** are conservative extensions of each of these systems (and of the Boolean extension **CR** of **R** as well). In the theorems and proofs of this section we presuppose the soundness and completeness results for **CNR** and **CR4** of sections V and VI below.

Extension Theorem for S4. *Let A be an* **S4** *formula. Then the following conditions are equivalent*:

(1) *A is a theorem of* **S4**
(2) *A is a theorem of* **CNR**
(3) *A is a theorem of* **CR4**.

Proof. Suppose that A is a theorem of **S4**. By A1, $T \rightarrow A$ is a theorem of **CNR**. Since, so evidently is T, A is provable in **CNR** by $\rightarrow E$. So (1) implies (2). Trivially, (2) implies (3). To show that (3) implies (1), suppose that A is a non-theorem of **S4**. Then there is a Kripke **S4**-model structure $\langle 0, K, S\rangle$ such that A is false on interpretation I (in our terminology at 0 on I), in view of the completeness result of Kripke (1963). Without loss of

12 Impossible worlds are impossible in the sense that one *may* make the negations of logical truths there on some valuation. Other valuations may have a happier effect. Adding impossible worlds, it is true, has been centrally important for relevant semantical enterprises; one wants (for reasons sufficiently set forth in other places) to take the rough with the smooth, and to consider theories deformed in various respects. But one wants also to be able to characterize the smooth; it is interesting that, in the present Boolean context, we are able to give a precise characterization of which worlds are possible and which worlds are not.

generality, we may assume $S0a$ for all $a \in K$. To show A a non-theorem of **CR4**, it will suffice (in view of the soundness theorem) to turn this structure into a CR4ms. Set $a^* = a$ for all $a \in K$. Set $Raaa$ for all $a \in K$, and otherwise let $Rabc$ fail. Set $P = K$. It is then easy to verify that $\mathbf{K} = \langle P, K, R, S,^* \rangle$ is an NRms and indeed a CR4ms. Accordingly, A is refuted in **K** on the same interpretation, and is a non-theorem of **CR4**. Contraposing, (3) implies (1), completing the proof.

Several previous studies have revealed special connections between **S4** and various relevant logics, especially **E** and **NR**. But, despite its triviality, the present result is the most satisfying. For example, the relations between **E** on the one hand and **S4** on the other explored in Meyer (1973) depend on translating the strict implication of **S4** as a certain kind of 'enthymematic' entailment. They do not render **S4** directly into **E** in the straightforward $\Box$, $\wedge$, $\sim$ vocabulary (for the simple reason that, on this translation, **S4** has more theorems than **E** does). Similar remarks apply to the relationship between **NR** and **S4**. Perhaps the fairest thing to say is that, read directly, the $\Box$ and $\wedge$ of **S4** are the $\Box$ and $\wedge$ of **NR**. But the *negation* of **S4** is Boolean, not De Morgan.

Or so one might think. In the **CNR** context, we may freely interchange occurrences of $\sim$ and $\neg$ in **S4** theorems. Specifically, let an **S4˜** tautology be any substitution instance of an S4˜ formula that is **S4** valid by identifying $\sim$ and $\neg$. In view of the equivalence of $\Diamond$ and $\Diamond$, it is evident that all **S4˜** tautologies are theorems of **CR4**. Somewhat more surprisingly, we have the

S4˜ Tautology Theorem. *All* **S4˜** *tautologies are theorems of* **CNR**.

Proof. Suppose that A is a formula in which no primitive particles occur but $\Box$, $\wedge$, $\neg$, and $\sim$. Suppose A is a non-theorem of **CNR**. We shall show it to be a non-theorem of **S4˜**. By the simple completeness theorem of section VI, there is a simple CNRms $\mathbf{K} = \langle P, K, R, S,^* \rangle$ which does in A. Specifically, there is a distinguished element 0 such that $P = \{x : S0x\}$ and an interpretation I such that A is false at 0 on I. Form a structure $\mathbf{P} = \langle 0, P, S_P \rangle$ by letting S_P be the restriction of S to P; since S is reflexive and transitive, it is a Kripke structure for **S4**. We define an interpretation I_P of **S4˜** in **P** by letting I_P agree with I on sentential variables in V, the truth values of compound formulas being determined by the Kripke truth conditions. By a simple induction, I_P agrees with I on all **S4˜** formulas at all worlds in P. This is trivial on inductive hypothesis for formulas of the form $B \wedge C$ and $\neg B$. For formulas of the form $\sim B$, it suffices to recall that, by PP, $p^* = p$ for all $p \in P$, thus in this domain T$\sim$ reduces to T$\neg$. For formulas of the form $\Box B$, it suffices to observe, by the S-P closure postulate P4, that if $\Box B$ is false on I at $p \in P$, there is some $q \in P$ such that Spq and B is false on I at q. On inductive hypothesis, $\Box B$ is false at p on I_P. If $\Box B$ is true on I, it is trivially true on I_P on inductive hypothesis. This suffices for the theorem.

Hitherto, Boolean systems such as **CR** have been conservative extensions of the corresponding non-Boolean systems. This relation does not hold between **NR**and **CNR**, since the **S4**˜ tautology

$$(\mathrm{K}^{\sim})\ {\sim}\Box({\sim} A \vee B) \vee {\sim}\Box A \vee \Box B,$$

which simply expresses distribution of $\Box$ over *De Morgan* material implication, is a non-theorem of **NR**. To show this, we exhibit an NRms in which (K˜) is invalid. Set $K = \{0, 1, 2\}$. Define R, S, and * in accordance with the following tables:[13 14]

	$*$
0	0
1	2
2	1

	S
0	01
1	01
2	012

	$\leq$
0	0
1	1
2	12

R	0	1	2
0	0		
1		1	12
2		12	12

It is readily verified that $\mathbf{K} = \langle 0, K, R, S, ^* \rangle$ is an NRms in the sense of [SEII]. To reject (K˜), make A true at 0 and 1 and B true at 0, and otherwise let these formulas be false. It turns out that all of the disjuncts of (K˜) are false at 0, whence (K˜) is a non-theorem of **NR**.

We turn now to the relation between **CR** and **CNR**. A CRms may be defined as a structure $\langle 0, K, R, ^* \rangle$ which satisfies the postulates P2, P5, P7, P8 and P=, *provided that* $a \leq b$ *is defined as* $R0ab$. Truth and validity conditions are as above (*mutatis mutandis*). It is shown in Meyer and Routley (1974) that **CR** is a conservative extension of **R**; in Meyer (1976) **CR** is proved complete, in the sense that A is a theorem of **CR** iff A is **CR**-valid. This will suffice for

Extension Theorem for CR. *Let A be a* **CR** *formula. The following conditions are equivalent:*

13 We are indebted to John Slaney's program MaGIC for a refutation of (K˜) in **E** which we were able to massage into the above **NR** refutation.

14 The relational tables are to be read as follows: Where X is a binary relation, the row entry for an element a correspond to the set of all b such that Xab. So, for example, the entry 01 in the 0 place for the relation S mean that both $S00$ and $S01$ hold, while $S02$ fails. Where X is a ternary relation, the entry in row a, column b, corresponds to the set of all elements c such that $Xabc$ holds. If one likes, this allows us to construe relations as 'many-valued functions'. (Brink, in particular, has looked at the relational semantics for relevant logics from this point of view. Cf. Brink (1989).)

(1) *A is a theorem of* **CR**
(2) *A is a theorem of* **CNR**
(3) *A is a theorem of* **CR4**.

Proof. It is again trivial that, if A is a theorem of **CR**, it is a theorem of **CNR** as well. So (1) implies (2). Since **CR4** is an extension of **CNR**, (2) implies (3). We complete the proof by showing that (3) implies (1). So suppose that A is not a theorem of **CR**. Let $\langle 0, K, R, ^* \rangle$ be a CRms and I an interpretation such that A is false at 0 on I. It will suffice to extend this model structure to a CR4ms $\mathbf{K} = \langle 0, K, R, S, ^* \rangle$ by defining S on K. Let Saa hold for all a in K, and otherwise let Sab fail. Note that, on this construction, 0 is the only possible world; accordingly, $\leq$ as defined on $\mathbf{K}$ by D$\leq$ agrees with its **CR** definition $R0ab$. The postulates on a CR4ms are then straightforwardly validated. The same interpretation I will now refute A in $\mathbf{K}$; whence by the Soundness Theorem for **CR4**, A is a non-theorem of **CR4**. Contraposing, this ends the proof.

Corollary. *Let A be an* **R** *formula. Then A is a theorem of* **CNR** *iff A is a theorem of* **R**.

Proof. By the extension theorem for **CR** and the fact that **CR** is a conservative extension of **R**.

IV

In this section we investigate CNRms further. We wish first to observe that any CNRms may be thought of as a union of component CRms, with the modal accessibility relation S defined among members of the union. Let $\mathbf{K} = \langle P, K, R, S, ^* \rangle$ be a CNRms. In virtue of P=, $\leq$ is reflexive; so, for every $x \in K$, there is a $p \in P$ such that $Rpxx$. We define the *galaxy* K_p determined by p as $\{x : Rpxx\}$. By the observation just made, K is the union of all the K_p for $p \in P$.

Galactic Lemma. *If $x \in K_p$, then $x^* \in K_p$. Moreover, if $Rxyz$, then if $x \in K_p$, then $y \in K_p$ and $z \in K_p$.*

Proof. Suppose $x \in K_p$. Then $Rpxx$. So Rpx^*x^*, by *-contraposition. This proves the first sentence of the lemma. Suppose moreover $Rxyz$. We wish to show that y and z are also in K_p. By definition, R^2pxyz. By Pasch, R^2pyxz. I.e. $Rpyy$, by P=, and $y \in K_p$. By *-contraposition, our initial assumption yields also Rxz^*y^*, whence by parity of reasoning z^* (and hence z) is in K_p. This concludes the verification of the lemma.

We now carry out our initial threat to break the Universe into Galaxies. For we shall dissolve a CNRms $\mathbf{K} = \langle P, K, R, S, ^* \rangle$ into non-overlapping component CRms's $\mathbf{K}_p = \langle 0_p, K_p, R_p, ^*{}_p \rangle$. For each $p \in P$, let $0_p = p$, let K_p be the galaxy determined by p, and let R_p, $^*{}_p$ be the restrictions of R and * respectively to K_p. We are ready for a lemma due originally to Brink (1989).

Galactic Theorem. *Let* $\mathbf{K} = \langle P, K, R, S, ^* \rangle$ *be a CNRms. Then* K *is the disjoint union of the* K_p*, for* $p \in P$*;* R *is the disjoint union of the* R_p*; and* * *is the disjoint union of the* $^*{}_p$*. Moreover, each Galaxy* $\mathbf{K}_p$ *is a CRms.*

Proof. We observed at the beginning of this section that K is the union of the K_p. Next we show that it is their *disjoint* union in the sense that each x in K belongs to exactly one of the K_p. Suppose otherwise. Then there are distinct $p, q \in P$ such that for some $x \in K$, we have both $Rpxx$ and $Rqxx$. By T1, $Rxqx$. By definition, R^2pxqx. By Pasch's Law, R^2pqxx. And so there is a y such that $Rpqy$. By definition of $\leq$, $q \leq y$. By P=, $q = y$. I.e. $Rpqq$. By T1, $Rqpq$, hence $p = q$. So, as claimed, each member of K belongs to exactly one galaxy K_p, where p is a possible world.

To show that * is the disjoint union of the $^*{}_p$, apply the Galactic Lemma to observe that if $x \in K_p$, then $x^* \in K_p$. To show the corresponding fact about the R_p, assume $Rxyz$. We must show that for some $p \in P$, R_pxyz. Let K_p be the galaxy containing x. Then, by the Galactic Lemma, $y \in K_p$ and $z \in K_p$. So R_pxyz, whence R is the union of all the R_p.

To conclude the proof, we show that every galaxy $\mathbf{K}_p = \langle 0_p, K_p, R_p, ^*{}_p \rangle$ is a CRms. Given the Galactic Lemma, the verifications of P2, P5, P7, and P8 are straightforward. We must verify P=; i.e. $R0_pab$ iff $a = b$. Suppose $R0_pab$. I.e. $Rpab$. So, by P= for $\mathbf{K}$, $a = b$. This concludes the verification that $\mathbf{K}_p$ is a CRms and the proof of the theorem.

Since the CNRms $\mathbf{K}$ is the disjoint union of the K_p, we may characterize the possible world p as the *centre* of the galaxy K_p. We may also say, where $x \in K_p$, that p is the centre of x.

The ternary relation R, as we have just seen relates only members of the same galaxy in a CNRms. We turn now to the relations that hold between worlds in different galaxies. Evidently, these must be S relations. But S relations between arbitrary worlds are largely parasitic on the S relations that hold between the possible worlds that determine their galaxies. Specifically, we have the

Intergalactic Lemma. *Let* $\mathbf{K} = \langle P, K, R, S, ^* \rangle$ *be a CNRms. Let* $p, q \in P$.

(α) *If* Spq*, then for every* $x \in K_p$ *there exist* $y, y^* \in K_q$ *such that* Sxy *and* Sx^*y^*.

(β) *Conversely, suppose* $x \in K_p$*,* $y \in K_q$*, and* Sxy*. Then* Spq*.*

Proof. Ad (α). Suppose Spq and $Rpxx$. Then Rxx^*p, by elementary CRms manipulations. By Hume's Dictum, there exist y and z such that Sxy, Sx^*z, and $Ryzq$. This can only happen in the CRms K_q if $z = y^*$, completing the verification of (α).

Ad (β). Suppose $Rpxx$, $Rqyy$, and Sxy. By Hume's Dictum, there exist p' and y' such that Spp', Sxy', and $Rp'y'y$. By the Galactic Lemma, $p', y' \in K_q$ (since $y \in K_q$). But, since P is S-closed, p' is a possible world. The only candidate is q; so $p' = q$. I.e. Spq as required, completing the proof of (β) and hence of the lemma.

Our Galactic Theorem broke a CNRms into component CRms, each clustering around a possible world p. The lemma just proved shows that the S-relations between members of different galaxies are largely induced by the S-relations that hold between possible worlds. Specifically, by the *centre* of a CNRms $\mathbf{K} = \langle P, K, R, S, ^* \rangle$ we mean the structure $\mathbf{P} = \langle P, S_P \rangle$, where S_P is the restriction of S to P. Evidently, the centre of $\mathbf{K}$ is composed of the centres of its component galaxies. We now prove an

Intergalactic Theorem. *Let* $\mathbf{K}$ *be a CNRms. Then the centre* $\mathbf{P}$ *of* $\mathbf{K}$ *is a Kripke model structure for* **S4** *(near enough). Moreover,* $\mathbf{P}$ *partially determines the* S*-relations that hold in* $\mathbf{K}$ *as a whole in the two senses corresponding to the lemma just proved. I.e., whenever* Sxy*, we also have* Spq *where* p *and* q *are the respective centres of* x *and* y*. Conversely, if* Spq *and* p *is the centre of* x*, then there exists a* y *such that* Sxy *and* Sx^*y^**, where* q *is the centre of* y*.*

Proof. That $\mathbf{P}$ is a Kripke model structure for **S4** was shown in the Extension Theorem for **S4**. The rest of the theorem follows from (β) and (α) respectively of the Intergalactic Lemma.

Since every CR4ms $\mathbf{K}$ is a CNRms, the theorems just proved apply immediately to all such $\mathbf{K}$. Observe also that in **CR4**, we have the following (slight) extension to the Intergalactic Lemma: Whenever Sxy, then by P12 Sx^*y^*. The lemma stated falls short of this conclusion. It requires only that when Spq and $x \in K_p$ there exists a z in K_q such that Sxz and Sx^*z^*. Not ruled out for **CNR** is that there exists *another* $y \in K_q$ such that Sxy but *not* Sx^*y^*.

The observations of this section impose a particularly smooth and pleasing structure on CNRms and CR4ms. Take a *horizontal* slice (so to speak) of a CNRms, and one has a CRms. Take a *vertical* slice (through the centres) of a CNRms, and one has a Kripke structure for **S4**. A further Kripke structure (on impossible worlds) is induced by any such central slice.

V

As promised above, in this section we show that **CNR** and **CR4** are sound relative to their suggested semantics. We adopt, in general, the strategy of [SEII]. That is, we show that when A is any instance of axioms A1–A8, $\Box A$ is **CNR**-valid. We show moreover that if $\Box A$ is **CNR**-valid, so are A and $\Box\Box A$. This suffices for all the axioms. To complete the proof of soundness, we must show that $\rightarrow E$ and $\wedge I$ preserve **CNR**-validity.

Lemma 5.1. *Let $A \rightarrow B$ be an instance of* A1–A10. *Then $A \Rightarrow B$ is* **CNR**-*valid.*

Proof. In each case, we appeal to SemEnt, assuming A true at an arbitrary world x on an arbitrary interpretation I and showing B true at x. This was done for A2–A8 in [SEII], to which we refer the reader. We consider A1, A9, A10. A1 may involve Boolean $\neg$. However, if $A \supset B$ is an S4 tautology, then at each world x at which A is true B is also true, in view of the **S4** semantics presented in Kripke (1963). This requires only reflexivity and transitivity of the modal accessibility relation S, and the straightforward truth conditions T$\wedge$, T$\neg$, and T$\Box$. So the same is true in **K**. So, by SemEnt, $A \Rightarrow B$ is verified always. For A9 and A10, it suffices by SemEnt to show that A and $\mathrm{t} \rightarrow A$ take the same values at each world x on an arbitrary interpretation I. Suppose $(\mathrm{t} \rightarrow A)x$. By reflexivity of $\leq$, there is a possible world a such that $Raxx$. By T1, $Rxax$; by Tt, $(\mathrm{t})a$; whence Ax by T$\rightarrow$. For the converse, suppose Ax. Suppose $(\mathrm{t})y$ and $Rxyz$. It will suffice to show Az, by T$\rightarrow$. Since $(\mathrm{t})y$, Py, by Tt. And, by T1, $Ryxz$; i.e. $x = z$, by D$\leq$ and P$=$. This concludes the proof of Lemma 5.1.

Lemma 5.2. *If $\Box A$ is* **CNR**-*valid, so are A and $\Box\Box A$.*

Proof. Suppose $\Box A$ is **CNR**-valid. By reflexivity of S, A is valid; given that P is S-closed, $\Box\Box A$ is valid by transitivity of S.

Lemma 5.3. **CNR**-*validity is preserved under $\rightarrow E$ and $\wedge I$.*

Proof. Suppose $(A \rightarrow B)a$ and Aa on I for all $a \in P$. Let a be an arbitrary member of P. By $Raaa$ and $T \rightarrow$, Ba. To verify $\wedge I$, just recall that the truths at a are closed under $\wedge$.

We conclude present business with the

Soundness Theorem for CNR. *Let A be a theorem of* **CNR**. *Then A is* **CNR**-*valid.*

Proof. By Lemmas 5.1, 5.2, and 5.3.

Soundness Theorem for CR4. *Let A be a theorem of* **CR4**. *Then A is* **CR4**-*valid.*

Proof. Lemmas 5.1, 5.2, and 5.3 being as above, *mutatis mutandis*, it will suffice to show that A12 is valid in all CR4ms. We leave this to the reader.

VI

In this section we prove that **CNR** and **CR4** are complete on their suggested semantics. We proceed, *mutatis mutandis*, as in [SEII]. A *CNR-theory* c (for the remainder of this section, simply a *theory*) is a set of formulas closed under adjunction and provable **CNR**-entailment; i.e., c is a theory iff whenever A, B are in c, $A \wedge B \in c$ and, moreover, if whenever $\vdash_{\mathbf{CNR}} A \Rightarrow B$ and $A \in c$, then $B \in c$. In addition, we call a theory c *regular* if $\mathrm{t} \in c$; *prime*, if whenever c contains $A \vee B$, it contains at least one of A, B; *maximal*[15], if the only proper supertheory of c is the set of all formulas; *De Morgan consistent*, if for no A does c contain both A and $\sim A$; *normal*, if c is regular, prime, and De Morgan consistent.

Let 0 be a regular theory. We call a theory c a 0-*theory*, if in addition to meeting other requirements, it is closed under 0-entailment; i.e., whenever $A \Rightarrow B \in 0$ and $A \in c$, then $B \in c$.

Let X now be the set of all 0-theories, where 0 is regular. Get a ternary relation R, a binary relation S, and a unary relation P on X, as follows for all theories a, b, c in X.

(R) $Rabc$ iff for all formulas A, B, if $A \in a$ and $B \in b$, then $A \circ B \in c$.

(S) Sab iff for all formulas A, if $\Box A \in a$, then $A \in b$.

(P) Pa iff $S0a$.

Let 0 be a maximal regular theory. The *canonical model structure* determined by 0 is a structure $\mathbf{K}_0 = \langle P_0, K_0, R_0, S_0, {}^*{}_0 \rangle$, where $P_0 = \{a : Pa\}$, K_0 is the set of all maximal 0-theories, R_0 and S_0 are defined on K_0 respectively by (R) and (S) above, and where $a^*{}_0 = \{A : \sim A \notin a\}$.[16] It is readily verified, by the methods of [SEII], that $\mathbf{K}_0$ is an NRms; to show moreover that it is a CNRms, it suffices to show that P= holds.

15 Since we are in a Boolean context, only the degenerate theories (the null set and the set of all formulas) are prime without being maximal. Our semantic interest here will be in maximal theories.

16 $^*{}_0$ is the usual Routley * operation on sets of formulas restricted to K_0. It is easy to verify that, so restricted, if $a \in K_0$, so does a^*; in the Boolean context, $a^*{}_0$ may be defined more simply as $\{A : {}^*A \in \mathrm{T}\}$.

Lemma 6.1. *Let* $\mathbf{K}_0$ *be as above and suppose* $a \leq b$ *for* $a, b \in K_0$. *Then* $a = b$.

Proof. By Lemma 3 of [SEII], we have at least $a \subseteq b$. But a is a maximal theory. So, by definition of maximality, either $b = a$ or b is the trivial theory. Whence, since $b \in K_0$, $a = b$.

We recall that a CNRms is *simple* if, roughly, it counts as an NRms in the sense of [SEII]. That is, a CNRms $\mathbf{K} = \langle P, K, R, S,^* \rangle$ is simple iff there exists an element 0 in P such that Pa iff $S0a$. And we call a formula A *simply* **CNR**-*valid* iff it is verified in all simple CNRms. We begin with a simple completeness theorem.

Simple Semantic Completeness Theorem for CNR. *Let A be simply* **CNR**-*valid. Then A is a theorem of* **CNR**.

Proof. Suppose that A is not a theorem of **CNR**. We shall, as is the fashion in these things, build a canonical model structure that rejects A at 0 on a canonical valuation.

First, we build in the usual way (i.e. by Lindenbaum's lemma adapted to this context) a regular maximal theory 0 in which A is a non-theorem. We let $\mathbf{K}_0 = \langle P_0, K_0, R_0, S^*_{0\,0} \rangle$ be the canonical model structure determined by 0 according to our definitions above, letting $P_0 = \{a : S0a\}$. As observed, and as readers with [SEII] at hand may amuse themselves by proving, $\mathbf{K}_0$ is a simple CNRms. The canonical valuation v is defined by setting $v(p, a) = \mathrm{T}$ iff $p \in a$. It is tedious but straightforward to show that $A_I a$ iff $A \in a$ on the associated interpretation I. Since 0 was constructed to lack A, A is false at 0 on I; but 0 is a possible world; hence A is **CNR**-invalid, completing the proof of the simple semantic completeness theorem.

Our snappy proof of completeness rested heavily on all the hard stuff having already been done in a closely related context. But, as a corollary we show that validity and simple validity coincide for **CNR**.

Corollary. *The following conditions are equivalent*:

(i) *A is a theorem of* **CNR**
(ii) *A is simply* **CNR**-*valid*
(iii) *A is* **CNR**-*valid.*

Proof. That (i) implies (iii) is the content of the soundness theorem of the last section. That (ii) implies (i) is the simple completeness theorem just proved. That (iii) implies (ii) is straightforward. By transitivity of *implies*, this completes the proof.

In analogy to the above, we call a formula simply **CR4**-*valid* iff it is valid in all simple CR4ms. We shall prove for **CR4** the appropriate simple completeness theorem. To give content to this theorem, we first insert an independence proof for A12 by exhibiting a CNRms **K** which refutes it. By the soundness theorem of the last section, evidently **K** will not be a CR4ms.

Independence Theorem. A12 *is not a theorem of* **CNR**.

Proof. Set $K = \{0, 1, 2, 3\}$. Let $P = \{0\}$. Let R, S, and * be determined by the following tables:

	$*$
0	0
1	1
2	3
3	2

	S
9	0
1	12
2	2
3	3

R	0	1	2	3
0	0	1	2	3
1	1	01	2	3
2	2	2	2	0123
3	3	3	0123	3

We leave it to the reader (and his computer) to show that these tables determine a CNRms. But they do not determine a CR4ms, since $S12$ holds but $S1^*2^*$ ($= S13$) fails. To refute A12, we shall make $\Box({\sim} p \vee q)$ true at 1, while making both ${\sim}\Box p$ and $\Box q$ false at 1. Set $v(p,1) = v(p,2) = v(q,1) = \mathrm{T}$ and otherwise make p and q false at the worlds in K. Enough said.

We now adapt the above completeness proof to **CR4**. To begin with, now that we have proved it complete, we note that there are simple semantic arguments to establish the **CNR**-theoremhood of the following:

N1. $\neg {\sim} A \leftrightarrow {\sim} \neg A$
N2. $(A \wedge B \rightarrow C) \supset (A \rightarrow \neg B \vee C)$
N3. $(A \rightarrow B \vee C) \supset (A \wedge \neg B \rightarrow C)$
N4. $\mathrm{T} \wedge A \leftrightarrow A$
N5. $\mathrm{F} \vee A \leftrightarrow A$
N6. $\mathrm{T} \leftrightarrow \Box \mathrm{T}$
N7. $\mathrm{F} \leftrightarrow \Diamond \mathrm{F}$
N8. $\mathrm{T} \leftrightarrow {\sim} \mathrm{F}$.

Since **CR4** is an extension of **CNR**, we may make free use of these principles therein. We also note that the set of theorems of **CNR** (and of **CR4**) is closed under *modus ponens* for $\supset$, under necessity introduction ($\Box$I), and replacement for provable equivalence. We may also appeal to double negation and De Morgan's laws (for either of ${\sim}$, $\neg$).

Next, we establish syntactically that our two versions of possibility are provably equivalent in **CR4** (though not of course in **CNR**).

Possibility Fact. $\vdash_{\mathbf{CR4}} \Diamond A \leftrightarrow \Diamonddot A$.

Proof sketch. Left to right:

$$\begin{array}{lll}
(1) & \mathrm{T} \to A \vee \neg A & \text{A1} \\
(2) & \mathrm{T} \to {\sim}{\sim} A \vee \neg A & \text{1, double neg.} \\
(3) & \mathrm{T} \to \Box({\sim}{\sim} A \vee \neg A) & \text{2, } \Box I\text{, A8, N6} \\
(4) & \mathrm{T} \to {\sim}\Box{\sim} A \vee \Box\neg A & \text{3, A2, A12} \\
(5) & \neg\Box\neg A \to {\sim}\Box{\sim} A & \text{4, N3, N4.}
\end{array}$$

I.e., by D$\Diamond$ and D$\Diamonddot$, $\Diamond A \to \Diamonddot A$.
Right to left:

$$\begin{array}{lll}
(1) & \mathrm{T} \to \neg{\sim} A \vee {\sim} A & \text{A1} \\
(2) & \mathrm{T} \to {\sim}\neg A \vee {\sim} A & \text{1, N1} \\
(3) & \mathrm{T} \to \Box({\sim}\neg A \vee {\sim} A) & \text{2, } \Box I\text{, A8, N6} \\
(4) & \mathrm{T} \to {\sim}\Box\neg A \vee \Box{\sim} A & \text{3, A12} \\
(5) & \Box\neg A \wedge {\sim}\Box{\sim} A \to F & \text{4, A6, N8} \\
(6) & {\sim}\Box{\sim} A \to \neg\Box\neg A & \text{5, N2, N5.}
\end{array}$$

I.e., by D$\Diamond$ and $D\Diamonddot$, $\Diamonddot A \to \Diamond A$.

Now, as promised, we give our

Simple Semantic Completeness Theorem for CR4. *Let A be simply* **CR4**-*valid. Then A is a theorem of* **CR4**.

Proof. As of the corresponding theorem for **CNR**. The only substantive point that requires comment is the demonstration that a canonical model structure built to reject an arbitrary non-theorem A of **CR4** satisfies the semantic postulate P12. Suppose then that S_0xy holds for arbitrary x and y in $\mathbf{K}_0 = \langle P_0, K_0, R_0, S_0, {}^*{}_0\rangle$, where 0 is a regular prime **CR4**-theory. We must show that $S_0x^*y^*$ holds as well. Let $\Box A \in x^*$. We must show that $A \in y^*$. Suppose not. Then ${\sim} A \in y$. Since S_0xy, $\Box{\sim} A \in x$. By the equivalence of the two kinds of possibility, $\Diamonddot{\sim} A \in x$. This contradicts $\Box A \in x^*$ and completes the argument that $\mathbf{K}_0$ is a CR4ms. All else being as before, the simple semantic completeness theorem for **CR4** is proved.

Corollary. *The following conditions are equivalent:*

(i) *A is a theorem of* **CR4**
(ii) *A is simply* **CR4**-*valid*
(iii) *A is* **CR4**-*valid.*

Proof. As above.

VII

In this section we conclude. Among open problems, the interesting one of conservative extension discussed above remains. Is **CR4** a conservative extension of **R4**? If not, what axioms should be added to **R4** to produce the Boolean-free fragment of **CR4**? For that matter, what is a semantics (if not the present one) on which **R4** is complete? On the **CNR** side, what is its Boolean-free fragment? All we know for sure about the non-Boolean fragment of **CNR** is that it is a *proper* extension of both **NR** and **S4**; and that the rule form of A12, namely,

$$\Box(\sim A \vee B) \vdash \sim\Box A \vee \Box B,$$

is admissible. At least, added to **NR**, this rule delivers all **S4** theorems (e.g., by translating the Gentzen system for **S4** of Curry (1963) into the **NR** vocabulary). Does this rule suffice?

More positively, there are pleasant echoes in the model theory of **CNR** of the Fine (1988) semantics for **RQ** and other first order relevant logics. For Fine, in order to accommodate quantifiers, had a 'galactic' picture of an RQms like that presented here for CNRms. In both cases, each horizontal slice through a model structure produces a model structure for the unenriched part of the logic (the propositional logic **R**, near enough). Vertical slices, for Fine, care for quantifiers; here, they care for modal operators. It would be interesting to combine the Fine semantics with the present ones to find a topological interpretation of the quantifiers, perhaps along the lines that provide topological understandings of intuitionist quantifiers on McKinsey-Tarski style analysis.

Finally, we hope the reader has found, as we have, that adding Boolean negation to relevant logics is useful and illuminating. This has certainly proved true at the technical level. For example, we now know that every non-theorem of **R** is refutable in a CRms (equivalently, a CRps, that is, a De Morgan monoid which is also a Boolean algebra). But we think it illuminating at a philosophical level as well. For, even if one rejects Boolean $\neg$ as an intuitive counterpart of 'not', preferring De Morgan $\sim$ instead, $\neg$ is still a connective to which we can assign a clear sense. If this makes sense in English (and many are prepared to die to defend that assertion that it does), fine! If not, as van Fraassen once said of relevant implication, it makes sense in English now![17]

---

17 We should like to thank various people with whom we have had conversations on these topics over the years. Among them are J. M. Dunn, N. D. Belnap Jr., John Slaney, Kit Fine, M. A. McRobbie, E. P. Martin, R. A. Girle, Alasdair Urquhart, Chris Brink, Richard Sylvan (ex Routley), Val Plumwood (ex Routley), Ross T. Brady, Paul

References

A.R. Anderson and N.D. Belnap Jr. (1975), *Entailment: The Logic of Relevance and Necessity, Vol. 1*, Princeton University Press, Princeton.

C. Brink (1989), '$R^{\neg}$-Algebras and $R^{\neg}$-Model Structures as Power Constructs.' *Studia Logica 48*, 85–109.

K. Fine (1988), 'Semantics for Quantified Relevant Logic.' *Journal of Philosophical Logic 17*, 27–59.

S. A. Kripke (1963), 'Semantical Analysis of Modal Logic I: Normal Modal Propositional Calculi.' *Zeitschrift für mathematische Logik und Grundlagen der Mathematik 9*, 113–116.

L. L. Maksimova (1973), 'A Semantics for the Calculus E of Entailment.' *Bulletin of the Section of Logic, Polish Academy of Sciences 2*, 18–21.

E. D. Mares and R. K. Meyer (1992), 'The Admissibility of γ in R4.' *Notre Dame Journal of Formal Logic 33*, 197–206.

R.K. Meyer (1968), 'Entailment and Relevant Implication.' *Logique et Analyse (N.S.) 11*, 472–479.

——— (1973), 'Intuitionism, Entailment, Negation.' in H. Leblanc (ed.) *Truth, Syntax and Modality*, North-Holland, Amsterdam, 168–198.

——— (1976), 'A Boolean-Valued Semantics for R.' Research Paper No. 4, Logic Group, Department of Philosophy (RSSS), The Australian National University.

——— (1979), 'Sentential Constants in R.' Research Paper No. 2, Logic Group, Department of Philosophy (RSSS), The Australian National University.

R.K. Meyer and R. Routley (1973), 'Classical Relevant Logic I.' *Studia Logica 32*, 51–68.

——— (1974), 'Classical Relevant Logic II.' *Studia Logica 33*, 183–194.

R. Routley and R.K. Meyer (1972), 'The Semantics of Entailment (II).' *Journal of Philosophical Logic 1*, 192–208.

——— (1973), 'The Semantics of Entailment (I).' in H. Leblanc (ed.) *Truth, Syntax and Modality*, North-Holland, Amsterdam, 199–243.

R. Routley, R. K. Meyer, V. Plumwood, and R. T. Brady (1982), *Relevant Logics and Their Rivals*, Ridgeview, Atascadero.

Thistlewaite, Jill LeBlanc, and John Bacon. (One or more of these people is responsible for any mistakes in this paper.) We should also like to thank the Automated Reasoning Project of The Australian National University, and the Research School of Social Sciences of which it is a part, for support for our research. Mares was financially supported by fellowship 456-89-0128 of The Social Sciences and Humanities Research Council of Canada.

Semantics for Substructural Logics

Hiroakira Ono
Department of Applied Mathematics
Faculty of Engineering, Hiroshima University

1 Introduction

In this paper, we will discuss algebraic and Kripke-type semantics for logics lacking some or all structural rules, which are now called *substructural logics* (for general information on substructural logics, see, for example, Došen [6], Ono [23] and Troelstra [33]). Various kinds of semantics for substructural logics have been introduced and studied in recent years; for instance, (intuitionistic) *phase structures* (Girard [9], Abrusci [1]), *pretopologies* (Sambin [31], [32]), *quantales* (Rosenthal [27], [28], Yetter [34]), Kripke-type semantics (Došen [4], [5], Ono-Komori [26], Ono [22]) and *coherence semantics* (Girard [9], also Troelstra [33]).

In a previous paper [25], we tried to clarify relations between these semantics and obtained some partial results. After completing [25], we learned of a paper by Yetter on quantale semantics [34], which became the starting point of the present study. The notion of quantale has been first introduced by Mulvey [20], as a generalization of *locale*, or complete Heyting algebra. It is easy to see that a (unital) quantale is nothing but a complete FL-algebra introduced in [25]. However, the simplicity of the definition has led us to a deeper understanding of relations between various semantics discussed in [25], by which we can grasp the essential meaning of completeness theorems with respect to these semantics. We have finally reached the conclusion that quantales, (intuitionistic) phase structures, pretopologies and complete so-monoids ([22] and [25]) are essentially the same thing. Moreover, we found that the Dedekind-MacNeille completion, which is a standard way of completing a given lattice, will give us a uniform way of establishing not only these results, but also completeness theorems for substructural logics with respect to (one of, and hence, all of) the semantics determined by these structures. (See [10] and [12] for the Dedekind-MacNeille completion.) This is the main result of the present paper (cf. [3], [27] and [28] for similar results on linear logic).

In §2, we will introduce the formal systems of substructural logics discussed in this paper. In §3, we will introduce the basic algebraic structures, called FL-algebras, which corresponds to the substructural logic FL given in §2. Then we will introduce also quantales and phase structures, which are special kinds of FL-algebras. They are the basis of quantale semantics and phase semantics respectively. We will show our main results (Lemma 4.5, Theorems 4.6 and 4.7) in §4. In fact, as corollaries of Lemma 4.5, we will show that every unital quantale is isomorphic to a phase structure (Theorem 4.6) and that every basic substructural logic is complete with respect to the phase semantics (Theorem 4.7). These results can be obtained by using the Dedekind-MacNeille completion and the close relation between complete lattices and closure systems. Then these results will be extended not only to *classical* substructural logics, like linear logic (§5), but also to substructural logics with the modal operators ! and ? (§6). In §7, we will explain first how Kripke-type semantics for substructural logics are related to algebraic semantics, and then, by using this explanation, we will introduce Kripke-type semantics for substructural propositional logics with modal operators, for which completeness theorems hold.

The first version of this paper was announced at the Conference on Logic and Computer Science (Marseille, June, 1990). After the talk, L. Esakia pointed out that our embedding results are closely related to the Dedekind-MacNeille completion of lattices. His suggestion helped us to refine our results and simplify proofs, and the paper was reorganized as [24]. Then the author delivered a talk based on [24] at the Conference on Logics with Restricted Structural Rules (Tübingen, October, 1990). This is the final version of our paper, which was revised once again after the conference. The author would like to express heartily gratitude to J. van Benthem, K. Došen, M. Dunn, L. Esakia, R. Hori, R.K. Meyer and M. Shimoda for their pertinent advices and helpful comments, and to I. Kubo for his invaluable help in completing the paper in the present form.

2 Formal Systems of Substructural Logics

In this section, we will introduce various logics that will be discussed in this paper. The basic system is called FL (*full Lambek logic*), which, roughly speaking, is the Gentzen-type sequent calculus obtained from the sequent calculus LJ for intuitionistic logic by deleting all the structural rules. In this paper, we call any extension of FL a *substructural logic*.

Next we will give a precise definition of sequent calculi for some basic substructural logics. (The description of formal systems given in [23] contains some errors; a corrected definition will be given below.) We assume that our language $\mathcal{L}$ consists of the propositional constants 1, 0, $\top$ and $\bot$, the connectives $\vee$, $\wedge$, $*$ (multiplicative conjunction), $\supset$ (and also $\supset'$), and the quantifiers $\forall$ and $\exists$. The negation $\sim\alpha$ of a formula α is defined as an abbreviation of $\alpha \supset 0$. (In this paper we will follow the notation for propositional constants of [33]. So, *readers must be careful with our constants* 0 *and* $\bot$, which are exactly opposite not only to Girard's original notation from [9] but also to our old notation from [22], [23] and [24].) Sometimes, we assume that the language $\mathcal{L}$ contains also two modal operators ! and ? (when we treat logics with these modal operators; see §6). On the other hand, we assume that $\mathcal{L}$ contains neither function symbols nor individual constants.

A sequent of FL is an expression of the form $\Gamma \to \theta$, where Γ is a finite sequence of formulas and θ is a formula. Both Γ and θ may be empty. In the sequel, capital Greek letters will denote finite (possibly empty) sequences of formulas. Initial sequents of FL are of the following forms:

(1) $\alpha \to \alpha$,

(2) $\quad \to 1$,

(3) $0 \to \quad$,

(4) $\Gamma \to \top$,

(5) $\Gamma,\ \bot,\ \Delta \to \theta$.

Rules of inference of FL are the following cut rule:

$$\frac{\Gamma \to \alpha \qquad \Delta, \alpha, \Sigma \to \theta}{\Delta, \Gamma, \Sigma \to \theta}\ (cut)$$

and rules for propositional constants, connectives and quantifiers:

$$\frac{\Gamma, \Delta \to \theta}{\Gamma, 1, \Delta \to \theta}\ (1w) \qquad \frac{\Gamma \to}{\Gamma \to 0}\ (0w)$$

$$\frac{\Gamma, \alpha, \Delta \to \theta \qquad \Gamma, \beta, \Delta \to \theta}{\Gamma, \alpha \vee \beta, \Delta \to \theta}\ (\vee \to)$$

$$\frac{\Gamma \to \alpha}{\Gamma \to \alpha \vee \beta}\ (\to \vee 1) \qquad \frac{\Gamma \to \beta}{\Gamma \to \alpha \vee \beta}\ (\to \vee 2)$$

$$\frac{\Gamma, \alpha, \Delta \to \theta}{\Gamma, \alpha \wedge \beta, \Delta \to \theta}\ (\wedge 1 \to) \qquad \frac{\Gamma, \beta, \Delta \to \theta}{\Gamma, \alpha \wedge \beta, \Delta \to \theta}\ (\wedge 2 \to)$$

$$\frac{\Gamma \to \alpha \qquad \Gamma \to \beta}{\Gamma \to \alpha \wedge \beta}\ (\to \wedge)$$

$$\frac{\Gamma, \alpha, \beta, \Delta \to \theta}{\Gamma, \alpha * \beta, \Delta \to \theta}\ (* \to) \qquad \frac{\Gamma \to \alpha \qquad \Delta \to \beta}{\Gamma, \Delta \to \alpha * \beta}\ (\to *)$$

$$\frac{\Gamma \to \alpha \qquad \Delta, \beta, \Sigma \to \theta}{\Delta, \alpha \supset \beta, \Gamma, \Sigma \to \theta}\ (\supset\to) \qquad \frac{\Gamma, \alpha \to \beta}{\Gamma \to \alpha \supset \beta}\ (\to\supset)$$

$$\frac{\Gamma, \alpha(x), \Delta \to \theta}{\Gamma, \exists z \alpha(z), \Delta \to \theta}\ (\exists \to) \qquad \frac{\Gamma \to \alpha(y)}{\Gamma \to \exists z \alpha(z)}\ (\to \exists)$$

$$\frac{\Gamma, \alpha(y), \Delta \to \theta}{\Gamma, \forall z \alpha(z), \Delta \to \theta}\ (\forall \to) \qquad \frac{\Gamma \to \alpha(x)}{\Gamma \to \forall z \alpha(z)}\ (\to \forall)$$

As usual, the variable x in $(\exists \to)$ and $(\to \forall)$ must satisfy the *eigenvariable* condition.

Since the calculus FL lacks the exchange rule, it may be natural, and sometimes it is convenient, to introduce another kind of implication, written here as $\supset'$, for which we assume the following rules:

$$\frac{\Gamma \to \alpha \qquad \Delta, \beta, \Sigma \to \theta}{\Delta, \Gamma, \alpha \supset' \beta, \Sigma \to \theta}\ (\supset'\to) \qquad \frac{\alpha, \Gamma \to \beta}{\Gamma \to \alpha \supset' \beta}\ (\to\supset')$$

The substructural logics FL_e, FL_c and FL_w are obtained from FL by adding the exchange rule $(e \to)$, the contraction rule $(c \to)$ and the weakening rules $(w \to)$ and $(\to w)$, respectively. These are the following rules:

$$\frac{\Gamma, \alpha, \beta, \Delta \to \theta}{\Gamma, \beta, \alpha, \Delta \to \theta}\ (e \to) \qquad \frac{\Gamma, \alpha, \alpha, \Delta \to \theta}{\Gamma, \alpha, \Delta \to \theta}\ (c \to)$$

$$\frac{\Gamma, \Delta \to \theta}{\Gamma, \alpha, \Delta \to \theta}\ (w \to) \qquad \frac{\Gamma \to}{\Gamma \to \theta}\ (\to w)$$

Similarly, we can get various substructural logics by adding some of these structural rules to FL. We will express that by adding the corresponding letters e, c and w to FL as subscripts. For instance, FL_{ec} is the logic obtained from FL by adding both exchange and contraction.

It is clear that in FL_w the rules $(1w)$ and $(0w)$ become redundant. Moreover, since 1 (and 0) is equivalent to $\top$ (and $\bot$, respectively) in FL_w (i.e., each of the sequents $1 \to \top$, $\top \to 1$, $0 \to \bot$ and $\bot \to 0$ is provable), we can dispense with the constants 1 and 0, and the initial sequents for them.

We say a substructural logic L (in the language $\mathcal{L}$ with $\supset'$) is *classical* if for every formula α both $(\alpha \supset' 0) \supset 0 \to \alpha$ and $(\alpha \supset 0) \supset' 0 \to \alpha$ are provable in it. When the exchange rule is admissible in L, this is equivalent to the condition that $\sim\sim\alpha \to \alpha$ is provable for every formula α.

We will now introduce some basic classical substructural logics. The logic GL (*Girard's linear logic* without modal operators) is defined as the Gentzen-type sequent calculus obtained from the sequent calculus LK for classical logic, by deleting both weakening and contraction rules. Of course, we must add initial sequents and rules for the propositional constants 1, 0, $\top$ and $\bot$, and the connective $*$. Moreover, we need to modify slightly initial sequents and rules of FL since in GL sequents of the form $\Gamma \to \Delta$ are allowed, where both Γ and Δ are arbitrary finite sequences of formulas. For example, $(0w)$ of GL becomes

$$\frac{\Gamma \to \Sigma}{\Gamma \to \Sigma, 0}$$

(for details, see e.g. [34]). It is easy to see that GL is equal to the logic obtained from FL_e by adding $\sim\sim\alpha \to \alpha$ as an initial sequent for every formula α. We remark that 1 and $0 \supset 0$ are equivalent in GL. Thus, we can eliminate the constant 1, its initial sequent and the rule $(1w)$.

In the same way, we can define the logic GL^- as the minimal classical substructural logic, i.e. the logic obtained from FL by adding both $(\alpha \supset' 0) \supset 0 \to \alpha$ and $(\alpha \supset 0) \supset' 0 \to \alpha$ as initial sequents. We can define GL^-_σ and GL_σ analogously, where σ is made of e, c and w. It is clear that $\mathrm{GL}^-_{e\sigma}$ is equal to GL_σ. Other well-known classical substructural logics are relevant logics. For instance, the relevant logic R (and R^+) is the logic obtained from GL_c (and the positive fragment of FL_{ec}, respectively) by adding the distributive law $\alpha\wedge (\beta \vee \gamma) \to (\alpha \wedge \beta) \vee (\alpha \wedge \gamma)$ (see e.g. [2] for details).

3 Full Lambek Algebras

Next we will introduce basic algebraic structures, called *full Lambek algebras*, which in the sequel we abbreviate to *FL-algebras*:

Definition 3.1. A structure $\mathbf{A} = \langle A, \to, \cup, \cap, *, 1, 0, \top, \bot\rangle$ is an *FL-algebra* if

(1) $\langle A, \cup, \cap, \top, \bot\rangle$ is a lattice with the least element $\bot$ and the greatest element $\top$ for which $\top = \bot \to \bot$ holds,

(2) $\langle A, *, 1\rangle$ is a monoid with the identity 1,

(3) $z * (x \cup y) * w = (z * x * w) \cup (z * y * w)$, for every $x, y, z, w \in A$,

(4) $x * y \leq z$ iff $x \leq y \to z$, for every $x, y, z \in A$,

(5) $0 \in A$.

We can show easily that $x * \bot \leq y$ holds for every x and y by using (1), (3) and (4). As the algebraic counterpart of the implication $\supset'$, we will sometimes introduce $\to'$ as an additional binary operation on FL-algebras, which must satisfy

(4′) $y * x \leq z$ iff $x \leq y \to' z$, for every $x, y, z \in A$.

Let $\mathbf{A}$ be an FL-algebra. Then we say that

(1) $\mathbf{A}$ is an *FL_e-algebra* if $\langle A, *, 1\rangle$ is a commutative monoid;

(2) $\mathbf{A}$ is an *FL_w-algebra* if (i) $0 = \bot$ and (ii) $x * y \leq x$ and $y * x \leq x$ for every x, $y \in A$;

(3) $\mathbf{A}$ is an *FL_c-algebra* if $x \leq x * x$ for every $x \in A$.

We can also define other kinds of FL-algebras in a trivial way. For example, $\mathbf{A}$ is an FL_{ec}-algebra if and only if it satisfies conditions for both FL_e-algebras and FL_c-algebras. Note that an FL_{ecw}-algebra is exactly a Heyting algebra, in which $x * y = x \cap y$ holds. We say that

(4) an FL-algebra $\mathbf{A}$ (with $\to'$) is *classical* if $(x \to 0) \to' 0 = x = (x \to' 0) \to 0$ for every $x \in A$.

For every element x of an FL-algebra, define $\sim x$ by $\sim x = x \to 0$. Then, when $\mathbf{A}$ is an FL_e-algebra, this last condition is equivalent to the condition that $\sim\sim x = x$ for every x. Sometimes, we call a classical FL-algebra (and a classical FL_e-algebra) a *GL^--algebra* (and a *GL-algebra*, respectively). The index σ denotes either the empty string, or one of e, c and w, or one of their combinations.

We remark that a GL_w-algebra is exactly a *latticed L^0-algebra* of [11] and that a GL_c-algebra satisfying the distributive law $a \cap (b \cup c) = (a \cap b) \cup (a \cap b)$ is a *De Morgan monoid*. Also, any distributive FL_{ec}-algebra (without assuming the existence of the least element $\bot$) is a *Dunn monoid*. (For more information, see [2], [7], [13], [14], [16], [17] and [18]; in particular, [14] and [18] for general discussions about various algebraic structures related to relevant logics.)

An FL-algebra $\mathbf{A}$ is said to be *complete* when $\langle A, \cup, \cap\rangle$ is a complete lattice, in which $y * (\bigcup_i x_i) * z = \bigcup_i (y * x_i * z)$ holds for every x_i, y, $z \in A$. It is easy to see that a complete FL_{ecw}-algebra is a complete Heyting algebra or a *locale* (for complete Heyting algebras, see e.g. [8]). Also, a complete distributive FL_{ec}-algebra is a complete Dunn monoid, and if it is moreover classical, it is a complete De Morgan monoid (see [19] and [29]). We will give here an alternative definition of complete FL-algebras, following Mulvey [20] and Niefield-Rosenthal [21].

Definition 3.2. A structure $\mathbf{B} = \langle B, \bigcup, * \rangle$ is a *quantale* if $\langle B, \bigcup \rangle$ is a complete lattice with the infinite join $\bigcup$, and $\langle B, * \rangle$ is a semigroup in which both $y * (\bigcup_i x_i) = \bigcup_i (y * x_i)$ and $(\bigcup_i x_i) * y = \bigcup_i (x_i * y)$ hold for every x_i, $y \in B$. A quantale $\langle B, \bigcup_i, * \rangle$ is *unital* if $\langle B, * \rangle$ has moreover an identity.

As mentioned in [20] and [21], there are many examples of quantales. For instance, the set of all left (right and two-sided) ideals of a given ring and the set of all left (right and two-sided) ideals of a given C^*-algebra are quantales. For more information on quantales, see [28].

It is easily shown that a complete FL-algebra is just a unital quantale with the greatest element $\top$, the least element $\bot$, the identity 1 and a designated element 0, in which $y \to z$ is defined by $y \to z = \bigcup\{x : x * y \leq z\}$ (and $y \to' z$ is defined by $y \to' z = \bigcup\{x : y * x \leq z\}$). We can prove that a complete GL-algebra is a commutative unital classical quantale, which amounts to saying that it is a commutative unital *Girard quantale* (for Girard quantales, see [28] and [34]). In the rest of this paper, we will concentrate mainly on unital quantales (with a fixed element 0), identifying them with complete FL-algebras. Sometimes we will call a complete FL_σ-algebra simply a *unital σ-quantale*.

Let f be a mapping from an FL-algebra $\mathbf{A}$ to another FL-algebra $\mathbf{B}$. Then f is an *FL-homomorphism* if f preserves the operations $\to$, $\cup$, $\cap$, $*$ and the constants 1, 0, $\top$, $\bot$ in $\mathbf{A}$. When both $\mathbf{A}$ and $\mathbf{B}$ are moreover unital quantales, we say that f is a *quantale homomorphism* if f preserves the operations $\bigcup$, $*$ and the constant 1. Clearly, any FL-homomorphism that preserves $\bigcup$ is a quantale homomorphism, but the converse doesn't hold always. On the other hand, any quantale isomorphism (preserving 0) is an FL-isomorphism that preserves $\bigcup$.

In [1], quantales of a special type, called *intuitionistic phase structures*, are introduced, which are intuitionistic forms of *phase structures* introduced by Girard [9]. They are generalized to FL phase structures in [25]. We will define these below.

For a given FL-algebra $\mathbf{A} = \langle A, \to, \cup, \cap, *, 1, 0, \top, \bot \rangle$, a mapping C from A to A is a *closure operation* on $\mathbf{A}$ if

(1) $x \leq Cx$,

(2) $x \leq y$ implies $Cx \leq Cy$,

(3) $CCx \leq Cx$,

(4) $Cx * Cy \leq C(x * y)$.

An element x of A is *C-closed* if $x = Cx$ holds. Let $\mathcal{C}(A)$ be the set of all C-closed elements of A. Then we can show that for every FL_σ-algebra $\mathbf{A} = \langle A, \rightarrow, \cup, \cap, *, 1, 0, \top, \bot \rangle$ and every closure operation C on $\mathbf{A}$, the set $\mathcal{C}(A)$ of all C-closed elements of A forms an FL_σ-algebra. More precisely, if $\mathbf{A}$ is an FL_σ-algebra, then $\mathcal{C}(\mathbf{A}) = \langle \mathcal{C}(A), \rightarrow, \cup_C, \cap, *_C, C1, C0, \top, C\bot \rangle$ is also an FL_σ-algebra, where $\cup_C$ and $*_C$ are defined by

$$x \cup_C y = C(x \cup y) \quad \text{and} \quad x *_C y = C(x * y).$$

Note that a similar result has been obtained already in [21], where a closure operation C on a given quantale is called a *quantic nucleus*. It is proved in [21, Theorem 2.1] that for every quantale $\mathbf{A}$ with a quantic nucleus C, the set $\mathcal{C}(\mathbf{A})$ of all C-closed elements of A forms a quantale with respect to $\bigcup_C$ and $*_C$, where $\bigcup_C x_i$ is defined by $\bigcup_C x_i = C(\bigcup x_i)$. We remark here that the following equations hold for every closure operation C:

(1) $C(\bigcup_i Cx_i) = C(\bigcup_i x_i)$,

(2) $C(\bigcap_i Cx_i) = \bigcap_i Cx_i$,

(3) $C(Cx * Cy) = C(x * y)$.

Suppose that a monoid $\langle M, *, \varepsilon \rangle$ with the identity ε is given. For every $X, Y \in \mathcal{P}(M)$, define $X \cdot Y$ and $X \Rightarrow Y$ by

$$X \cdot Y = \{x * y \in M : x \in X \text{ and } y \in Y\},$$
$$X \Rightarrow Y = \{z \in M : z * x \in Y, \text{ for every } x \in X\}.$$

Then it can be shown that for every monoid $\mathbf{M} = \langle M, *, \varepsilon \rangle$, the structure $\mathcal{P}(\mathbf{M}) = \langle \mathcal{P}(M), \bigcup, \cdot \rangle$ ($= \langle \mathcal{P}(M), \Rightarrow, \cup, \cap, \cdot, \{\varepsilon\}, O, M, \varnothing \rangle$ for a fixed O ($\subset M$)) forms a unital quantale. Combining this with the result above, we have that $\mathcal{C}(\mathcal{P}(\mathbf{M}))$ is a unital quantale for every monoid $\mathbf{M}$ and every closure operation C on $\mathcal{P}(\mathbf{M})$. When $\mathbf{M}$ is commutative, $\mathcal{C}(\mathcal{P}(\mathbf{M}))$ becomes an FL_e-algebra, called in [1] an *intuitionistic phase structure*. Following this usage, an FL-algebra (and, in general, an FL_σ-algebra) of the form above is called in [25] an *FL phase structure* (and *FL_σ phase structure*, respectively). It is noted in [25] that intuitionistic phase structures are essentially equivalent to the *pretopologies* introduced by Sambin [31].

4 Quantales and Phase Structures

In this section we will show two embedding results for FL-algebras (Lemma 4.5 and Theorem 4.6). Lemma 4.5 says that every FL-algebra **A** (under some conditions, which are satisfied in particular by the Lindenbaum algebra of the predicate logic FL) can be embedded into an FL phase structure. Theorem 4.6 says that when **A** is moreover complete, the embedding above becomes surjective and hence every unital quantale is isomorphic to an FL phase structure. From these two results, we obtain various types of completeness theorems for substructural logics (Theorem 4.7).

These results can be obtained by applying the Dedekind-MacNeille completion, which is a standard technique for completing a given lattice (for the Dedekind-MacNeille completion, see e.g. [10], [15] and [35]; see also the discussion in §3 of [3]).

It seems quite natural to state our results as extensions of results mentioned in §2.2 of [15], where relations between complete lattices and closure systems are discussed in a general setting. So we will refer to [15] for some definitions and theorems. Let K and L be any sets and let R be a subset of $K \times L$. We will define two functions $\rightarrow$ and $\leftarrow$, such that $\rightarrow$ is a function from $\mathcal{P}(K)$ to $\mathcal{P}(L)$ and $\leftarrow$ a function from $\mathcal{P}(L)$ to $\mathcal{P}(K)$, by

$$\begin{aligned} U^{\rightarrow} &= \{x \in L : uRx, \text{ for every } u \in U\}, \quad \text{for every } U \subset K, \\ X^{\leftarrow} &= \{v \in K : vRx, \text{ for every } x \in X\}, \quad \text{for every } X \subset L. \end{aligned}$$

Then we have the following proposition, which says that the pair $\rightarrow$ and $\leftarrow$ is a Galois connection (see Theorem 2.21 of [15] and also [3]):

Proposition 4.1. *Let K and L be sets and $R \subset K \times L$. Let $\rightarrow$ and $\leftarrow$ be functions defined as above. Then for every subset U and V of K and every subset X and Y of L:*

(1) $U \subset (U^{\rightarrow})^{\leftarrow}$ *and* $X \subset (X^{\leftarrow})^{\rightarrow}$,

(2.1) $U \subset V$ *implies* $V^{\rightarrow} \subset U^{\rightarrow}$,

(2.2) $X \subset Y$ *implies* $Y^{\leftarrow} \subset X^{\leftarrow}$,

(3) $U^{\rightarrow} = ((U^{\rightarrow})^{\leftarrow})^{\rightarrow}$ *and* $X^{\leftarrow} = ((X^{\leftarrow})^{\rightarrow})^{\leftarrow}$,

(4) $\rightarrow\leftarrow$ *and* $\leftarrow\rightarrow$ *are closure operations (in the usual sense) on K and on L, respectively (i.e. both $\rightarrow\leftarrow$ and $\leftarrow\rightarrow$ satisfy conditions* (1) *to* (3) *for closure operations stated in* §3).

Let L be a lattice, and take L for both L and K and $\leq$ for R in the proposition above. Then we can define an operation $*$ on $\mathcal{P}(L)$ by

$U^* = (U^{\rightarrow})^{\leftarrow}$ for every subset U of L. By the proposition above, $*$ is a closure operation. Moreover, we can show that U^* is an ideal of L. We say that a nonempty subset V of L is an *ideal* of L if for every a, $b \in V$ and every $c \in L$, $c \leq a \cup b$ implies $c \in V$. It is clear that the set $\{x \in L : x \leq c\}$ for an element c of L is an ideal of L. Such an ideal is called the *principal ideal* of L generated by c. We can show easily that if U is a principal ideal of L, then U is $*$-closed and that the converse holds if and only if L is complete. Let $C(L)$ be the set of all $*$-closed subsets of L and I_a the principal ideal of L generated by an element $a \in L$. Then $C(L)$ forms a complete lattice and the mapping f from L to $C(L)$ defined by $f(a) = I_a$, for every $a \in L$, is an embedding of L into $C(L)$ that preserves all existing infinite meets and joins in L. The complete lattice $C(L)$ is called the *Dedekind-MacNeille completion* of the lattice L. We can show that when L is a complete lattice, $C(L)$ is isomorphic to L by the mapping f. Thus we have shown the following (see Theorem 2.12 in [15]):

Proposition 4.2. *Every complete lattice is isomorphic to a lattice of all closed subsets for some closure operation (in the usual sense).*

In the sequel, we will follow the argument above, extending these results to FL_σ-algebras. Suppose that an FL-algebra $\mathbf{A} = \langle A, \rightarrow, \cup, \cap, *, 1, 0, \top, \bot \rangle$ is given. Then it is obvious that the reduct $\mathbf{A}^* = \langle A, *, 1 \rangle$ is a monoid, and therefore $\mathcal{P}(\mathbf{A}^*)$ forms a unital quantale, as shown in the previous section. Let us define a binary relation R between A and $A \times A$ (i.e. $R \subset A \times (A \times A)$) by

$$(1) \quad uR(x, y) \quad \text{if and only if} \quad x * u \leq y$$

for every u, x, $y \in A$. Then, by Proposition 4.1, $\rightarrow\leftarrow$ is a closure operation (in the usual sense) on the set $\mathcal{P}(\mathbf{A}^*)$. Moreover, we can show the following:

Lemma 4.3. *Let* $\mathbf{A}$ *be an FL-algebra. Then, for every subset* U *and* V *of* A, $((U^{\rightarrow})^{\leftarrow}) \cdot ((V^{\rightarrow})^{\leftarrow}) \subset (((U \cdot V)^{\rightarrow})^{\leftarrow})$ *holds. Therefore, the operation* M *on* $\mathcal{P}(\mathbf{A}^*)$ *defined by* $MU = (U^{\rightarrow})^{\leftarrow}$, *for every* $U \in \mathcal{P}(A)$, *is a closure operation on the FL-algebra* $\mathcal{P}(\mathbf{A}^*)$.

Proof. Suppose that $z \in ((U^{\rightarrow})^{\leftarrow}) \cdot ((V^{\rightarrow})^{\leftarrow})$. Then $z = u * v$ for some $u \in (U^{\rightarrow})^{\leftarrow}$ and $v \in (V^{\rightarrow})^{\leftarrow}$. Take any $(x, y) \in (U \cdot V)^{\rightarrow}$. By definition, $x * (u' * v') \leq y$ holds for every $u' \in U$ and $v' \in V$. Using the associativity of $*$, $(x * u', y) \in V^{\rightarrow}$ follows from this. Therefore, $(x * u') * v \leq y$ since $v \in (V^{\rightarrow})^{\leftarrow}$. Thus, $x * u' \leq v \rightarrow y$ for every $u' \in U$, and hence $(x, v \rightarrow y) \in U^{\rightarrow}$. Since $u \in (U^{\rightarrow})^{\leftarrow}$, $x * u \leq v \rightarrow y$ and thus $x * (u * v) = (x * u) * v \leq y$ for every $(x, y) \in (U \cdot V)^{\rightarrow}$. This means that $z = u * v \in ((U \cdot V)^{\rightarrow})^{\leftarrow}$.

Corollary 4.4. *Let* $\mathbf{A} = \langle A, \rightarrow, \cup, \cap, *, 1, 0, \top, \bot \rangle$ *be an* FL_σ*-algebra for* $\sigma \in \{e, c, w\}$ *and* M *the closure operation introduced in Lemma* 4.3. *Then* $\mathcal{M}(\mathcal{P}(\mathbf{A}^*))$ $(= \langle \mathcal{M}(\mathcal{P}(A)), \Rightarrow, \cup_M, \cap, \cdot_M, M\{1\}, M\{0\}, A, M\varnothing \rangle)$ *is an* FL_σ *phase structure.*

Proof. It suffices to show that if $\mathbf{A}$ is an FL_σ-algebra, then so is $\mathcal{M}(\mathcal{P}(\mathbf{A}^*))$ for every σ. It is clear that if $*$ is commutative then so is $\cdot_M$. Suppose that $\mathbf{A}$ is an FL_w-algebra. Then (1) $0 = \bot$ and (2) both $u * v \leq u$ and $u * v \leq v$ hold for every u and v. Since $x * \bot \leq y$ holds always, we can show that $M\varnothing = M\{\bot\}$ in every FL-algebra. Now, from $0 = \bot$ it follows immediately that $M\varnothing = M\{0\}$. Next we will show that $MU \cdot_M MV \subset MU$ and $MU \cdot_M MV \subset MV$ for every subset U and V of A. Take any $(x, y) \in U^{\rightarrow}$, any $u \in U$ and any $v \in V$. Then $x * u \leq y$ and hence $x * u * v \leq y$ since $u * v \leq u$. Thus, $(x, y) \in (U \cdot V)^{\rightarrow}$. Therefore, $U^{\rightarrow} \subset (U \cdot V)^{\rightarrow}$. Thus, by using Proposition 4.1 (2.2), $M(U \cdot V) \subset MU$. On the other hand, $MU \cdot_M MV \subset M(U \cdot V)$ follows from Lemma 4.3. Thus, $MU \cdot_M MV \subset MU$. Similarly, $MU \cdot_M MV \subset MV$. Thus, $\mathcal{M}(\mathcal{P}(\mathbf{A}^*))$ is also an FL_w-algebra. In the same way, we can show our corollary when σ is c.

The key element of the construction above is that though $\mathbf{A}^*$ is simply a monoid, it is the reduct of a partially-ordered monoid $\mathbf{A}$, and thus the Dedekind-MacNeille completion can be applied to it. The idea of using the binary relation R in (1) comes originally from the method developed in [26, §8] (see also [25]). We remark that if $\mathbf{A}$ is an FL_e-algebra, i.e., if $*$ is commutative, then for every subset U of A, the set MU is equal to the set U^* mentioned just below Proposition 4.1. Hence, $\mathcal{M}(\mathcal{P}(\mathbf{A}^*))$ is just the Dedekind-MacNeille completion of the lattice $\mathbf{A}$. This can be verified as follows. Let U be a subset of A. By definition, an element a of A belongs to U^* if and only if

(1) for every $z \in A$, if $u \leq z$ for every $u \in U$, then $a \leq z$,

and a belongs to MU if and only if

(2) for every x, $y \in A$, if $x * u \leq y$ for every $u \in U$, then $x * a \leq y$.

Suppose that (1) holds and that $x * u \leq y$ for every $u \in U$. Then $u * x \leq y$ and $u \leq x \rightarrow y$ for every $u \in U$. So, taking $x \rightarrow y$ for z in (1), we have $a \leq x \rightarrow y$. Thus, $x * a = a * x \leq y$. Conversely, (1) follows from (2) by taking 1 for x and z for y.

It will be easily seen from the argument above that when we have another implication $\rightarrow'$ in our language, we can replace our construction in Lemma 4.3 by the original Dedekind-MacNeille completion even in the case of noncommutative FL-algebras. We remark also that if we define a relation $\leq'$ between elements of $\mathbf{A}$ and subsets of $\mathbf{A}$ by

$$a \leq' U \quad \text{if and only if} \quad a \in MU,$$

then $\leq'$ is a *precover*, which determines a *pretopology* on $\mathcal{P}(\mathbf{A}^*)$ (see [31]).

The following lemma is an extension of Proposition 4.2:

Lemma 4.5. *Let* $\mathbf{A}$ *be an* FL_σ*-algebra and* M *the closure operation on* $\mathcal{P}(\mathbf{A}^*)$ *defined in Lemma* 4.3. *Define a mapping* h *from* $\mathbf{A}$ *to* $\mathcal{M}(\mathcal{P}(\mathbf{A}^*))$ *(in Corollary* 4.4) *by* $h(a) = M\{a\}$ $(= (\{a\}^{\rightarrow})^{\leftarrow})$, *for every* $a \in A$. *Then* h *is an FL-embedding that preserves all existing infinite meets and all existing infinite joins* $\bigcup a_i$ *in* A *satisfying* $x * (\bigcup a_i) = \bigcup (x * a_i)$ *for every* x *in* A. *If* $\mathbf{A}$ *is moreover complete, then every element of* $\mathcal{M}(\mathcal{P}(\mathbf{A}^*))$ *is a principal ideal of* $\mathbf{A}$, *and hence* h *is surjective.*

Proof. First, we will show that $h(a)$ is equal to the principal ideal I_a generated by a, for every $a \in A$. Suppose that $z \in h(a)$. By definition, this is equivalent to the condition that

(1) for every $x, y \in A$, if $x * a \leq y$, then $x * z \leq y$.

Taking 1 for x and a for y, $z \leq a$ follows from (1). Conversely, (1) follows from $z \leq a$. Thus, $z \in h(a)$ if and only if $z \leq a$. This means that $h(a) = I_a$. Now it is clear that the mapping h is injective.

Next we will show that h is an FL-homomorphism. To show that $h(a) \cdot_M h(b) = h(a * b)$, it is enough to prove that $M(MU \cdot MV) = M(U \cdot V)$ for every subset U and V of A. But, for every closure operation, this always holds, as shown in §3. We will show next that $h(a) \Rightarrow h(b) = h(a \rightarrow b)$. Of course, this is equivalent to $I_a \Rightarrow I_b = I_{a \rightarrow b}$. Now, $z \in I_a \Rightarrow I_b$ if and only if

(2) $z * x \leq b$, for every $x \leq a$.

Taking a for x in (2), we have $z \leq a \rightarrow b$. Conversely, if $z \leq a \rightarrow b$ and $x \leq a$, then $z * x \leq (a \rightarrow b) * a \leq b$. Thus $I_a \Rightarrow I_b = I_{a \rightarrow b}$ holds. It is easy to see that h satisfies the other conditions for FL-homomorphisms.

We will show that whenever $\bigcap a_i$ exists in A, $h(\bigcap a_i) = \bigcap h(a_i)$. For, $z \in \bigcap h(a_i) = \bigcap I_{a_i}$ if and only if $z \leq a_i$ for every i. But this holds if and only if $z \leq \bigcap a_i$, or equivalently $z \in h(\bigcap a_i)$. Finally, we will show that if $\bigcup a_i$ exists in A and satisfies $x * (\bigcup a_i) = \bigcup (x * a_i)$, then

(3) $h(\bigcup a_i) = \bigcup_M h(a_i)$.

As we noted in §3, $\bigcup_M MU_i = M(\bigcup U_i)$ for every set $\{U_i\}_i$ of subsets of A. Thus, $\bigcup_{M\, i\in I} h(a_i) = M(\{a_i : i \in I\})$. Now, $z \in M(\{a_i : i \in I\})$ if and only if

(4) for every $x,\ y \in A$, if $x * a_i \leq y$ for every $i \in I$, then $x * z \leq y$.

By taking 1 for x and $\bigcup a_i$ for y in (4), $z \leq \bigcup a_i$ follows. On the other hand, if $x * a_i \leq y$ for every $i \in I$, then $x * (\bigcup a_i) = \bigcup(x * a_i) \leq y$ by using our assumption. So, if moreover $z \leq \bigcup a_i$, then $x * z \leq y$. Thus, (3) holds.

Next, we will show that $\mathcal{M}(\mathcal{P}(\mathbf{A}^*))$ is the set of all principal ideals of $\mathbf{A}$ when $\mathbf{A}$ is complete. Recall that every element of $\mathcal{M}(\mathcal{P}(\mathbf{A}^*))$ is of the form $(U^{\rightarrow})^{\leftarrow}$ for a subset U of A. But,

$$\begin{aligned} (x,y) \in U^{\rightarrow} \quad & \text{if and only if} \quad x * u \leq y \text{ for every } u \in U, \\ & \text{if and only if} \quad x * (\textstyle\bigcup_{u\in U} u) = \bigcup_{u\in U}(x * u) \leq y, \end{aligned}$$

using the fact that $\mathbf{A}$ is a quantale. Therefore, $z \in (U^{\rightarrow})^{\leftarrow}$ if and only if

(5) for every $x,\ y \in A$, if $x * (\bigcup_{u\in U} u) \leq y$, then $x * z \leq y$,

which is equivalent to $z \in h(\bigcup_{u\in U} u)$ (= the principal ideal generated by the element $\bigcup_{u\in U} u$) (see (1) above). Thus, h is surjective when $\mathbf{A}$ is complete.

We remark that when either $*$ is commutative or we have the operation $\rightarrow'$ in our language, the condition that $x * (\bigcup a_i) = \bigcup(x * a_i)$ for every x becomes redundant. In fact, if $x * a_j \leq y$ for every j, then $a_j \leq x \rightarrow' y$ for every j, and therefore $\bigcup a_i \leq x \rightarrow' y$. Thus, $x * (\bigcup a_i) \leq y$. On the other hand, it is clear that $x * a_j \leq x * (\bigcup a_i)$ for every j. Hence, $x * (\bigcup a_i) = \bigcup(x * a_i)$ holds always.

Lemma 4.5 has two important consequences. The first is algebraic and the second logical.

Theorem 4.6. (cf. [28; Theorem 3.1.2]) *Every unital σ-quantale is FL-isomorphic to an FL_σ phase structure.*

Proof. Take a unital σ-quantale for $\mathbf{A}$ in Lemma 4.5. Then our theorem follows from Corollary 4.4 and Lemma 4.5.

The second consequence of Lemma 4.5 is the completeness theorem for substructural logics of the form FL_σ. To state our result precisely, we will give here some definitions. A *structure for FL_σ* is a pair $\langle \mathbf{A},\ V\rangle$ made of a unital σ-quantale $\mathbf{A}$ and a nonempty set V called the *domain*. For every element v of V, we will introduce a new individual constant $\bar{v}$ called the *name* of v. Let $\mathcal{L}[V]$ denote the language obtained from our original language by adding the names of elements of V and $CF[V]$ denote the set of all closed formulas in $\mathcal{L}[V]$. An *assignment* f of $\langle \mathbf{A}, V \rangle$ is a mapping from the set of all atomic formulas in $CF[V]$ to A satisfying $f(1) = 1$, $f(0) = 0$, $f(\top) = \top$ and $f(\bot) = \bot$ (we use the same symbols for both logical constants and specified elements of A). As usual, every assignment can be extended to a mapping from the set $CF[V]$ to A. For example, $f(\alpha \supset \beta) = f(\alpha) \to f(\beta)$ and $f(\alpha * \beta) = f(\alpha) * f(\beta)$ etc.

We say that a formula α is *true* in an assignment f of $\langle \mathbf{A}, V \rangle$ if $1 \leq f(\widetilde{\alpha})$ holds in $\mathbf{A}$, where $\widetilde{\alpha}$ is the universal closure of α. Let $\mathcal{K}$ be a class of unital σ-quantales. A formula α is *valid* in $\mathcal{K}$ if it is true in every assignment of every structure $\langle \mathbf{A}, V \rangle$ for FL_σ such that $\mathbf{A} \in \mathcal{K}$. For every sequent $\Gamma \to \theta$, by $\langle \Gamma \to \theta \rangle$ we denote the universal closure of the formula $\Gamma^* \supset \theta'$ where Γ^* is $\alpha_1 * \ldots * \alpha_m$ when Γ is $\alpha_1, \ldots, \alpha_m$ and 1 when Γ is empty, and θ' is θ when θ is a formula and 0 when θ is empty. We say that a sequent $\Gamma \to \theta$ is *valid* in $\mathcal{K}$ if the formula $\langle \Gamma \to \theta \rangle$ is valid in $\mathcal{K}$. A substructural logic FL_σ is *complete* with respect to a class $\mathcal{K}$ of unital σ-quantales if (1) and (2) below are equivalent for every sequent $\Gamma \to \theta$:

(1) $\Gamma \to \theta$ is provable in FL_σ,

(2) $\Gamma \to \theta$ is valid in $\mathcal{K}$.

Now, we have the following theorem:

Theorem 4.7. ([1], [25], [31]) *For every σ, the substructural logic FL_σ is complete with respect to the class of all FL_σ phase structures (and equivalently, the class of all FL_σ pretopologies and the class of all unital σ-quantales).*

Proof. It is easy to see that if a sequent is provable in FL_σ, then it is valid in the class of all unital σ-quantales. To show the converse, let us consider the Lindenbaum algebra $\mathbf{LA}_\sigma$ of the predicate logic FL_σ. Of course, $\mathbf{LA}_\sigma$ is an FL_σ-algebra. The algebra $\mathbf{LA}_\sigma$ can be constructed in the usual way. More precisely, define a binary relation $\sim$ on the set $CF[V]$, where V is a fixed countably infinite set, by

$$\alpha \sim \beta \quad \text{if and only if both } \alpha \to \beta \text{ and } \beta \to \alpha \text{ are provable in } \mathrm{FL}_\sigma.$$

Then $\sim$ is an equivalence relation on $CF[V]$ and the quotient set $CF[V]/{\sim}$ forms naturally an FL_σ-algebra, which is denoted by $\mathbf{LA}_\sigma$. Remark that (1) the order $\leq$ of $\mathbf{LA}_\sigma$ is defined by

$$[\alpha] \leq [\beta] \quad \text{if and only if } \alpha \to \beta \text{ is provable in } \mathrm{FL}_\sigma,$$

where $[\gamma]$ denotes the equivalence class to which a formula γ belongs, and (2) $\alpha * \exists x\beta(x) \sim \exists x(\alpha * \beta(x))$ holds always when α has no free occurrence of x.

By Lemma 4.5, $\mathbf{LA}_\sigma$ can be embedded into an FL_σ phase structure $\mathbf{B}$ by some mapping h that preserves all necessary infinite meets and joins in $\mathbf{LA}_\sigma$. Now, let g be the canonical mapping from $CF[V]$ to $\mathbf{LA}_\sigma$, i.e. $g(\alpha) = [\alpha]$ for every $\alpha \in CF[V]$. Define a mapping f from $CF[V]$ to $\mathbf{B}$ by $f = h \circ g$. Then the restriction f_0 of f to the set of all atomic formulas in $CF[V]$ can be regarded as an assignment of the structure $\langle \mathbf{B}, V\rangle$ for FL_σ. Moreover, it can be shown that the extension of f_0 to the domain $CF[V]$ is equal to f, since h is an FL-homomorphism that preserves all existing infinite meets and joins (satisfying the condition in Lemma 4.5). Now suppose that a sequent $\Gamma \to \theta$ is not provable in FL_σ. Then the sequent $1 \to \langle \Gamma \to \theta\rangle$ is not provable. We will show that $\Gamma \to \theta$ is not true in the assignment f_0 of $\langle \mathbf{B}, V\rangle$. Suppose otherwise. Then $1_{\mathbf{B}} = h([1]) \leq f(\langle \Gamma \to \theta\rangle) = h([\langle \Gamma \to \theta\rangle])$. Since h is injective, $[1] \leq [\langle \Gamma \to \theta\rangle]$ holds in $\mathbf{LA}_\sigma$. But this contradicts the fact that the sequent $1 \to \langle \Gamma \to \theta\rangle$ is not provable in FL_σ. Thus, our theorem holds.

It is well-known that the Dedekind-MacNeille completion of a distributive lattice is not necessarily distributive (see e.g. p. 98 and p. 123 of [10]). On the other hand, when a given FL-algebra $\mathbf{A}$ is a Heyting algebra, i.e. an FL_{ecw}-algebra, $\mathcal{M}(\mathcal{P}(\mathbf{A}^*))$ is also a Heyting algebra by Corollary 4.4 and hence it is distributive. This follows from the fact that (1) $a*b = a\cap b$ holds in every FL_{ecw}-algebra, and thus in particular $MU \cdot_M MV = MU \cap MV$ holds in $\mathcal{M}(\mathcal{P}(\mathbf{A}^*))$, and (2) the distributivity of $*$ over $\cup$ holds in every FL-algebra.

So we cannot expect that the Dedekind-MacNeille completion will work well in general when we deal with FL-algebras with the distributive law, such as Dunn monoids, which are related to relevant logics. Thus it is necessary to modify our method in these cases.

For an FL-algebra $\mathbf{A}$, let $\mathcal{I}(A)$ be the set of all ideals of $\mathbf{A}$. Then it is easy to see the following:

Lemma 4.8. *Let* $\mathbf{A}$ *be an FL-algebra and* M *the operation on* $\mathcal{P}(\mathbf{A}^*)$ *defined in Lemma* 4.3. *Then, for every subset* U *of* A, MU *is an ideal of* $\mathbf{A}$. *Thus,* $M(\mathcal{P}(A^*)) \subset \mathcal{I}(A) \subset \mathcal{P}(A)$.

For every subset U of A, IU denotes the ideal generated by the set U, i.e.,

$$IU = \{x : x \leq u_1 \cup \ldots \cup u_m \text{ for some } u_1, \ldots, u_m \in U\}.$$

Of course, IU belongs to $\mathcal{I}(A)$. For every subset U and V of A, define $U \circ V$ and $U \vee V$ by $U \circ V = I(U \cdot V)$ and $U \vee V = I(U \cup V)$ (for the definition of $\cdot$, see the last paragraph of §3). Then we have the following (for complete Dunn monoids, see [19]):

Lemma 4.9. (1) *Let* $\mathbf{A}$ *be an FL-algebra. For every subset* U *and* V *of* A, *if* U *and* V *belong to* $\mathcal{I}(A)$, *then* $U \Rightarrow V$, $U \vee V$, $U \cap V$ *and* $U \circ V$ *belong to* $\mathcal{I}(A)$.

(2) *If* $\mathbf{A}$ *is moreover a Dunn monoid, then* $\mathcal{I}(\mathbf{A}) = \langle \mathcal{I}(A), \Rightarrow, \vee, \cap, \circ, I\{1\}, I\{\bot\}, A, I\{\bot\}\rangle$ *is a complete Dunn monoid, i.e. a Dunn monoid in which* $\langle \mathcal{I}(A), \vee, \circ\rangle$ *forms a quantale and satisfies the following infinite distributive law:*

$$U \cap (\vee_i V_i) = \vee_i (U \cap V_i).$$

The following result is a weak analogue of Lemma 4.5 (cf. an embedding result for De Morgan monoids in [29, Corollary 9.2]).

Theorem 4.10. *Let* $\mathbf{A}$ *be a Dunn monoid and* h *a mapping from* $\mathbf{A}$ *to* $\mathcal{I}(\mathbf{A})$ *defined by* $h(a) = I_a$ *(the principal ideal generated by* a*). Then* h *is an FL-embedding of* $\mathbf{A}$ *into* $\mathcal{I}(\mathbf{A})$. *Thus, every Dunn monoid is embeddable into a complete Dunn monoid.*

Note that in this case h preserves infinite meets existing in $\mathbf{A}$, but it does not always preserve infinite joins.

5 Classical Substructural Logics

Logics we have discussed so far are intuitionistic substructural logics. In this section, we show how our previous results can be extended to classical substructural logics. For every σ, we call an FL_σ phase structure a *GL_σ^- phase structure* if it is also a classical quantale. More precisely, a GL_σ^- phase structure $\mathcal{C}(\mathcal{P}(\mathbf{M}))$ is an FL_σ phase structure in which $(U \Rightarrow O) \Rightarrow' O = U = (U \Rightarrow' O) \Rightarrow O$ always holds, where $\Rightarrow'$ is defined by

$$X \Rightarrow' Y = \{z \in M : x * z \in Y, \text{ for every } x \in X\}.$$

Sometimes, we abbreviate $\mathrm{GL}_{e\sigma}^-$ phase structures to *GL_σ phase structures*. We remark that if $CU = (U \Rightarrow O) \Rightarrow O$ holds for every U in a GL phase structure $\mathcal{C}(\mathcal{P}(\mathbf{M}))$, then it is exactly a *phase structure* in Girard's sense [9]. The following is a classical version of Lemma 4.5:

Theorem 5.1. *Let* σ *be either empty or one of* c, w *and* cw. *For every* GL_σ*-algebra* $\mathbf{A}$, *there exist a* GL_σ *phase structure* $\mathcal{C}(\mathcal{P}(\mathbf{M}))$ *and an FL-embedding* h *of* $\mathbf{A}$ *into* $\mathcal{C}(\mathcal{P}(\mathbf{M}))$ *satisfying that*

(1) *for every subset U of M, U is C-closed if and only if* $U = (U \Rightarrow O) \Rightarrow O$,

(2) *h preserves all existing infinite meets and joins in A,*

(3) *h is surjective when* $\mathbf{A}$ *is complete.*

Proof. The proof is quite similar to that of Lemma 4.5, taking $\mathcal{M}(\mathcal{P}(\mathbf{A}^*))$ for $\mathcal{C}(\mathcal{P}(\mathbf{M}))$, i.e., $\mathcal{C} = \mathcal{M}$ and $\mathbf{M} = \mathbf{A}^*$. In the present case, we will assume moreover that $\sim\sim x = x$ holds for every $x \in A$, where $\sim x$ is defined as $x \to 0_A$ for a fixed element $0_A \in A$. Recall that the zero element O of $\mathcal{C}(\mathcal{P}(\mathbf{M}))$ is given by $O = M\{0_A\}$ ($= h(0_A)$), where 0_A is the zero element of $\mathbf{A}$ (for the definitions of M and h, see Lemmas 4.3 and 4.5, respectively). Thus, to show our theorem, it suffices to prove that $(U \Rightarrow O) \Rightarrow O = MU = (U^{\rightarrow})^{\leftarrow}$ for every $U \subset A$. Let $V = (U \Rightarrow O) \Rightarrow O$. Then $v \in V$ iff

(1) for every $z \in A$, $v * z \leq 0_A$ if $z * u' \leq 0_A$ for every $u' \in U$.

But, as shown below, (1) is equivalent to

(2) for every $w \in A$, $v \leq w$ if $u' \leq w$ for every $u' \in U$,

which means that $v \in (U^{\rightarrow})^{\leftarrow}$ (see the paragraph just below Corollary 4.4). So, it remains to show that (1) is equivalent to (2). Suppose that (1) holds and moreover that $u' \leq w$ for every $u' \in U$. Then $(\sim w)*u' = u'*(\sim w) \leq 0_A$ for every $u' \in U$. Thus, $v * (\sim w) \leq 0_A$ by (1), but this is equivalent to $v \leq w$, using the fact that $\sim\sim w = w$. Conversely, if $z * u' \leq 0_A$ for every $u' \in U$, then $u' \leq \sim z$ for every $u' \in U$, and hence $v \leq \sim z$ by (2). Thus, $v * z \leq 0_A$. So, (2) implies (1). Therefore, $U \in \mathcal{M}(\mathcal{P}(\mathbf{A}^*))$ if and only if $(U \Rightarrow O) \Rightarrow O = U$.

By Theorem 5.1, we have immediately the following result (see [3], [27] and [28]):

Corollary 5.2. *Every commutative unital classical σ-quantale is FL-isomorphic to a GL_σ phase structure for every $\sigma \in \{c, w, cw\}$.*

By modifying slightly the argument above, we can also treat the non-commutative case (cf. [27]). Thus we have the following:

Corollary 5.3. *Every (noncommutative) unital classical σ-quantale is FL-isomorphic to a GL_σ^- phase structure for every $\sigma \in \{c, w, cw\}$.*

With Corollaries 5.2 and 5.3, we can demonstrate completeness theorems for classical substructural logics similarly to Theorem 4.7:

Theorem 5.4. ([9], [27], [28], [34]) *For every $\sigma \in \{c, w, cw\}$, the classical substructural logic GL_σ is complete with respect to the class of all GL_σ phase structures (and equivalently, the class of all commutative unital classical σ-quantales). This holds also for every noncommutative substructural logic GL_σ^-.*

By using results obtained in §4 and §5, we will show a syntactic embedding theorem for the substructural propositional logics FL_e and GL, i.e. intuitionistic and classical linear propositional logic, into extensions of propositional FL_e with the distributive law. In the rest of this section, we will discuss only propositional logics.

First, we will introduce some extensions of FL_e with the distributive law. Let DFL_e be the sequent calculus obtained from FL_e by adding the following distributive law

(1) $\alpha \wedge (\beta \vee \gamma) \to (\alpha \wedge \beta) \vee (\alpha \wedge \gamma)$

as an initial sequent. Next, we will add a modal operator $\Diamond$ to our language $\mathcal{L}$, and define $\mathrm{DFL}_e\Diamond$ to be the calculus obtained from DFL_e by adding the following initial sequents:

(2) $\alpha \to \Diamond\alpha$,

(3) $\Diamond\Diamond\alpha \to \Diamond\alpha$,

and the rule for $\Diamond$:

$$\frac{\Gamma \to \theta}{\Diamond\Gamma \to \Diamond\theta}.$$

Here, Γ may be empty but θ may not. The calculus $\mathrm{DFL}_e\Diamond^+$ is defined as $\mathrm{DFL}_e\Diamond$ save that θ can be empty in the application of the rule for $\Diamond$.

Next we will introduce algebraic structures corresponding to these logics. A pair $\langle \mathbf{A}, C\rangle$ is called a *closure FL-algebra* if $\mathbf{A}$ is an FL-algebra and C is a closure operation on $\mathbf{A}$. Of course, every assignment f of $\langle \mathbf{A}, C\rangle$ must satisfy $f(\Diamond\alpha) = Cf(\alpha)$ for every formula α. By considering Lindenbaum algebras for these logics and using the proof of Corollary 4.4 and Lemma 4.5, we can show the following:

Lemma 5.5. (1) *A sequent is provable in FL_e if and only if it is valid in every closure FL_e-algebra, and also if and only if it is valid in every FL_e phase structure $C(\mathcal{P}(\mathbf{M}))$ with a commutative monoid $\mathbf{M}$ as a closure FL_e-algebra, i.e. a closure FL_e-algebra $\langle \mathcal{P}(\mathbf{M}), C\rangle$.*

(2) *A sequent is provable in $DFL_e\Diamond$ if and only if it is valid in every closure FL-algebra $\langle \mathbf{A}, C\rangle$ with a distributive FL_e-algebra $\mathbf{A}$.*

(3) *A sequent is provable in $DFL_e\Diamond^+$ if and only if it is valid in every closure FL-algebra $\langle \mathbf{A}, C\rangle$ with a distributive FL_e-algebra $\mathbf{A}$ satisfying $C0 = 0$.*

We define inductively a modal formula $\alpha^\dagger$ for every (nonmodal) propositional formula α:

(1) $p^\dagger = \Diamond p$, if p is either a propositional variable or any of 1, 0 and $\bot$,

(2) $\top^\dagger = \top$,

(3) $(\beta \circ \gamma)^\dagger = \beta^\dagger \circ \gamma^\dagger$, if $\circ$ is either $\supset$ or $\wedge$,

(4) $(\beta \circ \gamma)^\dagger = \Diamond(\beta^\dagger \circ \gamma^\dagger)$, if $\circ$ is either $\vee$ or $*$.

For a sequent $\Gamma \rightarrow \theta$ with $\Gamma = \alpha_1, \ldots, \alpha_m$, $\Gamma^\dagger \rightarrow \theta^\dagger$ denotes the sequent $\alpha_1^\dagger, \ldots, \alpha_m^\dagger \rightarrow \theta^\dagger$. We have the following theorem:

Theorem 5.6. *The following three conditions are equivalent for every (nonmodal) sequent* $\Gamma \rightarrow \theta$:

(1) $\Gamma \rightarrow \theta$ *is provable in* FL_e,

(2) $\Gamma^\dagger \rightarrow \theta^\dagger$ *is provable in* $DFL_e\Diamond$,

(3) $\Gamma^\dagger \rightarrow \theta^\dagger$ *is provable in* $DFL_e\Diamond^+$.

Proof. First, we will show that (1) implies (2). By using Lemma 5.5, it suffices to show that for every closure FL_e-algebra $\langle \mathbf{A}, C\rangle$ with a distributive $\mathbf{A}$ and every formula α, $\alpha^\dagger$ is valid in $\langle \mathbf{A}, C\rangle$ if α is valid in every FL_e-algebra. (Note that the formula $(\Gamma^* \supset \theta)^\dagger$ is valid in an FL-algebra if and only if $(\Gamma^\dagger)^* \supset \theta^\dagger$ is valid in it. For the definition of Γ^*, see §4.) As mentioned in §3, $\mathcal{C}(\mathbf{A})$ is also an FL_e-algebra. For every assignment f of $\langle \mathbf{A}, C\rangle$, define an assignment g of $\mathcal{C}(\mathbf{A})$ by $g(p) = Cf(p)$ for every propositional variable p. Then by induction we can show that $g(\beta) = f(\beta^\dagger)$ for every formula β. Now suppose that α is valid in every FL_e-algebra. Then $1_\mathbf{A} \leq C1_\mathbf{A} = 1_{\mathcal{C}(\mathbf{A})} \leq g(\alpha) = f(\alpha^\dagger)$. Thus, $\alpha^\dagger$ is valid in $\langle \mathbf{A}, C\rangle$.

It is clear that (2) implies (3). So we will show that (3) implies (1). By using Lemma 5.5, it suffices to show that for every FL_e phase structure $\mathcal{C}(\mathcal{P}(\mathbf{M}))$ with a commutative $\mathbf{M}$ and every formula α, α is valid in $\mathcal{C}(\mathcal{P}(\mathbf{M}))$ if $\alpha^\dagger$ is valid in every closure FL_e-algebra satisfying condition (3) of Lemma 5.5. Now suppose otherwise. Then there exists an assignment g of $\mathcal{C}(\mathcal{P}(\mathbf{M}))$ such that $1_{\mathcal{C}(\mathcal{P}(\mathbf{M}))} \not\leq g(\alpha)$. Let us consider the structure $\langle \mathcal{P}(\mathbf{M}), C\rangle$. It is easy to see that $\mathcal{P}(\mathbf{M})$ is a distributive FL-algebra. Moreover, $\mathcal{P}(\mathbf{M})$ is an FL_e-algebra since $\mathbf{M}$ is commutative. Define an assignment f of $\langle \mathcal{P}(\mathbf{M}), C\rangle$ by $f(p) = g(p)$ for every propositional variable p. Since $g(p) \in \mathcal{C}(\mathcal{P}(M))$, $g(p) = Cg(p) = Cf(p)$. Thus, as in the argument above, we can show that $g(\beta) = f(\beta^\dagger)$ for every formula β. Thus, if $1_{\mathcal{P}(\mathbf{M})} \leq f(\alpha^\dagger)$ then $1_{\mathcal{C}(\mathcal{P}(\mathbf{M}))} = C(1_{\mathcal{P}(\mathbf{M})}) \leq Cf(\alpha^\dagger) = Cg(\alpha) = g(\alpha)$.

But this contradicts our assumption. Hence α is valid in $\mathcal{C}(\mathcal{P}(\mathbf{M}))$. Note here that in the discussion above the closure operation C may not satisfy $C0 = 0$. In such a case, we have only to take $C0$, instead of 0, as the new zero element. Then $C(C0) = C0$ and hence condition (3) of Lemma 5.5 will be satisfied.

Next we will show an embedding theorem for classical linear logic by using an analogous argument. Similarly to Lemma 5.5, we can show the following:

Lemma 5.7. (1) *A sequent is provable in GL if and only if it is valid in every GL-algebra, and also if and only if it is valid in every GL phase structure* $\mathcal{C}(\mathcal{P}(\mathbf{M}))$ *with a commutative monoid* $\mathbf{M}$ *and a closure operation* C *satisfying* $CU = {\sim\sim}U$ *for every* $U \subset M$.

(2) *A sequent is provable in* DFL_e *if and only if it is valid in every distributive* FL_e*-algebra* $\mathbf{A}$.

It is easy to see that double negation $\sim\sim$ can be regarded as a modal operator satisfying the two initial sequents above and the rule for $\diamondsuit$. For every formula α, let $\alpha^{\sim\sim}$ be the formula obtained from $\alpha^{\dagger}$ by replacing every $\diamondsuit$ by $\sim\sim$. Now we have the following:

Theorem 5.8. *The following two conditions are equivalent for every (non-modal) sequent* $\Gamma \to \theta$;

(1) $\Gamma \to \theta$ *is provable in GL*,

(2) $\Gamma^{\sim\sim} \to \theta^{\sim\sim}$ *is provable in* DFL_e.

Proof. The proof is similar to that of Theorem 5.6. Suppose first that (1) holds. By Lemma 5.7, it suffices to show that for every distributive FL_e-algebra $\mathbf{A}$ and every formula α, $\alpha^{\sim\sim}$ is valid in $\mathbf{A}$ if α is valid in every GL-algebra. Let us define a closure operation C on $\mathbf{A}$ by $Cx = {\sim\sim}x$ for every $x \in A$. Then $\mathcal{C}(\mathbf{A})$ is a GL-algebra. As in the proof of Theorem 5.6, we can show that $\alpha^{\sim\sim}$ is valid in $\mathbf{A}$ by using the fact that α is valid in $\mathcal{C}(\mathbf{A})$. Conversely, suppose that (2) holds. By Lemma 5.7, it suffices to show that if $\mathcal{C}(\mathcal{P}(\mathbf{M}))$ is a GL phase structure with a commutative monoid $\mathbf{M}$ and a closure operation C satisfying $CU = {\sim\sim}U$ for every $U \subset M$, then for every formula α, α is valid in $\mathcal{C}(\mathcal{P}(\mathbf{M}))$ when $\alpha^{\sim\sim}$ is valid in every distributive FL_e-algebra. Suppose otherwise. Then there exists an assignment g of $\mathcal{C}(\mathcal{P}(\mathbf{M}))$ such that $1_{\mathcal{C}(\mathcal{P}(\mathbf{M}))} \not\leq g(\alpha)$. It is easy to see that $\mathcal{P}(\mathbf{M})$ is a distributive FL_e-algebra. Define an assignment f of $\mathcal{P}(\mathbf{M})$ by $f(p) = g(p)$ for every propositional variable p. Since $g(p) \in \mathcal{C}(\mathcal{P}(\mathbf{M}))$, $g(p) = {\sim\sim}f(p)$. Thus, as in the proof of Theorem 5.6, $g(\beta) = f(\beta^{\sim\sim})$ holds for every β, and this leads to a contradiction.

We can easily verify the following, which, with Theorems 5.6 and 5.8, will show the difference between intuitionistic and classical linear logic through the translation introduced in this section:

Lemma 5.9. (1) *The logic $DFL_e\Diamond^+$ is a conservative extension of DFL_e.*
(2) *In $DFL_e\Diamond^+$, $\Diamond\alpha \to \sim\sim\alpha$ is provable for every formula α.*

Therefore, double negation is the maximum modal operator satisfying the conditions above, and classical linear logic can be determined by the double-negation translation, while intuitionistic linear logic can be determined by a modal translation. We don't know whether similar results will hold for other logics FL_σ.

6 Substructural Logics with Modal Operators

We will extend the embedding results obtained in §4 and §5 to results between FL-algebras with *modal operations*, which correspond to the logical connectives ! and ? (see e.g. [1] and [9]). The basic calculus FL^M is obtained from the calculus FL by adding the following rules for the modal operators ! and ?:

$$\frac{\Gamma, \alpha, \Delta \to \theta}{\Gamma, !\alpha, \Delta \to \theta}\ (! \to) \qquad \frac{!\Gamma \to \alpha}{!\Gamma \to !\alpha}\ (\to !)$$

$$\frac{\Gamma, \Delta \to \theta}{\Gamma, !\alpha, \Delta \to \theta}\ (!w) \qquad \frac{\Gamma, !\alpha, !\alpha, \Delta \to \theta}{\Gamma, !\alpha, \Delta \to \theta}\ (!c)$$

$$\frac{\Gamma, !\alpha, \beta, \Delta \to \theta}{\Gamma, \beta, !\alpha, \Delta \to \theta}\ (!e) \qquad \frac{\Gamma \to}{\Gamma \to ?\alpha}\ (?w)$$

$$\frac{!\Gamma, \alpha, !\Delta \to ?\theta}{!\Gamma, ?\alpha, !\Delta \to ?\theta}\ (? \to) \qquad \frac{\Gamma \to \alpha}{\Gamma \to ?\alpha}\ (\to ?)$$

Here, $!\Gamma$ denotes the sequence of formulas $!\gamma_1, \ldots, !\gamma_m$ if Γ is $\gamma_1, \ldots, \gamma_m$. The extension of FL_σ (and GL_σ) obtained by adding these rules will be denoted by FL_σ^M (and GL_σ^M, respectively). The following definition of modal FL-algebras will naturally arise from these syntactic properties of ! and ? in FL^M (see also [3] and [23]). Note that we don't assume here the commutativity of $*$ (see also [34]).

Definition 6.1. A *modal FL-algebra* is an FL-algebra with additional unary operations ! and ? satisfying:

(1) $!a \leq a$,

(2) $!1 = 1$,

(3) $!a \leq !!a$,

(4) $!(a \cap b) = !a * !b$,

(5) $!a * b = b * !a$,

(6) $!(a \to b) \leq ?a \to ?b$,

(7) $a \leq ?a$,

(8) $?0 = 0$,

(9) $??a \leq ?a$,

(10) $0 \leq ?a$.

Clearly, when an FL-algebra is commutative, condition (5) is redundant. The following can be shown easily (see e.g. [2]):

Lemma 6.1. *In every modal FL-algebra, the following holds:*

(1) $!a \leq 1$,

(2) $!a \leq !a * !a$,

(3) $!a * !b \leq !(!a * !b)$,

(4) $a \leq b$ *implies* $!a \leq !b$,

(5) $a \leq b$ *implies* $?a \leq ?b$,

(6) $!a * b \leq ?c$ *implies* $!a * ?b \leq ?c$.

We remark here that, in every FL-algebra, condition (4) in Definition 6.1 is equivalent to the conjunction of (1), (2), (3) and (4) of Lemma 6.1, under the assumptions (1) to (3) (of Definition 6.1).

We note also that in every modal FL-algebra (1) $?a \leq \sim!\sim a$ holds, and (2) $\sim!b = \sim'!b$ and $!\sim b = !\sim' b$ hold where $\sim' c$ is an abbreviation for $c \to' 0$. Then it is easy to see that in every GL$^-$-algebra, if an operation ! satisfies conditions (1) to (5) of Definition 6.1, then the operation $?'$, which is defined by $?'x = \sim!\sim x$, for every x, satisfies the other conditions for ? of Definition 6.1. Therefore, when we discuss modal *classical* substructural logics we usually consider only the operation ! satisfying conditions (1) to (5).

The following lemma says when a given FL-algebra can be extended to a modal FL-algebra. Similar results have been proved by various authors (see e.g. [1], [2], [7], [17] and [20]):

Lemma 6.2. *Let* **A** *be a unital quantale and let* R *and* S *be subsets of* A *satisfying the following conditions:*

(1) R *is closed under* $*$,

(2) $x * x = x$, *for every* $x \in R$,

(3) $x * u = u * x$, *for every* $x \in R$ *and every* $u \in A$,

(4) 1 *is the greatest element of* R,

(5) 0 *is the least element of* S,

(6) *if* $x \in R$ *and* $y \in S$, *then* $x \to y \in S$.

Define $!a$ *and* $?a$ *by* $!a = \bigcup\{x \in R : x \leq a\}$ *and* $?a = \bigcap\{x \in S : a \leq x\}$ *for every* $a \in A$. *Then* **A** *with* $!$ *and* $?$ *is a modal unital quantale.*

Proof. First, we note that when **A** is commutative, condition (3) above is redundant. It suffices to show that both $!$ and $?$ satisfy the conditions in Definition 6.1. Here we will give a proof of (i) $!(a \cap b) = !a * !b$, and (ii) $!(a \to b) \leq ?a \to ?b$. By the definition of $!$, $!(a \cap b) = \bigcup\{x \in R : x \leq a \cap b\}$. Suppose that $x \in R$ and $x \leq a \cap b$. Then both $x \leq a$ and $x \leq b$ hold. Thus, $x \leq !a$ and $x \leq !b$ by the definition of $!$, and hence $x = x * x \leq !a * !b$, by condition (2) above. Therefore, $!(a \cap b) \leq !a * !b$. On the other hand,

$$\begin{aligned} !a * !b &= (\bigcup\{x \in R : x \leq a\}) * (\bigcup\{y \in R : y \leq b\}) \\ &= \bigcup\{x * y : x, y \in R,\ x \leq a \text{ and } y \leq b\}. \end{aligned}$$

Since $y \in R$, $y \leq 1$ holds, and hence $x * y \leq x * 1 = x \leq a$. Similarly, $x * y \leq b$. Thus, $x * y \leq a \cap b$. Note that $x * y \in R$, since R is closed under $*$. Hence,

$$!a * !b \leq \bigcup\{z \in R : z \leq a \cap b\} = !(a \cap b).$$

Next, we will show $!(a \to b) * ?a \leq ?b$, which implies $!(a \to b) \leq ?a \to ?b$. It is easy to see that

$$\begin{aligned} !(a \to b) * ?a &= (\bigcup\{x \in R : x \leq a \to b\}) * ?a \\ &= \bigcup\{x * ?a : x \in R \text{ and } x \leq a \to b\}. \end{aligned}$$

Suppose that $b \leq y$ and $y \in S$. If $x \in R$ and $x \leq a \to b$, then $a * x = x * a \leq b \leq y$. Therefore, $a \leq x \to y$. On the other hand, $x \to y \in S$ by condition (6) above. Thus, $?a \leq x \to y$ by the definition of $?$, and hence $x * ?a = ?a * x \leq y$ for every $x \in R$ such that $x \leq a \to b$ and every $y \in S$ such that $b \leq y$. This means that $!(a \to b) * ?a \leq \bigcap\{y \in S : b \leq y\} = ?b$.

We have assumed in the lemma above that **A** is a quantale. But, if either **A** is commutative or $\rightarrow'$ is defined in **A**, then it is enough to assume the existence of the join $\cup\{x \in R : x \leq a\}$ and the meet $\cap\{x \in S : a \leq x\}$ for every $a \in A$ in order to show the result above. We note that if $\bigcup_i x_i$ exists, then for every y, $\bigcup_i (x_i * y)$ exists and is equal to $(\bigcup_i x_i) * y$. A similar result holds for $\bigcup_i (y * x_i)$ when we have $\rightarrow'$ in **A**.

We remark also that when **A** in Lemma 6.2 is classical and a subset R of A satisfies conditions (1) to (4) (of Lemma 6.2), then the set S defined by

$$x \in S \quad \text{if and only if} \quad {\sim}x \in R,$$

satisfies the remaining conditions (5) and (6). Moreover, if we define $!a$ and $?a$ in the same way as in Lemma 6.2 by using these R and S, then $?a = {\sim}!{\sim}a$ holds.

We can verify the following lemma, which says that the converse of Lemma 6.2 holds without assuming that FL-algebras are complete:

Lemma 6.3. *For every modal FL-algebra* **A** *with* $!$ *and* $?$, *define* $R = \{!x : x \in A\}$ *and* $S = \{?x : x \in A\}$. *Then,* R *and* S *satisfy conditions* (1) *to* (6) *of Lemma* 6.2. *Moreover, both* $\cup\{u \in R : u \leq a\}$ *and* $\cap\{u \in S : a \leq u\}$ *exist and are equal to* $!a$ *and* $?a$, *respectively, for every* $a \in A$.

A mapping f from a modal FL-algebra **A** to another modal FL-algebra **B** is a *modal FL-homomorphism* if f is an FL-homomorphism which preserves ! and ?. In the following lemma, we will consider FL_e-algebras, i.e. commutative FL-algebras:

Lemma 6.4. *Suppose that* f *is an FL-embedding of a modal* FL_e*-algebra* **A** *into a commutative unital quantale* **B**. *Then, for some operations* $!^*$ *and* $?^*$ *on* **B**, **B** *is a modal* FL_e*-algebra with modal operations* $!^*$ *and* $?^*$, *and* f *is a modal FL-embedding of* **A** *into* **B**.

Proof. Define the subsets R and S of B by

$$R = \{f(!x) : x \in A\} \text{ and } S = \{f(?x) : x \in A\}.$$

Then we can show easily that R and S satisfy conditions (1) to (6) of Lemma 6.2. Note that (3) holds since **B** is commutative. Now define $!^*$ and $?^*$ by

$$!^*b = \cup\{y \in R : y \leq b\} \text{ and } ?^*b = \cap\{y \in S : b \leq y\}.$$

Then the unital quantale **B** with $!^*$ and $?^*$ is a modal FL_e-algebra by Lemma 6.2. It remains to show that $f(!a) = !^*f(a)$ and $f(?a) = ?^*f(a)$ for every $a \in A$. Since f is order-preserving, $f(!a) \leq f(a)$. Clearly, $f(!a) \in R$ and hence $f(!a) \leq !^* f(a)$. Suppose next that $y \leq f(a)$ for $y \in R$. Then there exists an $x \in A$ such that $y = f(!x)$. Since f is an order-isomorphism, $!x \leq a$, and therefore $!x \leq !a$. Thus, $y = f(!x) \leq f(!a)$. Hence, $!^* f(a) = \cup\{y \in R : y \leq f(a)\} \leq f(!a)$. Similarly, we can show that $f(?a) = ?^* f(a)$.

When **A** is a noncommutative FL-algebra, R above will not satisfy condition (3) of Lemma 6.2 in general. Thus, it seems to be difficult to establish the corresponding result for noncommutative FL-algebras. But, this can be done in the special case where **B** is the FL phase structure $\mathcal{M}(\mathcal{P}(\mathbf{A}^*))$ and f is the FL-embedding h defined by $h(a) = M\{a\}$ (see Lemma 4.5), by which we can obtain the required result:

Lemma 6.5. *Suppose that* **A** *is a modal FL-algebra and h an FL-embedding from* **A** *to $\mathcal{M}(\mathcal{P}(\mathbf{A}^*))$, defined by $h(a) = M\{a\}$. Then, for some operations $!^*$ and $?^*$ on $\mathcal{M}(\mathcal{P}(\mathbf{A}^*))$, $\mathcal{M}(\mathcal{P}(\mathbf{A}^*))$ is a modal FL phase structure, and h is a modal FL-embedding of* **A** *into $\mathcal{M}(\mathcal{P}(\mathbf{A}^*))$.*

Proof. The proof is analogous to the proof of Lemma 6.4. In the present case, it suffices to show that R satisfies condition (3) of Lemma 6.2; that is, for every $z \in A$ and every $U \subset A$, $M(M\{!z\} \cdot MU) = M(MU \cdot M\{!z\})$. But, $\{!z\} \cdot U = U \cdot \{!z\}$ by Definition 6.1 (5), and hence $M(M\{!z\} \cdot MU) = M(\{!z\} \cdot U) = M(U \cdot \{!z\}) = M(MU \cdot M\{!z\})$.

By Lemmas 4.5, 6.4 and 6.5, we have the following theorem which is a modal version of Lemma 4.5:

Theorem 6.6. *For every modal FL_σ-algebra* **A**, *there exists a modal FL_σ phase structure* **B** *and a modal FL-embedding f of* **A** *into* **B** *such that:*

(1) *f preserves all existing infinite meets, and all existing infinite joins $\cup a_i$ in A satisfying $x * (\cup a_i) = \cup(x * a_i)$ for every x in A,*

(2) *f is surjective when* **A** *is complete.*

We can obtain similar results for modal classical FL-algebras with the help of the remark following Lemma 6.2. Note that if we define the subsets R and S as in Lemma 6.4 in a modal classical FL-algebra **A**, then it can be shown that for every $y \in B$, $y \in S$ if and only if $\sim y \in R$.

Thus, we have modal versions of Theorem 4.6, Corollaries 5.2 and 5.3, and completeness theorems for modal substructural logics that are extensions of Theorems 4.7 and 5.4:

Corollary 6.7. (1) *Every modal unital σ-quantale is modal FL-isomorphic to a modal FL_σ phase structure.*

(2) *Every modal unital classical σ-quantale is modal FL-isomorphic to a modal GL_σ^- phase structure.*

Corollary 6.8. (cf. [1], [15], [17], [20]) *For every σ, the modal substructural logic FL_σ^M (and $(GL_\sigma^-)^M$) is complete with respect to the class of all modal FL_σ phase structures (and the class of all modal GL_σ^- phase structures). Thus, it is also complete with respect to the class of all modal unital σ-quantales (and to the class of all modal unital classical σ-quantales, respectively).*

7 Kripke-Type Semantics

In this section, we will consider Kripke-type semantics for substructural logics. Došen [5] and Ono-Komori [26] introduced a Kripke-type semantics based on semilattice-ordered monoids, which seems to be a natural extension of the usual Kripke-type semantics for intuitionistic logic. To explain the idea in [5] and [26], we will repeat here the definition.

Definition 7.1. A structure $\mathbf{D} = \langle D, \cap, \cdot, \varepsilon, \omega\rangle$ is a *semilattice-ordered monoid* (abbreviated to *so-monoid*) if

(1) $\langle D, \cap\rangle$ is a meet-semilattice with the greatest element ω,

(2) $\langle D, \cdot, \varepsilon\rangle$ is a monoid with the identity ε that satisfies $x \cdot \omega = \omega$ for every $x \in D$,

(3) $z \cdot (x \cap y) \cdot w = (z \cdot x \cdot w) \cap (z \cdot y \cdot w)$, for every $x, y, z, w \in D$.

For every so-monoid $\mathbf{D}$, a subset A of D is a *filter* of $\mathbf{D}$ if for every $x, y \in A$ and every $z \in D$, $x \cap y \leq z$ implies $z \in A$. Let $\mathcal{F}(D)$ be the set of all nonempty filters of $\mathbf{D}$. A *Kripke model* (for propositional substructural logics) is defined as a pair $\langle \mathbf{D}, g\rangle$ of an so-monoid $\mathbf{D}$ and a mapping g from the set of all propositional variables with the constant 0 to $\mathcal{F}(D)$. For a given Kripke model $\langle \mathbf{D}, g\rangle$, the relation $x \models \alpha$ for an element $x \in D$ and a formula α is defined inductively as follows:

(1) $x \models p$ iff $x \in g(p)$, if p is a propositional variable or 0,

(2) $x \models 1$ iff $x \geq \varepsilon$,

(3) $x \models \top$, for every x,

(4) $x \models \bot$ iff $x = \omega$,

(5) $x \models \alpha \supset \beta$ iff, for every y and z, if $y \models \alpha$ and $x \cdot y \leq z$, then $z \models \beta$,

(6) $x \models \alpha \vee \beta$ iff, for some y and z such that $y \cap z \leq x$, ($y \models \alpha$ or $y \models \beta$) and ($z \models \alpha$ or $z \models \beta$),

(7) $x \models \alpha \wedge \beta$ iff $x \models \alpha$ and $x \models \beta$,

(8) $x \models \alpha * \beta$ iff, for some y and z such that $y \cdot z \leq x$, $y \models \alpha$ and $z \models \beta$.

Sometimes, $\models$ is called a *valuation* on $\mathbf{D}$ (determined by g). It is easy to see that the set $\{x \in D : x \models \alpha\} \in \mathcal{F}(D)$ for every formula α. We say that a formula α is *true* in a Kripke model $\langle \mathbf{D}, g\rangle$ if $x \models \alpha$ for every $x \geq \varepsilon$, and α is *valid* in a class $\mathcal{K}$ of Kripke models if α is true in every Kripke model in $\mathcal{K}$. As before, a sequent $\Gamma \to \theta$ is *valid* in $\mathcal{K}$ if the formula $\langle \Gamma \to \theta\rangle$ is valid in $\mathcal{K}$. Then we have the following:

Proposition 7.1. ([5], [26]) *The propositional logic FL is complete with respect to the class of all Kripke models.*

Similar results can be shown for other propositional logics of the form FL_σ with respect to certain subclasses of Kripke models (see [5], [26]). For example, $\langle \mathbf{D}, g\rangle$ is a Kripke model for FL_e if $\mathbf{D}$ is a commutative so-monoid, it is a Kripke model for FL_w if (1) $g(0) = \{\omega\}$ and (2) $x \leq x \cdot y$ and $x \leq y \cdot x$, and it is a Kripke model for FL_c if $x \cdot x \leq x$ holds. (Compare this with Kripke-type semantics for relevant logics, in e.g. [14], [29] and [30].)

In [22] and [25], we tried to extend our Kripke-type semantics to a semantics for predicate substructural logics, by using *complete so-monoids*. An so-monoid $\mathbf{D} = \langle D, \cap, \cdot, \varepsilon, \omega\rangle$ is complete if $\langle D, \cap\rangle$ is a complete meet-semilattice (and hence *it is in fact a complete lattice*) satisfying for every $x_i, z, w \in D$, $z \cdot (\cap_i x_i) \cdot w = \cap_i (z \cdot x_i \cdot w)$. This time, g must be a mapping from the set of all atomic formulas (in $CF[V]$ for a given domain V) to the set of all *principal filters*. We say that a subset A of D is a principal filter if it is a set of the form $\{x \in D : c \leq x\}$ for some element $c \in D$. In this case, every valuation $\models$ must satisfy:

(9) $x \models \exists x \alpha(x)$ iff there exists a subset $\{z_i\}_{i \in I}$ of D and a subset $\{v_i\}_{i \in I}$ of V such that $\cap_{i \in I} z_i \leq x$ and $z_j \models \alpha(\bar{v}_j)$ for every $j \in I$,

(10) $x \models \forall x \alpha(x)$ iff, for every $v \in V$, $x \models \alpha(\bar{v})$.

Now we have the following:

Proposition 7.2. ([22], [25]) *The predicate logic FL is complete with respect to the class of all Kripke models based on complete so-monoids.*

However, the introduction of the notion of quantale will reveal that *a complete so-monoid is just a unital quantale with the reverse order*. So, in particular, every principal filter of a given complete so-monoid is a principal ideal of a unital quantale, and vice versa. Therefore, the completeness results mentioned above become trivial after we know the completeness results of the previous sections. This discovery may be disappointing, since it says that our Kripke-type semantics gives us nothing new.

On the other hand, if we concentrate only on propositional logics, then there still seems to be room for further study of Kripke-type semantics for substructural logics. For, in the propositional case, it is not necessary to assume that so-monoids are complete and filters are principal. This observation will help us in developing Kripke-type semantics.

In this way we reached the definition given below of valuation in a Kripke model for formulas of the form $!\alpha$ and $?\alpha$. Let $\mathbf{D} = \langle D, \cap, \cdot, \varepsilon, \omega \rangle$ be an so-monoid with a fixed element $\mu \in D$, and let F and G be binary relations on D satisfying the following conditions:

(1) if $\varepsilon \leq x$, then $\varepsilon\ F\ x$,

(2) if $y\ F\ x$ and $x \leq x'$, then $y\ F\ x'$,

(3) if $y\ F\ x$, then $\varepsilon \leq y$ and $y \cdot y \leq x$,

(4) if $y\ F\ x$, then $y\ F\ u$ and $u\ F\ x$ for some u,

(5) if $y\ F\ x$, then there exists a w such that
(i) $y\ F\ w$ and (ii) for every u, $w \cdot u \leq u \cdot x$ and $u \cdot w \leq x \cdot u$,

(6) if $y\ F\ x$ and $y'\ F\ x'$, then $y \cap y'\ F\ x \cap x'$ and $y \cdot y'\ F\ x \cdot x'$,

(7) if $y\ G\ x$ and $x \leq x'$, then $y\ G\ x'$,

(8) $y\ G\ x$ for every y if and only if $\mu \leq x$,

(9) if $y\ G\ x$ and $y\ G\ x'$, then $y\ G\ x \cap x'$,

(10) for every x and y, there exists a z such that if $x\ F\ u$ for some u, then
(i) $y\ G\ x \cdot z$ and (ii) for every w, $y\ G\ x \cdot w$ iff $z\ G\ w$.

Remark that (i) $y\ F\ x$ implies $\varepsilon \leq y \leq x$ (by (3)), and (ii) (5) is redundant when $\mathbf{D}$ is commutative. Such a pair $\langle F, G \rangle$ is called *a pair of accessibility relations on* $\mathbf{D}$. A Kripke model for the modal propositional substructural logic FL_σ^M is a quintuple $\langle \mathbf{D}, F, G, g, \mu \rangle$ where $\langle \mathbf{D}, g \rangle$ is a Kripke model for FL_σ, μ an element of D and $\langle F, G \rangle$ a pair of accessibility relations on $\mathbf{D}$. (In this case, we take $g(0) = \{x : \mu \leq x\}$). Of course, we must extend the definition of valuations to formulas containing modal operators. This can be done in the following way:

(1) $x \models\ !\alpha$ iff $z \models \alpha$ for some z such that $z\ F\ x$,

(2) $x \models\ ?\alpha$ iff, for every y, if $z \models \alpha$ implies $y\ G\ z$ for every z, then $y\ G\ x$.

We can verify easily that the set $\{x \in D : x \models \alpha\}$ is a filter of **D** when α contains modal operators.

To give an intuitive idea of these accessibility relations F and G, let us take subsets R and S of D that satisfy the following conditions (cf. the conditions in Lemma 6.2):

(11) R is closed under $\cap$ and $\cdot$,

(12) $x \cdot x \leq x$ for every $x \in R$,

(13) $x \cdot u = u \cdot x$ for every $x \in R$ and $u \in D$,

(14) ε is the least element of R,

(15) μ is the greatest element of S,

(16) for every $x \in R$ and every $y \in S$, the least element z_0 of $\{z \in D : y \leq z \cdot x\}$ exists and belongs to S.

Define the binary relations F and G by:

$$\begin{array}{lll} y\ F\ x & \text{if and only if} & y \leq x \text{ and } y \in R, \\ y\ G\ x & \text{if and only if} & y \in S \text{ implies } y \leq x. \end{array}$$

Then we can show that F and G satisfy conditions (1) to (10). In this case, the pair $\langle F, G\rangle$ is said to be a pair of accessibility relations *induced by the binary relations R and S*. Now, we will show the following theorem:

Theorem 7.3. *For every σ, the modal substructural propositional logic FL_σ^M is complete with respect to the class of all Kripke models for FL_σ^M.*

Proof. First we will show that if a sequent is provable in FL_σ^M, then it is true in every Kripke model for FL_σ^M. As usual, this can be proved by showing that every initial sequent of FL_σ^M is true and that if the upper sequent(s) of a given rule is true, then so is the lower sequent. We will show this for the rules $(!w)$ and $(? \rightarrow)$. For the sake of brevity, we will consider the following cases:

$$\frac{\gamma, \delta \rightarrow \theta}{\gamma, !\alpha, \delta \rightarrow \theta}\ (!w) \qquad \frac{!\gamma, \alpha, !\delta \rightarrow ?\theta}{!\gamma, ?\alpha, !\delta \rightarrow ?\theta}\ (? \rightarrow).$$

1) Suppose that $x \models \gamma * !\alpha * \delta$. Then, for some y, z, w such that $y \cdot z \cdot w \leq x$, $y \models \gamma$, $z \models\ !\alpha$ and $w \models \delta$. Since $z \models\ !\alpha$, there exists a u such

that $u\ F\ z$ and $u \models \alpha$. But $u\ F\ z$ implies $\varepsilon \leq z$. Thus, $y \cdot w \leq y \cdot z \cdot w \leq x$, $y \models \gamma$ and $w \models \delta$ hold. Hence, $x \models \theta$ holds by the hypothesis that the upper sequent of $(!w)$ is true.

2) Suppose that $x \models\ !\gamma * ?\alpha * !\delta$. Then, for some y, z, w such that $y \cdot z \cdot w \leq x$, $y \models\ !\gamma$, $z \models\ ?\alpha$ and $w \models\ !\delta$. By $w \models\ !\delta$, there exists a w_0 such that $w_0\ F\ w$ and $w_0 \models \delta$. By using condition (5) of F, there exists a w' such that $w_0\ F\ w'$ and $w' \cdot u \leq u \cdot w$ for every u. Taking z for u, we have:

(1) $y \cdot w' \cdot z \leq y \cdot z \cdot w \leq x$.

On the other hand, from $y \models\ !\gamma$, it follows that $y_0\ F\ y$ and $y_0 \models \gamma$ for some y_0. By applying condition (4) of F to $y_0\ F\ y$ and $w_0\ F\ w'$, we have that there exist y_1 and w_1 such that $y_0\ F\ y_1$, $y_1\ F\ y$, $w_0\ F\ w_1$ and $w_1\ F\ w'$. Hence,

(2) $y_1 \models\ !\gamma$, $y_1 \leq y$ and $w_1 \leq w'$.

By using condition (5) once again, there exists a w_2 such that $w_0\ F\ w_2$, for which $u \cdot w_2 \leq w_1 \cdot u$ holds for every u. To show that $x \models\ ?\theta$, we assume now that

(3) $u \models \theta$ implies $v\ G\ u$, for every u.

Moreover, suppose that $t \models \alpha$. Since $y_1 \models\ !\gamma$ and $w_2 \models\ !\delta$,

(4) $y_1 \cdot t \cdot w_2 \models\ !\gamma * \alpha * !\delta$.

Then $y_1 \cdot w_1 \cdot t \models\ ?\theta$, by using the assumption that the upper sequent of $(? \rightarrow)$ is true and the fact that $y_1 \cdot t \cdot w_2 \leq y_1 \cdot w_1 \cdot t$. But, this implies that, for every v', if $u \models \theta$ implies $v'\ G\ u$ for every u, then $v'\ G\ y_1 \cdot w_1 \cdot t$. Taking v for v' and using (3), we have $v\ G\ y_1 \cdot w_1 \cdot t$. On the other hand, $y_1 \cdot w_1\ F\ y \cdot w'$ holds by using condition (6) of F, and therefore, by condition (10):

(5) for some w^*, $v\ G\ y_1 \cdot w_1 \cdot w^*$ and, for every w^+, $v\ G\ y_1 \cdot w_1 \cdot w^+$ if and only if $w^*\ G\ w^+$.

Let $w^+ = t$ in (5). Then we have $w^*\ G\ t$. Thus we have shown that

(6) $t \models \alpha$ implies $w^*\ G\ t$, for every t.

Since $z \models\ ?\alpha$, it follows from (6) that $w^*\ G\ z$. Taking z for w^+ in (5), we have that $v\ G\ y_1 \cdot w_1 \cdot z$ under assumption (3). This means that $y_1 \cdot w_1 \cdot z \models\ ?\theta$. Combining the fact that $y_1 \cdot w_1 \cdot z \leq y \cdot w' \cdot z$ with (1), we have proved that $x \models\ ?\theta$.

Next we will show completeness. Suppose that $\alpha = \langle \Gamma \rightarrow \theta \rangle$ is not provable in FL_σ^M. By Corollary 6.8, there exists a modal unital σ-quantale

A with its assignment f such that $1 \not\leq f(\alpha)$ in **A**. Let $\mathbf{A}^\dagger$ be a (complete) so-monoid obtained by turning **A** upside down, with the order $\leq^\dagger$. That is, the order $\leq^\dagger$ of $\mathbf{A}^\dagger$ is defined by:

$$x \leq^\dagger y \quad \text{if and only if} \quad y \leq x \quad (\text{in } \mathbf{A}).$$

Let R and S be subsets of A defined by $R = \{!x : x \in A\}$ and $S = \{?x : x \in A\}$, and let $\langle F, G\rangle$ be a pair of accessibility relations on $\mathbf{A}^\dagger$ induced by R and S. Furthermore, let g be a valuation on $\mathbf{A}^\dagger$ defined by $g(p) = \{x : f(p) \leq^\dagger x\}$ (here we let $\mu = f(0) = 0$). Then it is easy to see that $\langle \mathbf{A}^\dagger, F, G, g, \mu\rangle$ is a Kripke model for FL_σ^M. By induction we can show that for every $x \in A$ and every formula β:

$$x \models \beta \quad \text{if and only if} \quad f(\beta) \leq^\dagger x, \quad \text{if and only if} \quad x \leq f(\beta).$$

In particular, we have that $1 \not\models \alpha$. Thus, we have shown completeness.

References

[1] V. M. Abrusci, Sequent calculus for intuitionistic linear propositional logic, in: *Mathematical Logic, Proceedings of the Summer School and Conference 'Heyting '88'*, P. P. Petkov ed., Plenum Press, 1990, 223–242.

[2] A. R. Anderson and N. D. Belnap Jr., *Entailment: The Logic of Relevance and Necessity, Vol. 1*, Princeton University Press, 1975.

[3] A. Avron, The semantics and proof theory of linear logic, *Theoretical Computer Science 57* (1988), 161–184.

[4] K. Došen, Sequent systems and groupoid models. I, *Studia Logica 47* (1988), 353–385.

[5] ———, Sequent systems and groupoid models. II, *Studia Logica 48* (1989), 41–65.

[6] ———, A brief survey of frames for the Lambek calculus, *Zeitschrift für mathematische Logik und Grundlagen der Mathematik 38* (1992), 179–187.

[7] J. M. Dunn, Relevance logic and entailment, in: *Handbook of Philosophical Logic, Vol. III*, D. Gabbay and F. Guenthner eds., D. Reidel Publishing Co., 1986, 117–224.

[8] M. P. Fourman and D. S. Scott, Sheaves and logic, Applications of Sheaf Theory to Algebra, Analysis and Topology, M. P. Fourman, C. J. Mulvey and D. S. Scott eds., *Lecture Notes in Mathematics*, Springer-Verlag, 1979, 302–401.

[9] J.Y. Girard, Linear logic, *Theoretical Computer Science 50* (1987), 1–102.

[10] G. Grätzer, *General Lattice Theory*, Academic Press, 1978.

[11] V.N. Grishin, Predicate and set-theoretic calculi based on logic without contractions, *Math. USSR Izvestiya 18* (1982), 41–59.

[12] H.M. MacNeille, Partially ordered sets, *Transactions of the American Mathematical Society 42* (1937), 416–460.

[13] L.L. Maksimova, An interpretation and separation theorems for the logical systems E and R, *Algebra and Logic 10* (1971), 232–241.

[14] ———, Implication lattices, *Algebra and Logic 12* (1973), 445–467.

[15] R.N. McKenzie, G.F. McNulty and W.F. Taylor, *Algebras, Lattices, Varieties, Vol. I*, Wadsworth & Brooks/Cole, 1987.

[16] R.K. Meyer, Conservative extension in relevant implication, *Studia Logica 31* (1972), 39–46.

[17] R.K. Meyer, J.M. Dunn and H. Leblanc, Completeness of relevant quantification theories, *Notre Dame Journal of Formal Logic 15* (1974), 97–121.

[18] R.K. Meyer and R. Routley, Algebraic analysis of entailment. I, *Logique et Analyse 15* (1972), 407–428.

[19] ———, Classical relevant logic. I, *Studia Logica 32* (1973), 51–66.

[20] C.J. Mulvey, &, Second Topology Conference (Taormina, 1984), *Rendiconti del Circolo Matematico di Palermo, Ser. 2, Supplement No 12* (1986), 94–104.

[21] S.B. Niefield and K.I. Rosenthal, Constructing locales from quantales, *Mathematical Proceedings of the Cambridge Philosophical Society 104* (1988), 215–234.

[22] H. Ono, Semantical analysis of predicate logics without the contraction rule, *Studia Logica 44* (1985), 187–196.

[23] ———, Structural rules and a logical hierarchy, in: *Mathematical Logic, Proceedings of the Summer School and Conference 'Heyting '88'*, P.P. Petkov ed., Plenum Press, 1990, 95–104.

[24] ———, Phase structures and quantales, Technical Report No 4, The Division of Mathematical & Information Sciences, Faculty of Integrated Arts & Sciences, Hiroshima University, 1990.

[25] ———, Algebraic aspect of logics without structural rules, in: *Contemporary Mathematics 3, Proceedings of the International Conference on Algebra Honoring A. Mal'cev*, Novosibirsk, 1989, L. A. Bokut, Yu. L. Ershov, O. H. Kegel and A. I. Kostrikin eds., American Mathematical Society, 1992, 601–621.

[26] H. Ono and Y. Komori, Logics without the contraction rule, *The Journal of Symbolic Logic 50* (1985), 169–201.

[27] K. I. Rosenthal, A note on Girard quantales, *Cahiers de Topologie et Géometrie Différentielle Catégoriques 31* (1990), 3–11.

[28] ———, *Quantales and their Applications*, Longman Scientific & Technical, 1990.

[29] R. Routley and R. K. Meyer, The semantics of entailment. I, in: *Truth, Syntax, Modality*, H. Leblanc ed., North-Holland, 1973, 194–243.

[30] ———, The semantics of entailment. III, *Journal of Philosophical Logic 1* (1972), 192–208.

[31] G. Sambin, Intuitionistic formal spaces and their neighbourhood, in: *Logic Colloquium '88*, R. Ferro, C. Bonotto, S. Valentini and A. Zanardo eds., North-Holland, 1989, 261–285.

[32] ———, Pretopologies and completeness proofs, Manuscript, 1989.

[33] A. S. Troelstra, *Lectures on Linear Logic*, Center for the Study of Language and Information, University of Stanford, 1992.

[34] D. N. Yetter, Quantales and (noncommutative) linear logic, *The Journal of Symbolic Logic 55* (1990), 41–64.

[35] B. A. Davey and H. A. Priestley, *Introduction to Lattices and Order*, Cambridge University Press, 1990.

The Semantics of Pretopologies

Giovanni Sambin

Department of Pure and Applied Mathematics
University of Padova

The notion of pretopology, introduced in [S2], may be used as a basic tool to obtain a complete semantics for several substructural logics. The present paper is a short and partial survey of what is known at the moment on pretopologies, with particular attention to the completeness proofs. Some suggestions for further work are also given.

1 Origins and Definition of Pretopologies

By the summer of 1986, P. Martin-Löf and the author had reached a satisfactory formulation of the first notions of an intuitionistic[1] point-free, or formal, approach to topology (cf. [S1]). Roughly speaking, the definition of formal topology is obtained by assuming all the formal properties of the collection ΩX of open subsets of an arbitrary topological space X which can be expressed without mentioning its points $x, y, \ldots \in X$; of course, one has to choose carefully a stock of primitives, and the best choice seems to be to consider the operation of intersection, the relation of covering (in the sense of inclusion in the union) and the property of being inhabited.[2] Moreover, one assumes that opens are given by means of a base (which may well happen to be a set even when ΩX is a proper category, in the sense of [ITT]). We thus have formally:

— a set S, of formal basic neighbourhoods $a, b, c, \ldots$, which is supposed to be closed under a binary operation $\cdot$ and to contain a distinguished element 1,

1 Which here more specifically means: within the framework of intuitionistic type theory as developed by Martin-Löf, see e.g. [ITT].

2 Note that the choice of primitives in the standard approach to point-free topology is a bit different: the property of being inhabited is usually not considered, and one has either an operation giving directly arbitrary joins or a covering relation (called coverage in [J]), which differs from ours since it is interpreted through equality.

— an infinitary relation $a \lhd U$ (read 'a is covered by U') between basic neighbourhoods and subsets of S, i.e. coverings,

— a predicate $Pos(a)$ defined for any $a \in S$.

Now one writes down as many as possible of the properties which hold under the interpretation in ΩX: S as a base for ΩX, $\cdot$ as intersection, 1 as the whole X, $a \lhd U$ as $a \subseteq \bigcup\{b : b \in U\}$, $Pos(a)$ as $(\exists x \in X)\, x \in a$. In a certain sense (to be specified below), one cannot do better than the following:

Definition. A *formal topology* is a structure $\mathcal{A} = (S, \cdot, 1, \lhd, Pos)$ where

— $(S, \cdot, 1)$ is a commutative monoid;

— $\lhd$ is a *cover*, i.e., for any $a \in S$, $U \subseteq S$, it satisfies:

reflexivity $\dfrac{a \in U}{a \lhd U}$

transitivity $\dfrac{a \lhd U \qquad (\forall b \in U)\, b \lhd V}{a \lhd V}$

$\cdot$-right $\dfrac{a \lhd U \qquad a \lhd V}{a \lhd \{b \cdot c : b \in U, c \in V\}}$

$\cdot$-left $\dfrac{a \lhd U}{a \cdot b \lhd U}$

— *Pos* is a *positivity predicate*, i.e., it satisfies:

monotonicity $\dfrac{a \lhd U \qquad Pos(a)}{(\exists b \in U) Pos(b)}$

positivity $\dfrac{Pos(a) \to a \lhd U}{a \lhd U}$

It is convenient to adopt systematically a few abbreviations: $U \lhd V$ for $(\forall b \in U)\, b \lhd V$, $a \lhd b$ for $a \lhd \{b\}$ (and then one sees that $\lhd$ gives a preorder on S), $U \cdot V$ for $\{b \cdot c : b \in U, c \in V\}$ and finally $Pos(U)$ for $(\exists b \in U) Pos(b)$. With this notation, any visible reference to quantification and set formation in the definition above disappears.

Starting with the notion above, it is possible to reconstruct formal, or point-free, counterparts of the usual topological notions. First of all, what is a (formal) open subset? In ΩX, two sets of basic neighbourhoods U, V

determine the same open subset if they have equal union, i.e., if they cover each other; we thus put

$$U =_{\mathcal{A}} V \equiv (U \triangleleft V \text{ and } V \triangleleft U).$$

Since $=_{\mathcal{A}}$ is an equivalence relation on the powerset $\mathcal{P}S$, we may consider $\mathcal{P}S/_{=_{\mathcal{A}}}$; by the assumptions on $\triangleleft$, the equivalence class of $U \cdot V$ and $\bigcup_{i \in I} U_i$ defines finite meet and arbitrary join, respectively, on $\mathcal{P}S/_{=_{\mathcal{A}}}$, which then turns out to be a locale[3], called *Open*$(\mathcal{A})$. A (formal) open is thus an equivalence class with respect to $=_{\mathcal{A}}$. Conversely, any locale Ω is presentable in this way: if S is a base for Ω, that is, $u = \bigvee\{a \in S : a \leq u\}$ for any $u \in \Omega$ (cf. e.g. [TvD], p. 719), then put $a \triangleleft U \equiv a \leq \bigvee U$ for any $a \in S$, $U \subseteq S$.

In this sense, one cannot find better conditions on $\triangleleft$. Also, at this point, by analogy with the completeness of the topological interpretation, a close connection between covers and intuitionistic logic is to be expected: in fact, we will obtain it as a corollary.

Secondly, what is a (formal) point? Again, writing down as many as possible of the properties satisfied by a generic point in a topological space, we reach the following characterization: a point α over $\mathcal{A}$ is a subset of S which satisfies

$$1 \in \alpha, \qquad a \cdot b \in \alpha \text{ iff } a \in \alpha \text{ and } b \in \alpha$$

$$\frac{a \triangleleft U \qquad a \in \alpha}{(\exists b \in U)\, b \in \alpha}$$

$$a \in \alpha \rightarrow Pos(a), \quad \text{i.e.} \quad \alpha \subseteq Pos.$$

The collection of points over $\mathcal{A}$ is denoted by $Pt(\mathcal{A})$. It can be shown easily that they correspond biunivocally to completely prime filters of *Open*$(\mathcal{A})$, which is the way in which points are reconstructed in the standard approach to point-free topology (see e.g. [J], p. 41).

Starting with the definitions above, an intuitionistic approach to point-free topology can be developed (see [S1], [BS]; a full-scale monograph is planned).

Apart from its motivation in a foundational choice, the notion of cover has revealed to be solid and flexible enough technically to be adapted to different situations, and in particular to logics in which structural rules are kept under control.

Structural rules are, so to speak, built both in the intuition and in the formal definition of cover. So, if the basic idea — interpreting a formula or proposition as a formal open — is kept, the first task becomes that of separating, among the conditions defining covers, those which are necessary

3 Some authors would call it a frame (e.g. [J]) or a complete Heyting algebra.

to define the quotient on $\mathcal{P}S$ and its operations $\cdot$ and $\vee$, from those which correspond to structural rules.

For that, we must see in more detail how the quotient on $\mathcal{P}S$ is defined. The most convenient way to do it is to start with the biunivocal correspondence between infinitary relations like $\triangleleft$ and operators on subsets, i.e. functions $\mathcal{C} : \mathcal{P}S \to \mathcal{P}S$, which is defined as follows: given $\triangleleft$, for any $U \subseteq S$ we put

$$\mathcal{C}U \equiv \{a \in S : a \triangleleft U\},$$

and conversely, given $\mathcal{C}$ we put

$$a \triangleleft U \equiv a \in \mathcal{C}U.$$

It is easy to see that $\triangleleft$ satisfies reflexivity and transitivity iff the corresponding $\mathcal{C}$ is a closure operator; i.e., $U \subseteq \mathcal{C}U$, $\mathcal{C}\mathcal{C}U \subseteq \mathcal{C}U$, $U \subseteq V \to \mathcal{C}U \subseteq \mathcal{C}V$ (note that the second and third condition together can equivalently be replaced by $U \subseteq \mathcal{C}V \to \mathcal{C}U \subseteq \mathcal{C}V$, which is more easily seen to correspond to the transitivity of $\triangleleft$). Then, putting $U =_{\mathcal{C}} V \equiv U \triangleleft V \,\&\, V \triangleleft U$ as we did above for covers, we have $U =_{\mathcal{C}} V$ iff $\mathcal{C}U = \mathcal{C}V$, so that it is immediate that $=_{\mathcal{C}}$ is an equivalence relation on $\mathcal{P}S$; we write $[U]_{\mathcal{C}}$ for the equivalence class of U.

Moreover, since any closure operator $\mathcal{C}$ satisfies $\mathcal{C}(\bigcup_{i \in I} U_i) = \mathcal{C}(\bigcup_{i \in I} \mathcal{C}U_i)$ for any family of subsets $(U_i)_{i \in I}$, it is also immediate that $=_{\mathcal{C}}$ respects arbitrary unions, or equivalently that putting $\bigvee_{i \in I} [U_i] = [\bigcup_{i \in I} U_i]$ is a good definition, and that it gives a supremum on $\mathcal{P}S/_{=_{\mathcal{C}}}$.

In other words, the conditions of reflexivity and transitivity on $\triangleleft$ are sufficient to define $\mathcal{P}S/_{=_{\mathcal{C}}}$ and to equip it with the structure of a complete lattice. Are they necessary? Yes. In fact, the equivalence class $[U]_{\mathcal{C}}$ has a privileged representative, namely $\mathcal{C}U$, which is the greatest in the class, since $V =_{\mathcal{C}} U$ implies $V \subseteq \mathcal{C}U$. So, given *any* equivalence relation θ on $\mathcal{P}S$, we extract an operator $\mathcal{C}$ by putting, for any $U \subseteq S$, $\mathcal{C}U \equiv \bigcup\{V : V \,\theta\, U\}$ or, more constructively, $a \in \mathcal{C}U \equiv (U \cup \{a\} \,\theta\, U)$; with a little work on definitions, one can see that θ is an equivalence which respects arbitrary unions iff the operator $\mathcal{C}$ so associated is a closure operator. As a partial conclusion, the following notions are just different formulations of one and the same concept:

(1) reflexive and transitive $\triangleleft$,

(2) closure operator on $\mathcal{P}S$,

(3) equivalence relation on $\mathcal{P}S$ respecting arbitrary unions, i.e. congruence on $(\mathcal{P}S, \cup)$.

Now it is easier to determine which additional condition we should put on $\triangleleft$ in order to lift the operation $\cdot$ to the quotient $\mathcal{P}S/_{=_\mathcal{C}}$. In fact, the operation $[U]\cdot[V] \equiv [U\cdot V]$ is well defined iff $U =_\mathcal{C} U'$ & $V =_\mathcal{C} V' \Rightarrow U\cdot V =_\mathcal{C} U'\cdot V'$, which is equivalent to $U' \triangleleft U$ & $V' \triangleleft V \Rightarrow U'\cdot V' \triangleleft U\cdot V$, and hence also to $a \in \mathcal{C}U$ & $b \in \mathcal{C}V \Rightarrow a\cdot b \in \mathcal{C}(U\cdot V)$; i.e.

$$\text{stability} \qquad \frac{a \triangleleft U \qquad b \triangleleft V}{a\cdot b \triangleleft U\cdot V}$$

or equivalently, in terms of the operator $\mathcal{C}$, $\mathcal{C}U\cdot\mathcal{C}V \subseteq \mathcal{C}(U\cdot V)$.

We say that $\triangleleft$ is a *precover* if it is reflexive, transitive and stable; and that $\mathcal{C}$ is compatible with $\cdot$ if $\mathcal{C}U\cdot\mathcal{C}V \subseteq \mathcal{C}(U\cdot V)$. We thus have that:

(1) precover,

(2) closure operator compatible with $\cdot$,

(3) congruence on $(\mathcal{P}S, \cdot, \cup)$

are three different formulations of the same mathematical content.

In other words, precovers embody in a simple form the information which is necessary to generate congruences, i.e. to define a quotient on $\mathcal{P}S$ and to provide it with operations $\cdot$ and $\vee$, which is exactly half of what we were looking for (the other half is to identify structural rules, which is done in the next section). We thus put:

Definition. A *pretopology* is a structure $\mathcal{F} = (S, \cdot, 1, \triangleleft_\mathcal{F})$ where $(S, \cdot, 1)$ is a commutative monoid, and $\triangleleft_\mathcal{F}$ a precover on it.

We reserve the letter $\mathcal{F}$ for pretopologies and for the closure operator associated to a precover. As one can expect from the discussion above, just as all locales can be presented as *Open*$(\mathcal{A})$ for some formal topology $\mathcal{A}$, it turns out that all quantales (or better, all unital commutative quantales, see e.g. [R] or [T] for definitions) can be presented as *Open*$(\mathcal{F})$ for some pretopology $\mathcal{F}$ (for details, see [BS]). But then, since quantales give a complete algebraic semantics for basic linear logic (cf. [T]), it is also to be expected that linear logic can be interpreted directly in a pretopology; we will see that in the next section.

2 The Semantics of Pretopologies and its Uniform Completeness

Independently of its topological origin, the notion of pretopology is simple enough to be given a direct intuitive interpretation. In fact, we may look at the set S as a universe of objects, or of atomic data, or of tokens of information, as long as we conceive them as actually or concretely given, and thus bound to their occurrence. Then $a \cdot b$ is the combination of a with b, or the result of putting together a with b, and 1 is an object which costs nothing to be produced, or a token carrying no information. With respect to formal topologies, the interpretation is somehow shifted to a lower level of abstraction: the elements of S are now considered so concrete to be indivisible, which makes them so similar to points, rather than neighbourhoods. This is due to the fact that we assume no relation to hold beetween $a \cdot b$ and a, b (in particular, neither $a = a \cdot a$ nor $a \cdot b \lhd a$).

The task of the relation $\lhd$, or of its associated operator $\mathcal{F}$, is to express the 'logic' on objects: $a \lhd U$ means that a is forced to fall under the 'logically' correct concept $\mathcal{F}U$ generated by the crude set U, or to belong to the least 'well-behaved' subset containing U. In this sense, $\lhd$ can be seen as a generalization of the crude membership relation $\in$.

Given the above intuitive justification of pretopologies, the link with logic is then at hand: the interpretation of a formula is a 'well-behaved' concept of the universe of objects, i.e. a subset of the form $\mathcal{F}U$ for some U. Since $U \lhd V$ iff $\mathcal{F}U \lhd \mathcal{F}V$, the relation $U \lhd V$ expresses the fact that the concept (generated by) V is a consequence of the concept (generated by) U; in particular, $\mathcal{F}U$ is true if it is a consequence of the token carrying no information, or if it includes the object with no cost, i.e., if $1 \lhd U$. So $\lhd$ lends itself to be the interpretation of the derivability relation $\vdash$.

What about the interpretation of connectives? Let us consider the collection $Sat(\mathcal{F})$ of all $\mathcal{F}$-saturated subsets, or $\mathcal{F}$-subsets, i.e. subsets of the form $\mathcal{F}U$ for some U. Recalling that $\mathcal{F}U$ is the greatest in the equivalence class $[U]_{\mathcal{F}}$, the mapping $[U] \mapsto \mathcal{F}U$ is a bijection from the quotient $\mathcal{P}S/_{=_{\mathcal{F}}}$ onto $Sat(\mathcal{F})$, and hence all the operations of $\mathcal{P}S/_{=_{\mathcal{F}}}$ are immediately transferred to $Sat(\mathcal{F})$. The result is that $Sat(\mathcal{F})$ has a complete lattice structure, with $\bigvee_{i \in I} \mathcal{F}U_i \equiv \mathcal{F}\big(\bigcup_{i \in I} \mathcal{F}U_i\big)$ and $\bigwedge_{i \in I} \mathcal{F}U_i \equiv \bigcap_{i \in I} \mathcal{F}U_i$, which is what is needed to interpret additives,[4] connectives & and $\oplus$ and the quantifiers $\forall$ and $\exists$. The result of transferring $\cdot$ from subsets to $\mathcal{F}$-subsets is the operation $\mathcal{F}U \cdot_{\mathcal{F}} \mathcal{F}V \equiv \mathcal{F}(\mathcal{F}U \cdot \mathcal{F}V)$, which will be used to interpret multiplicative conjunction $\otimes$.

We now have to define an operation $\to_{\mathcal{F}}$ corresponding to linear implication $\multimap$. Since $\multimap$-rules are to be valid, we must have $W \cdot U \lhd V$ iff $W \lhd U \to_{\mathcal{F}} V$ for any W, U, V; hence, in particular, $a \cdot U \lhd V$ iff $a \lhd U \to_{\mathcal{F}} V$, for any a; but then nothing prevents us from taking this as the definition, and we put $U \to_{\mathcal{F}} V \equiv \{a \in S : a \cdot U \lhd V\}$.

Finally, any pretopology has also three distinguished $\mathcal{F}$-subsets, the whole universe S and the concepts $\mathcal{F}\emptyset$ and $\mathcal{F}1$, generated by $\emptyset$ and $\{1\}$

4 We keep Girard's notation and terminology, as in [G].

respectively: they will be the interpretation of the constant atomic formulas $\top$, 0 and 1 respectively.

Note that the subset *Pos* of positive neighbourhoods does not appear in the definition of pretopology; this corresponds to the fact that no assumption is made on the constant $\bot$ in basic linear logic. In fact, we may think of the subset in which $\bot$ is interpreted as the subset of impossible objects (like a magic wand, or a proof of consistency of ZF, ...), and hence the complement of *Pos*; the behaviour of $\bot$ will be determined by how impossible objects are conceived. In any case, in line with Brouwer's spirit, the negation of a concept $\mathcal{F}U$ is defined to consist of all objects which, combined with an object in $\mathcal{F}U$, give an impossible object.

It is now almost routine to turn the remarks above into a precise definition of interpretation. Let us do it first for the propositional fragment. Given any assignment of truth values in $\mathcal{F}$ to propositional variables P_1, P_2,..., i.e. any function $V : \{P_1, P_2, \ldots\} \to Sat(\mathcal{F})$, and the value of $V(\bot)$, we extend it to all propositional formulas by means of the following inductive clauses:

$$\begin{aligned}
V(A \otimes B) &\equiv \mathcal{F}(V(A) \cdot V(B)) \\
V(A \& B) &\equiv V(A) \cap V(B) \\
V(A \oplus B) &\equiv \mathcal{F}(V(A) \cup V(B)) \\
V(A \multimap B) &\equiv V(A) \to_{\mathcal{F}} V(B) \\
V(\top) \equiv S, \qquad & V(1) \equiv \mathcal{F}1, \qquad V(0) \equiv \mathcal{F}\emptyset.
\end{aligned}$$

To check validity, it is convenient to adopt a two-sided sequent calculus with exactly one formula at the right-hand side, called BL for 'basic linear'; the reader can find it spelled out in [S2] and [S3], or obtain it from ILL in [T] by suppressing exponentials and their rules (and paying attention to different notation).

We say that a sequent $B_1, \ldots, B_n \vdash A$ is valid in a pretopology $\mathcal{F}$ for a valuation V if $V(B_1) \cdot \ldots \cdot V(B_n) \lhd V(A)$ holds; in particular, this means that a formula A is valid if $1 \lhd V(A)$, or equivalently $1 \in V(A)$ since $V(A)$ is $\mathcal{F}$-saturated. We say that $\Gamma \vdash A$ is valid if it is valid for every valuation in every pretopology. It is easy to check that all rules of BL are valid; a detailed proof is given in [S3], but the reader who prefers to do it by himself will experience directly the close similarity of $\vdash_{BL}$ with $\lhd$.

To prove completeness, the idea is, as usual, to build up from syntax a pretopology in which exactly the derivable formulas are valid. So, we define as usual an equivalence relation $=_{BL}$ on the set Frm of formulas by putting $(A =_{BL} B) \equiv (A \vdash B \text{ and } B \vdash A)$, and define an operation $\otimes$ on the quotient $Frm/_{=_{BL}}$ by putting $[A] \otimes [B] \equiv [A \otimes B]$; by the rules for $\otimes$ and 1, the structure $(Frm/_{=_{BL}}, \otimes, [1])$ is a commutative monoid, called the Lindenbaum base (since it is a reduct of the full Lindenbaum algebra). From now on, it is absolutely harmless to forget about equivalence classes and suppress all square brackets, since any notion depends only on $\vdash_{BL}$, which respects $=_{BL}$.

We now wish to provide the Lindenbaum base with a precover $\lhd_{\mathrm{Pr}}$, expressing its 'logic', which is to be as close as possible to the logic given by $\vdash_{BL}$. So we start by putting $A \lhd_{\mathrm{Pr}} B$ iff $A \vdash_{BL} B$; i.e., define $\mathrm{Pr}\, B$ to be $\{A : A \vdash_{BL} B\}$, the downward cone generated by B in the order induced by $\vdash_{BL}$. The minimal closure operator is then obtained by closing the family of all generated cones under arbitrary intersections; so, for any set of formulas Σ, we put $\mathrm{Pr}\,\Sigma \equiv \bigcap\{\mathrm{Pr}\, C : \Sigma \subseteq \mathrm{Pr}\, C\}$. It can be proved easily that, because of properties of the implication $\multimap$, the associated infinitary relation $A \lhd_{\mathrm{Pr}} \Sigma \equiv A \in \mathrm{Pr}\,\Sigma$, that is, $A \lhd_{\mathrm{Pr}} \Sigma$ iff $(\forall C)(\Sigma \subseteq \mathrm{Pr}\, C \rightarrow A \lhd_{\mathrm{Pr}} C)$, satisfies stability; hence $\lhd_{\mathrm{Pr}}$ is a precover. We thus have a pretopology Pr_{BL}, called the BL-canonical pretopology; in more abstract terms, $Sat(\mathrm{Pr}_{BL})$ is exactly the Dedekind-MacNeille completion of the Lindenbaum base (cf. [O] or [T]).

The next step is to find a valuation V on Pr such that A is valid under V exactly when $\vdash_{BL} A$ is provable. Assigning to the propositional variable P_i the Pr-subset $\mathrm{Pr}\, P_i$ is a simple guess, but it works:

Lemma on the canonical valuation. *If* $V(P_i) = \mathrm{Pr}\, P_i$ *and* $V(\bot) = \mathrm{Pr}\,\bot$, *then* $V(A) = \mathrm{Pr}\, A$ *for any formula* A.

An easy proof, by induction, is given in detail in [S3]; we review here only two sample cases. By the induction hypothesis and the definition of valuation, to prove $V(A\&B) = \mathrm{Pr}\,(A\&B)$ it is enough to show that $\mathrm{Pr}\, A \cap \mathrm{Pr}\, B = \mathrm{Pr}\,(A\&B)$, that is, $C \vdash A\&B$ iff $C \vdash A$ and $C \vdash B$, for any formula C, which is immediate by &-rules. Similarly, for $\oplus$, we have to prove $\mathrm{Pr}\,(\mathrm{Pr}\, A \cup \mathrm{Pr}\, B) = \mathrm{Pr}\,(A \oplus B)$, which, by the characterization of Pr, follows from $\mathrm{Pr}\, A \cup \mathrm{Pr}\, B \subseteq \mathrm{Pr}\, C$ iff $A \oplus B \in \mathrm{Pr}\, C$, that is, $A \vdash C$ and $B \vdash C$ iff $A \oplus B \vdash C$, for any formula C, which is also immediate, by $\oplus$-rules.

Completeness is now an easy corollary: if A is valid, then, in particular, it is valid for the canonical valuation, which by the lemma means that $1 \lhd_{\mathrm{Pr}} A$, i.e. $1 \vdash_{BL} A$, from which also $\vdash_{BL} A$, since $\vdash_{BL} 1$.

The same method of proof is immediately extended to predicate logic, once the correct notion of interpretation is defined. In fact, since $Sat(\mathcal{F})$ has arbitrary joins and meets, which will serve for the interpretation of quantifiers, the structure $Sat(\mathcal{F})$ of truth-values is left unchanged. But then a model is obtained from a standard model, by substituting the classical structure of truth values, true and false or 1 and 0, by the richer structure $Sat(\mathcal{F})$. Thus we say that $\mathcal{D} = (D, f^{\mathcal{D}}, ..., R^{\mathcal{D}}, ..., \mathcal{F})$ is an $\mathcal{F}$-structure if $(D, f^{\mathcal{D}}, ...)$ is a domain of interpretation for terms as usual (so

D is a set, and for any n-ary function sign f, $f^{\mathcal{D}}$ is a function $D^n \to D$) and, for any relation sign R, $R^{\mathcal{D}}$ gives the truth-value of R on any n-tuple $(d_1, \ldots, d_n)$, that is, $R^{\mathcal{D}}$ is a function $D^n \to Sat(\mathcal{F})$. Then the valuation of atomic formulas is determined as usual by an assignment σ to individual variables in D, in the sense that we put $V^\sigma(Rt_1 \ldots t_n) \equiv R^{\mathcal{D}} t_1^\sigma \ldots t_n^\sigma$, where t^σ is the value of the term t under σ. Writing $V^\sigma(Ad)$ as a shorthand for $V^{\sigma_d}(Ax)$, σ_d differing from σ only on x and $\sigma_d(x) = d$, the previous inductive clauses are supplemented by:

$$\begin{aligned} V^\sigma(\exists x Ax) &\equiv \mathcal{F}\Big(\bigcup_{d \in D} V^\sigma(Ad)\Big) \\ V^\sigma(\forall x Ax) &\equiv \bigcap_{d \in D} V^\sigma(Ad). \end{aligned}$$

Validity of the rules for quantifiers is easy to check. To prove completeness, in the style of Henkin, we consider the domain T formed by all terms, and adjoin it to the BL-canonical pretopology Pr_{BL}. To complete the definition of a Pr-structure $\mathcal{T}$, the new trick is to put, again self-referentially, $R^{\mathcal{T}}(t_1, \ldots, t_n) \equiv \mathrm{Pr}\,(R(t_1, \ldots, t_n))$. Then by definition, choosing the identity assignment $\iota : x \mapsto x$,

$$V^\iota(R(t_1, \ldots, t_n) \equiv R^{\mathcal{T}}(t_1^\iota, \ldots, t_n^\iota) \equiv \mathrm{Pr}\,(R(t_1, \ldots, t_n)).$$

So the assumption of the lemma on the canonical valuation is satisfied, and to complete its proof it is enough to check the inductive steps for quantifiers (with some care about names of variables; see [S3]). The conclusion of the completeness proof is exactly as before. We thus have:

Basic completeness theorem. *Basic linear predicate logic BL is complete with respect to the semantics given by $\mathcal{F}$-structures.*

A nice peculiarity of pretopologies, and of the method of proof above, is that they can be adjusted with very little work to give complete semantics for extensions of BL too. In fact, since no step in the proof makes use of negative information, exactly the same definitions and proofs apply to any sequent calculus L extending BL, and thus give a Pr_L-structure for which the lemma on the canonical valuation holds. Then to prove completeness of L it is enough, first, to specify with respect to which subclass of pretopologies completeness is to be proved (this subclass is determined simply by the conditions making L valid), and, second, to show that Pr_L is in that subclass.

A few examples may illustrate how simple the method is. Let us consider BLw, the extension of BL obtained by adding the rule **w** of weakening

$$\frac{\Gamma \vdash A}{\Gamma, B \vdash A}$$

By an automatic translation, we see that **w** is valid in a pretopology iff $W \lhd Z \Rightarrow W \cdot U \lhd Z$ holds for all W, U, Z. A tiny bit of creative work is needed to see that this condition is equivalent to $W \cdot U \lhd W$, hence to $a \cdot b \lhd a$ and finally to $b \lhd 1$ for any b; thus we restrict ourselves to pretopologies satisfying this condition. Now it is clear that the canonical structure Pr_{BLw} is a pretopology satisfying $b \lhd 1$ for any b, since $B \vdash_{BL} 1$ indeed holds by weakening. So BLw is complete with respect to pretopologies in which any object b is covered by 1.

Quite similarly, the rule **c** of contraction

$$\frac{\Gamma, B, B \vdash A}{\Gamma, B \vdash A}$$

is valid if $W \cdot U \cdot U \lhd Z \Rightarrow W \cdot U \lhd Z$, which is equivalent to $U \lhd U \cdot U$, and also to $a \lhd a \cdot a$ for any a. Clearly, Pr_{BLc} satisfies $a \lhd a \cdot a$, since $A \vdash A \otimes A$ follows by contraction from $A, A \vdash A \otimes A$; so BLc is complete with respect to pretopologies in which any object a is covered by $a \cdot a$.

We have thus fully accomplished also the task of identifying, among the properties of covers, those which correspond to structural rules. In fact it is easy to see that, assuming stability, $\cdot$-left is equivalent to $a \lhd 1$ and $\cdot$-right to $a \lhd a \cdot a$ and that conversely stability follows from $\cdot$-rules. So covers may alternatively be defined as precovers satisfying $a \lhd 1$ and $a \lhd a \cdot a$ for any a.

Usual intuitionistic logic may be reformulated as BL extended by **w**, **c** and the axiom $\bot \vdash A$, or equivalently $\bot \vdash 0$; thus, putting together the arguments above, which is possible since they use only positive information, we have that intuitionistic logic is complete with respect to the semantics of covers and valuations with $V(\bot) = V(0)$.[5]

The method is general enough to be applied also to other extensions of BL; a noteworthy example is Girard's linear logic (without exponentials), which is obtained by adding the axiom **dn** of double negation $\neg\neg A \vdash A$. A valuation V on a pretopology $\mathcal{F}$ satisfies **dn** iff $(U \to V(\bot)) \to V(\bot) \lhd U$ for any U; putting $\bot \equiv V(\bot)$ and $U^{\bot} \equiv U \to \bot$, it is easy to see that then $\mathcal{F}U$ must be equal to $U^{\bot\bot}$, and so $\mathcal{F}$ is nothing but the phase space

5 Note that this is not a proof of completeness with respect to formal topologies, because of the absence of *Pos*; actually, it is still unknown to me whether the canonical pretopology for intuitionistic logic can be provided constructively with a positivity predicate making it a formal topology.

(cf. [G]) on the base $(S, \cdot, 1)$ and with dualizer $\bot$. Since conversely any phase space is a pretopology, it comes out that phase spaces are exactly those pretopologies in which **dn** is valid. Also, for any set Σ of formulas, $A \in (\Sigma \to \Pr\bot) \to \Pr\bot \equiv (\Sigma \to \Pr\bot) \otimes A \subseteq \Pr\bot \equiv (\forall C)((\forall B \in \Sigma)$ $B \otimes C \vdash \bot \Rightarrow A \otimes C \vdash \bot)$ iff, by **dn**, $(\forall C)((\forall B \in \Sigma)\, B \vdash C \Rightarrow A \vdash C)$ $\equiv A \in \Pr\Sigma$; that is, $\Pr_{BLdn}$ is a phase space, with dualizer $\Pr\bot$; this proves completeness of BLdn with respect to phase spaces (for more details on this proof, see [S3]).

Note that again all arguments are positive, and thus we automatically have completeness also for all logics obtained by extending BL by any combination of **w**, **c**, **dn** and $\bot \vdash 0$ (so, in particular, classical logic).

Finally, such a method for proving completeness looks uniform enough to suggest that the examples above could be instances of a more general and abstract theorem; in other words, it becomes quite possible to develop a correspondence theory, between formulas and conditions on pretopologies, analogous to the well-known correspondence theory, between modal formulas and first order conditions on Kripke frames (cf. e.g. [vB]).

3 Some Results and Problems Arising from the Comparison with Formal Topologies

Once the completeness proofs have given a *raison d'être* to pretopologies, it is natural to ask how much of the theory of formal topologies, or more generally of point-free topology, can be extended to the more general setting of pretopologies.

It has already been mentioned that, as any locale can be presented as $Open(\mathcal{A})$, or $Sat(\mathcal{A})$, for some formal topology $\mathcal{A}$, so any quantale can be presented as $Sat(\mathcal{F})$, for some pretopology $\mathcal{F}$; actually, the method of proof is exactly the same (cf. [BS]). Quite analogously, the notion of morphism between formal topologies is extended to pretopologies with no change. A morphism $f : \mathcal{A} \to \mathcal{B}$ is defined (cf. [S1]) to be a function $f : S \to \mathcal{P}T$, from the base S of $\mathcal{A}$ into subsets of the base T of $\mathcal{B}$, which satisfies

$$f1 =_{\mathcal{B}} 1, \qquad f(a \cdot b) =_{\mathcal{B}} fa \cdot fb,$$
$$a \lhd_{\mathcal{A}} U \Rightarrow fa \lhd_{\mathcal{B}} \bigcup\{fb : b \in U\}.$$

Any such f induces a morphism of locales $\bar{f} : Sat(\mathcal{A}) \to Sat(\mathcal{B})$, where $\bar{f}(U) \equiv \bigcup\{fb : b \in U\}$; conversely, it is easy to show that, since a morphism of locales preserves arbitrary joins, any morphism from $Sat(\mathcal{A})$ into $Sat(\mathcal{B})$ is induced in this way by a morphism $f : \mathcal{A} \to \mathcal{B}$ (which corresponds to the fact that the inverse of a continuous function between topological spaces is determined by its behaviour on a base).

Exactly the same proof shows that, if we choose to adopt the above definition of morphism also for pretopologies, we obtain a biunivocal correspondence between morphisms $\mathcal{F} \to \mathcal{G}$ and quantale morphisms $Sat(\mathcal{F}) \to Sat(\mathcal{G})$; this fact of course confirms the choice.

But then the first problem arises as follows. In fact, a basic result in point-free topology is the link between locales and topological spaces, which in our framework takes the following form. The collection $Pt(\mathcal{A})$ of points (see section 1) of a formal topology $\mathcal{A}$ is provided with a topology $\Omega Pt(\mathcal{A})$ by taking $\phi a \equiv \{\alpha \in Pt(\mathcal{A}) : a \in \alpha\}$, for any $a \in S$, as basic neighbourhoods. Putting $\phi U \equiv \bigcup\{\phi b : b \in U\}$, the mapping $\phi : Open(\mathcal{A}) \to \Omega Pt(\mathcal{A})$ becomes a surjective homomorphism of locales, hence an isomorphism when $\mathcal{A}$ has 'enough points', namely when $\phi a \subseteq \phi U$ iff $a \triangleleft U$.

The problem is to see whether and how this can be extended to pretopologies, and the following approach seems to be useful. It is easy to see that points of $\mathcal{A}$ can be identified with positive morphisms $\mathcal{A} \to \mathcal{P}1$, where $\mathcal{P}1$ is the trivial formal topology (i.e. the degenerate monoid with only one element 1, with cover $a \triangleleft U \equiv a \in U$ and with $Pos(1)$) and a morphism f is positive if $Pos(fa)$ implies $Pos(a)$ for any a.

Of course, $\mathcal{P}1$ is also the trivial pretopology, but not informative enough: if we repeat on pretopologies the above construction with no change, i.e., if $Pt(\mathcal{F})$ is the collection of morphisms $\mathcal{F} \to \mathcal{P}1$ and $\phi : Open(\mathcal{F}) \to \Omega Pt(\mathcal{F})$ is defined as above, then $\phi a \cdot \phi b = \phi a \cap \phi b$, so that, in any case, $\Omega Pt(\mathcal{F})$ is a locale, and thus, far from being isomorphic to $Open(\mathcal{F})$. This is due to the fact that $\mathcal{P}1$ is 'too small', and so the question is: is it possible to find the least 'significative' pretopology, let us call it $\mathcal{P}$, and give to $Hom(\mathcal{F}, \mathcal{P})$ the structure of a quantale which, under some conditions, is isomorphic to $Sat(\mathcal{F})$?

More generally, the interplay between formal topologies or pretopologies and some notions with categorical flavour seems to be a promising topic of research. For instance, it has been shown by John L. Bell (in his lectures in Padova, June, 1991) how to adapt the definition of precover to a topos (by generalizing stability to uniformity, i.e., $fa \triangleleft U$ iff $a \triangleleft f^{-1}[U]$ for any morphism f) so that it becomes an alternative characterization of modalities or of universal closure operators (which, by the way, is a formal rendering of the intuitive idea that $\triangleleft$ generalizes $\in$, see [B], ch. 5).

Also, it is straightforward to reformulate the notion of Ω-set in terms of formal topologies, obtaining that of $\mathcal{A}$-set, simply by replacing Ω by $Sat(\mathcal{A})$. This does not seem to give much, but still, passing to pretopologies, i.e. looking for the notion of $\mathcal{F}$-set, could help to solve the intriguing problem of finding a useful definition of quantale-valued sets, or Q-sets.

The approach of formal topology allows to include in a simple way as particular cases both Stone's representation theory and the theory of Scott domains. A formal topology $\mathcal{A}$ is said to be Stone, or compact (resp. Scott, or 1-compact) when $a \lhd U$ only if $a \lhd u$ for some finite subset $u \subseteq U$ (resp. for a subset $u \subseteq U$ with at most one element). A property of closure operators, familiar to algebraists, is that of being algebraic: $\mathcal{A}$ is algebraic if for any subset U, $\mathcal{A}U = \bigcup\{\mathcal{A}u : u \subseteq_\omega U\}$ (where we write $u \subseteq_\omega U$ for: u is a finite subset of U, and similarly $u \subseteq_1 U$ for u with at most one element). Simply by unfolding definitions, it is seen that a formal topology $\mathcal{A}$ is Stone iff the associated operator $\mathcal{A}$ is algebraic. Quite similarly, $\mathcal{A}$ is Scott iff $\mathcal{A}$ is 1-algebraic, i.e., $\mathcal{A}U = \bigcup\{\mathcal{A}u : u \subseteq_1 U\}$ for any U.

A standard result is that a closure operator is algebraic iff it is inductive, i.e., the collection of saturated subsets is closed under directed unions. It is not difficult to supply an intuitionistic proof of this fact, and to see that it extends immediately to a proof of: $\mathcal{A}$ is 1-algebraic iff the collection of saturated subsets is closed under arbitrary unions.

All of this is extended to pretopologies with no change; again, the problems come when truly topological notions are involved. If $\mathcal{A}$ is Stone, a subset α of S satisfies the second condition on points, namely $a \lhd U$ & $a \in \alpha \Rightarrow (\exists b \in U)\, b \in \alpha$, iff for any $b_1, \ldots, b_n \in S$ the following is satisfied:

$$\frac{a \lhd \{b_1, \ldots, b_n\} \qquad a \in \alpha}{(\exists i \leq n)\, b_i \in \alpha}$$

So points of a Stone formal topology $\mathcal{A}$ correspond biunivocally to prime filters of $Open(\mathcal{A})$, and then the condition of having 'enough points' becomes, by classical logic, just another formulation of the prime filter principle. This is enough to obtain Stone's representation theorem for distributive lattices as a special case of the isomorphism $Open(\mathcal{A}) \cong \Omega Pt(\mathcal{A})$; however, like in the general case, it seems problematic to extend this to pretopologies.

The connection with Scott domains, though stemming from a similar remark, is more rewarding. In fact, as above in the Stone case, it is easy to see that points of a Scott formal topology $\mathcal{A}$ correspond biunivocally to (proper) filters of $Open(\mathcal{A})$; in particular, $\{b : a \lhd b\}$ is a point for every positive basic neighbourhood a, and this is enough to prove, completely constructively, that $Open(\mathcal{A}) \cong \Omega Pt(\mathcal{A})$ when $\mathcal{A}$ is Scott. The link with Scott domains is not only that $Pt(\mathcal{A})$, equipped with the inclusion ordering, is a Scott domain when $\mathcal{A}$ is Scott, but also, conversely, that any Scott domain can be represented in this way[6]. Moreover, since a Scott formal topology is completely determined by the structure it induces on its base, by applying Ockham's razor, one can reduce everything to a very elementary notion, called information base in [SVV]: a set S with an associative binary

6 It is by working out the proof of this theorem that one realizes the need of the assumptions on the predicate *Pos*.

operation $\cdot$, a preorder $\lhd$ with positive top element 1 and a property Pos, which satisfy:

$$\frac{Pos(a) \quad a \lhd b}{Pos(b)} \qquad\qquad \frac{Pos(a) \to a \lhd b}{a \lhd b}$$

$$\frac{a \lhd b}{a \cdot c \lhd b} \qquad \frac{a \lhd b}{c \cdot a \lhd b} \qquad\qquad \frac{a \lhd b \quad a \lhd c}{a \lhd b \cdot c}$$

Consequently, the notion of point acquires a simpler characterization too: a point, or concept, is a nonempty subset α of positive elements which is closed under $\cdot$ and upward under $\lhd$. Thirdly, given two information bases $\mathcal{S}$ and $\mathcal{T}$, a morphism $\mathcal{S} \to \mathcal{T}$, called a translation, is a relation $R \subseteq S \times T$ such that $\{b : aRb\}$ is a point for any positive a in S, and which respects $\lhd$ and Pos, that is, $a \lhd c \;\&\; cRb \Rightarrow aRb$ and $(Pos(a) \to aRb) \Rightarrow aRb$ respectively.

The category formed by information bases and translations is equivalent, via the functor Pt sending $\mathcal{S}$ to $Pt(\mathcal{S})$ and R to $\alpha \mapsto \{b : (\exists a \in \alpha)\, aRb\}$, to the category of Scott domains (and thus it plays the same role as, for instance, the category of information systems in [S]). And this is so leisurely simple (at least, it looks so for us after we worked it out, cf. [SVV]) that one would like to repeat it for pretopologies and linear logic. Now, the analogue of the category of Scott domains in the case of linear logic is Girard's category of coherent spaces and stable functions (cf. [G]). While coherent spaces are concrete presentations of a subclass of Scott domains, the notion of morphism is changed, i.e., stable functions are intrinsically different from morphisms of domains; so the first problem is: find a notion, expressible in terms of formal topology, corresponding to stable functions. However, maybe it is better to follow a different approach, namely to keep the notion of morphism unchanged, and instead generalize that of object. The problem in this case is to find a notion of 'Scott' *pre*topology leading to a full subcategory (of the category of pretopologies) which can play the role of the category of coherent spaces.

References

[vB] J. van Benthem, Correspondence theory, in: *Handbook of Philosophical Logic, Vol. II*, D.M. Gabbay and F. Guenthner eds, Reidel, 1984, 167–247.

[BS] G. Battilotti, G. Sambin, Presentation of locales and quantales by means of covers and precovers, (to appear).

[B] J.L. Bell, *Toposes and Local Set Theories: An Introduction*, Clarendon Press, 1988.

[G] J.-Y. Girard, Linear logic, *Theor. Computer Sc. 50* (1987), 1–102.

[J] P. T. Johnstone, *Stone Spaces*, Cambridge University Press, 1982.

[ITT] P. Martin-Löf, *Intuitionistic Type Theory*, Bibliopolis, 1984.

[O] H. Ono, Semantics for substructural logics, (in this book).

[R] K. I. Rosenthal, *Quantales and their Applications*, Longman, 1990.

[S1] G. Sambin, Intuitionistic formal spaces — a first communication, in: *Mathematical Logic and its Applications*, D. Skordev ed., Plenum, 1987, 187–204.

[S2] ———, Intuitionistic formal spaces and their neighbourhood, in: *Logic Colloquium '88*, R. Ferro et al. eds, North-Holland, 1989, 261–285.

[S3] ———, Pretopologies and completeness proofs, (to appear).

[SVV] G. Sambin, S. Valentini, P. Virgili, Constructive domain theory as a branch of intuitionistic pointfree topology, Dip. di Matematica Pura e Appl., Univ. di Padova, Preprint, August 1992.

[S] D. S. Scott, Domains for denotational semantics, in: *Automata, Languages and Programming*, M. Nielsen and E. M. Schmidt eds, Springer, 1982, 577–613.

[T] A. S. Troelstra, *Lectures on Linear Logic*, Center for the Study of Language and Information, Stanford, 1992.

[TvD] A. S. Troelstra, D. van Dalen, *Constructivism in Mathematics: An Introduction, Vol. 2*, North-Holland, 1988.

Lambda Terms with Functional Symbols and Decidability in Certain Closed Categories

Sergei V. Soloviev

St Petersburg Institute for Informatics and Automatization

Introduction

There is a close connection between substructural logics, weak systems of lambda terms and closed categories of different types. Normalization-like procedures on lambda terms provide decision procedures for equality of morphisms in free closed categories [1,2,7]. A decision procedure for cartesian closed (CC) categories was obtained by using ordinary typed lambda terms related to the (&, $\Rightarrow$)-fragment of the intuitionistic propositional calculus with the constant I (truth). Some weak systems of lambda terms (cf. [4,5]) provide decision procedures for symmetric monoidal closed (SMC) categories, non-symmetric monoidal closed categories etc. (see [7]). It would be interesting to apply this apparatus to categories that are not free, i.e. categories where some morphisms do not belong to the closed structure and/or satisfy additional equations.

We consider here the following problem. Given a symmetric monoidal (SM) category V, not closed, take the free closed category CL(V) on V. In general CL(V) is only relatively free. Now, supposing equality in V is decidable, try to obtain a decision procedure for equality of morphisms in CL(V).[1]

We show that this problem is logically equivalent to the problem of decidability of an equivalence relation on lambda terms. This part depends strongly on the technique of Mints and Babaev (see [1,2,7]; cf. also the work of Lambek [5]).

After that, we consider the particular case when V is the SM category of finite ordinals and all set-theoretical functions (with sum as tensor),

1 This problem was suggested to the author by J. Bénabou.

which is interesting to algebraic topology. As an easy corollary we obtain a decision procedure for the SMC category on the SM subcategory of V containing only permutations as morphisms.

These results may also have some application in theoretical computer science, taking into account the role of the typed lambda calculus and its models in the category of finite sets (cf. [3,8]). On the other hand, they may also be used in the study of substructural logics enriched by axioms related to those morphisms that do not belong to the closed structure.

1 Symmetric Monoidal and Symmetric Monoidal Closed Categories

An SM category consists of the following:

(i) a category V,

(ii) $\mathrm{I} \in \mathrm{Ob}(\mathrm{V})$,

(iii) a bifunctor $\otimes : \mathrm{V} \times \mathrm{V} \to \mathrm{V}$ (tensor),

(iv) natural isomorphisms

$$\begin{array}{l} a = a_{ABC} : A \otimes (B \otimes C) \to (A \otimes B) \otimes C \\ b = b_A : A \otimes \mathrm{I} \to A \\ c = c_{AB} : A \otimes B \to B \otimes A \end{array}$$

and their inverses a^{-1}, b^{-1}, c^{-1}.

The following equations must be satisfied (f, g, h denoting arbitrary morphisms of V, and 1 being the appropriate unit; we include here also the ordinary identities expressing that V is a category, since these equations will be used in the construction of a free category):

(0.1) $f1 = 1f = f$; (0.2) $(fg)h = f(gh)$
(these equations express that V is a category);

(1) $aa^{-1} = 1$; $a^{-1}a = 1$; $bb^{-1} = 1$; $b^{-1}b = 1$; $c_{AB}c_{BA} = 1$
(the last equation shows that there is no need for c^{-1});

(2) $a_{A(B\otimes C)D}a_{AB(C\otimes D)} = (a_{ABC} \otimes 1_D)a_{A(B\otimes C)D}(1_A \otimes a_{BCD})$;

(3) $a_{CAB}c_{(A\otimes B)C}a_{ABC} = (c_{AC} \otimes 1_B)a_{ACB}(1_A \otimes c_{BC})$;

(4) $b_{A\otimes B}a_{AB\mathrm{I}} = 1_A \otimes b_B$

(these are the so-called Mac Lane coherence conditions for a, b, c [6]);

(5) $(fg) \otimes (f'g') = (f \otimes f')(g \otimes g')$,
where $f : A \to B$, $g : B \to C$, $f' : A' \to B'$, $g' : B' \to C'$;
$1_{A \otimes B} = 1_A \otimes 1_B$
(these conditions express the functoriality of $\otimes$);

(6)
$$\begin{aligned}((f \otimes 1_B) \otimes 1_C)a_{ABC} &= a_{A'BC}(f \otimes 1_{B \otimes C});\\ ((1_A \otimes g) \otimes 1_C)a_{ABC} &= a_{AB'C}(1_A \otimes (g \otimes 1_C));\\ ((1_{A \otimes B} \otimes h)a_{ABC} &= a_{ABC'}(1_A \otimes (1_B \otimes h));\\ fb_A &= b_{A'}(f \otimes 1_{\mathrm{I}});\\ (1_A \otimes f)c_{AB} &= c_{A'B}(f \otimes 1_B);\\ (g \otimes 1_B)c_{AB} &= c_{AB'}(1_A \otimes g),\end{aligned}$$

where $f : A \to A'$, $g : B \to B'$, $h : C \to C'$
(these conditions express the naturality of a, b, c).

If the category V is not free, the morphisms and objects of V satisfy some additional equalities; for example, $f \otimes g = h$, $A \otimes (B \otimes C) = B$. Below we shall suppose that the equality relation in V is decidable.

For an SMC category W the following must be added to (i)–(iv) above (with W instead of V):

(v) a bifunctor $\Rightarrow :\ \mathrm{W}^{op} \times \mathrm{W} \to \mathrm{W}$,

(vi) natural transformations

$$e = e_{AB} : (A \Rightarrow B) \otimes A \to B \quad \text{and} \quad d = d_{BA} : B \to A \Rightarrow (B \otimes A).$$

Of course, some additional equalities must be satisfied:

(7) $(1 \Rightarrow e_{AC})d_{(A \Rightarrow C)A} = 1_{A \Rightarrow C}$; $e_{(B \otimes A)A}(d_{BA} \otimes 1_A) = 1_{B \otimes A}$
(coherence conditions for e and d);

(8) $e_{DC}(1_{D \Rightarrow C} \otimes f) = e_{AC}((f \Rightarrow 1_C) \otimes 1_A)$;
$(1_A \Rightarrow 1_B \otimes f)d_{BA} = (f \Rightarrow 1_{B \otimes D})d_{BD}$,
where $f : A \to D$;
$(1_A \Rightarrow g \otimes 1_A)d_{BA} = d_{B'A}g$, $ge_{AB} = e_{AB'}((1_A \Rightarrow g) \otimes 1_A)$,
where $g : B \to B'$
(naturality conditions for e and d);

(9) $(f'f \Rightarrow gg') = (f \Rightarrow g)(f' \Rightarrow g')$,
where $f : A \to A'$, $f' : A' \to A''$, $g' : B'' \to B'$, $g : B' \to B$;
$1_A \Rightarrow 1_B = 1_{A \Rightarrow B}$
(these two conditions express the functoriality of $\Rightarrow$).

2 Free Symmetric Monoidal Closed Category on a Given Symmetric Monoidal Category

Let O(V) and M(V) denote the sets of constants, one for every $A \in \mathrm{Ob}(V)$ and $f \in \mathrm{Mor}(V)$ respectively. (For simplicity we shall suppose that V is small; the interesting examples are countable.) If no confusion is likely, we denote the constants from O(V) and M(V) by the same letters as the corresponding objects and morphisms.

Let us consider the set FO(V) of all terms built from the elements of O(V) using $\Rightarrow$, $\otimes$ and brackets, in the ordinary way. The elements of FO(V) will be called formal objects on V. Let T be a term that does not contain $\Rightarrow$. Then $\#T$ will denote the object of V obtained by replacing the constants and the symbol $\otimes$ by the corresponding objects and functor of V. Now let $\sim$ be the smallest equivalence relation on FO(V) such that:

i) $T \sim T'$ for every term T and T' that do not contain $\Rightarrow$ and $\#T = \#T'$ in V;

ii) if $T \sim T'$ and $S \sim S'$, then $T \otimes S \sim T' \otimes S'$ and $T \Rightarrow S \sim T' \Rightarrow S'$.

Note that if equality of objects in V is decidable, then $\sim$ on FO(V) is also decidable.

Let us define the set FM(V). This definition is more complicated than the definition of FO(V) and we shall give it in detail.

The elements of FM(V) will be called formal morphisms. Every formal morphism will have a type of the form $S \to T$ where S and T are formal objects. If φ is a formal morphism of the type $S \to T$, we shall often write $\varphi : S \to T$.

(1) For every $R, S, T \in \mathrm{FO}(V)$ the symbols 1_R, a_{RST}, a^{-1}_{RST}, b_R, b^{-1}_R, c_{RS}, e_{RS}, d_{SR} are formal morphisms of types

$$R \to R, R \otimes (S \otimes T) \to (R \otimes S) \otimes T, \ldots, S \to R \Rightarrow (S \otimes R)$$

respectively.

(2) If $f : A \to B \in \mathrm{Mor}(V)$, then the corresponding constant symbol f is a formal morphism of type $A \to B$.

(3) If $\varphi : S \to T$ and $\psi : S' \to T'$ are formal morphisms then $\varphi \otimes \psi : S \otimes S' \to T \otimes T'$ and $\varphi \Rightarrow \psi : T \Rightarrow S' \to S \Rightarrow T'$ are formal morphisms. If $\varphi : R \to S$, $\psi : S' \to T$ are formal morphisms and $S \sim S'$, then $(\psi\varphi) : R \to T$ is a formal morphism. (We may write these rules in a more familiar form:

$$\frac{\varphi : S \to T \qquad \psi : S' \to T'}{\varphi \otimes \psi : S \otimes S' \to T \otimes T'}$$

$$\frac{\varphi : S \to T \quad \psi : S' \to T'}{\varphi \Rightarrow \psi : T \Rightarrow S' \to S \Rightarrow T'}$$

$$\frac{\varphi : R \to S \quad \psi : S' \to T \quad S \sim S'}{\psi\varphi : R \to T}$$

Note that the third of these rules is a cut rule.)

Formal morphisms are exactly those that may be obtained from (1) and (2) by the rules (3).

By *formal* V*-morphisms* we understand those formal morphisms that do not contain the symbols $\Rightarrow$, e and d. Now let us define the relation $\sim$ on formal morphisms. If φ is a formal V-morphism, then we may replace all formal objects T (in indexes) by $\#T$, all symbols $a, b, a^{-1}, b^{-1}, c, 1$ and functional constants by corresponding morphisms of V, and $\otimes$ by the tensor in V. The resulting morphism of V will be denoted by $\#\varphi$. Now we define $\sim$ on formal morphisms as the smallest equivalence relation such that:

(i) if φ, ψ are formal V-morphisms and $\#\varphi = \#\psi$ in V, then $\varphi \sim \psi$ (for formal V-morphisms we shall often write $\varphi = \psi$ instead of $\varphi \sim \psi$);

(ii) all the conditions (1)–(9) for SMC categories above are fulfilled (with formal objects instead of objects of the category and formal morphisms of appropriate types instead of morphisms);

(iii) if $\varphi \sim \psi$ and $\varphi' \sim \psi'$, then $\varphi \otimes \psi \sim \varphi' \otimes \psi'$ and $\varphi \Rightarrow \psi \sim \varphi' \Rightarrow \psi'$; if, moreover, the compositions $\varphi\varphi'$ and $\psi\psi'$ exist, then $\varphi'\varphi \sim \psi'\psi$.

It is easily checked that taking equivalence classes of objects as objects and equivalence classes of formal morphisms as morphisms, we obtain the free SMC category on V. It will be denoted by CL(V).

3 Central Isomorphisms

Remark 1. An important role is played by so-called central (iso)morphisms in CL(V). We call $\varphi \in$ FM(V) *central* if it is built from a, b, c, their inverses and 1's by applications of $\otimes$. (The difference between central morphisms and V-morphisms is that the first don't contain symbols of the morphisms of V except a, b, c and 1; at the same time, the indexes of a, b, c in central morphisms may be arbitrary formal objects.) Let $\varphi, \psi : S \to T$ be central morphisms. Then $\varphi \sim \psi$ (see [6]). (This may fail for $\varphi : S \to T$ and $\psi : S' \to T'$ with $S \sim S'$ and $T \sim T'$.) Let $\Sigma = S_1, \ldots, S_n$ and $\Gamma = T_1, \ldots, T_m$ and let some elements of Σ be identical with some elements of Γ. Let all members which belong only to Γ (only to Σ) be I. Let S and T be formal objects obtained from $S_1 \otimes \ldots \otimes S_n$ and $T_1 \otimes \ldots \otimes T_m$

respectively by any allowable distribution of brackets. Then there exists a central morphism $\Theta : S \to T$ with S_i and T_j as indexes that is unique in the sense described above.

4 Morphisms and λ-Terms

Even if equality of objects and morphisms in V is decidable, the relation $\sim$ on FM(V) may be undecidable. Now we shall describe a system of λ-terms, a correspondence between formal morphisms and these λ-terms and an equivalence relation on them such that two terms are equivalent iff the corresponding formal morphisms are equivalent.

The content of this section is a generalization of [1,2,7] to the case when functional symbols are involved.

The language of our λ-terms includes the constant I, the variables x^T, y^T, $z^T, \ldots$ for every $T \in$ FO(V), the functional symbols f_A^B for every functional constant $f : A \to B \ \in$ M(V), the symbols l_{ST}, r_{ST} for every $S, T \in$ FO(V), λx^T for every variable x^T and three types of brackets $\langle , \rangle$; $\{ , \}$; $(,)$. Some variables will be considered as bound (in the ordinary way) by λ-operators. We consider identical λ-terms obtained from each other by renaming of bound variables. By $t_x[s]$ we denote the result of substituting a term s for all free occurrences of the variable x in t, with renaming of bound variables if collisions arise (we follow the same convention not only in the case when t is a term, but for some other expressions, quasi-terms, where the notion of bound variable makes sense).

Every term t will have as its type $T \in$ FO(V). The expressions $t \in S$ and t^S mean that t is a term of type S. The terms are defined inductively (t and s below are supposed to have no common free variable):

(1) I is a term of type I; x^S is a term of type S;

(2.1) if $t \in T$, $s \in S$, then $\langle t, s \rangle \in T \otimes S$;

(2.2) if $t \in S \Rightarrow T$, $s \in S'$, $S \sim S'$, then $(t, s) \in T$;

(2.3) $\{t^T, s^{\mathrm{I}}\}, \{s^{\mathrm{I}}, t^T\} \in T$;

(2.4) $\lambda x^S t^T \in S \Rightarrow T$;

(2.5) $t^R_{x,y}[l_{ST} s^{S\otimes T},\ r_{ST} s^{S\otimes T}] \in R$, where $x \in S$, $y \in T$;

(2.6) if $t \in A$, where $A \in$ O(V), then $f_A^B(t) \in B$.

The system of terms described above is connected with a system of natural deduction with restricted structural rules. Namely:

(1) corresponds to the axioms $\to \mathrm{I}$ and $S \to S$,

(2.1) corresponds to the following rule ($\bullet$ plays the role of tensor):

$$\frac{\Gamma \to T \qquad \Sigma \to S}{\Gamma\Sigma \to T \bullet S}$$

(2.2) corresponds to:

$$\frac{\Gamma \to S \Rightarrow T \qquad \Sigma \to S}{\Gamma\Sigma \to T}$$

(2.3) corresponds to restricted weakening:

$$\frac{\Gamma \to T \qquad \Sigma \to \mathrm{I}}{\Gamma\Sigma \to T}$$

(2.4) corresponds to:

$$\frac{\Gamma S \to T}{\Gamma \to S \Rightarrow T}$$

(2.5) corresponds to:

$$\frac{\Gamma \to S \bullet T \qquad S, T, \Sigma \to R}{\Gamma\Sigma \to R}$$

(2.6) corresponds to:

$$\frac{f : A \to B \qquad \Gamma \to A}{\Gamma \to B}$$

Here Γ and Σ are lists of formulas.

We add this comment to make clearer the connections with substructural logics. Below we shall prefer the language of terms.

We shall often omit the subscripts of l and r (in fact, they are needed only when S and T are constants corresponding to objects of V ($S, T \in$ O(V)) and it is possible that $S \otimes T \sim S' \otimes T'$ while it is not the case that $S \sim S'$ and $T \sim T'$.

Note that terms are closed under substitution in the sense that if $t \in T$, $s \in S$ are terms and $x \in S'$, where $S \sim S'$, is a variable, then $t[s]$ is a term of type $T' \sim T$. This is easily checked by induction on the construction of t.

By *quasi-terms* we understand the expressions obtained in a similar way but such that (2.5) in the definition above is replaced by (2.5') $l_{ST}t \in S$,

$r_{ST}t \in T$ where $t \in S \otimes T$ (and 'term' in the modified definition means 'quasi-term').

Note that when we apply (2.5) some subterms of t may become quasi-terms (because this rule is not 'local').

Lemma 1. *A quasi-term t is a term iff for any presentation of t in the form $t'_x[\delta s]$, we also have*

$$t = t''_{x,y}[\delta s, \delta^* s],$$

where t', t'', s are quasi-terms, δ is l or r; $l^* = r$ and $r^* = l$.

A term t will be called a V-*term* if the types of all variables in t belong to O(V) (i.e., they are constants corresponding to the objects of V) and t does not contain sub(quasi)terms of the forms λxs and (t, s).

Let $\Gamma = S_1, \dots S_n$ be the list of types of the free variables in a term $t \in T$ (possibly permuted). We call the expression $\Gamma \to T$ the *full type* of the term t and write $t : \Gamma \to T$.

To every formal morphism $\varphi : S \to T$ corresponds a term $t(\varphi) : S \to T$ defined by recursion (we omit the obvious indexes of l, r):

$$t(1 : S \to S) = x^S; \;\; t(f : A \to B) = f_A^B(x^A);$$

$$t(a_{RST}) = \langle\langle lx^{R\otimes(S\otimes T)}, lrx^{R\otimes(S\otimes T)}\rangle, rrx^{R\otimes(S\otimes T)}\rangle;$$

$$t(a_{RST}^{-1}) = \langle llx^{(R\otimes S)\otimes T}, \langle rlx^{(R\otimes S)\otimes T}, rx^{(R\otimes S)\otimes T}\rangle\rangle;$$

$$t(b_S) = \{lx^{S\otimes \mathrm{I}}, rx^{S\otimes \mathrm{I}}\}; \; t(b_S^{-1}) = \langle x^S, \mathrm{I}\rangle;$$

$$t(c_{ST}) = \langle rx^{S\otimes T}, lx^{S\otimes T}\rangle;$$

$$t(d_{TS}) = \lambda x^S\langle x^S, x^T\rangle; \;\; t(e_{ST}) = (lx^{(S\Rightarrow T)\otimes S}, rx^{(S\Rightarrow T)\otimes S});$$

$$t(\varphi \Rightarrow \psi) = \lambda x^{R'} t(\psi)_{x^S}[(x^{R\Rightarrow S}, t(\varphi))],$$
where $\varphi : R' \to R$ and $\psi : S \to S'$;

$$t(\varphi \otimes \psi) = \langle t(\varphi), t(\psi)\rangle_{x^S, x^{S'}}[lx^{S\otimes S'}, rx^{S\otimes S'}],$$
where $\varphi : S \to T$ and $\psi : S' \to T'$;

$$t(\varphi\psi) = t(\varphi)_{x^{S'}}[t(\psi)], \text{ for } \psi : R \to S, \;\; \varphi : S' \to T, \text{ where } S \sim S'.$$

Remark 2. The terms corresponding to formal V-morphisms are V-terms. Let us consider an example. Take the usual natural isomorphism (currying)

$$\pi = \pi_{RST} : \mathrm{Hom}(R \otimes S, T) \to \mathrm{Hom}(R, S \Rightarrow T).$$

Let $\varphi : R \otimes S \to T$ and $\psi : R \to S \Rightarrow T$. The morphism $\pi_{RST}(\varphi)$ may be defined as $(1_S \Rightarrow \varphi)d_{RS}$. To simplify matters, let $\varphi = f : A \otimes B \to C$. We have:

$$t(f) = f^C_{A\otimes B}(x^{A\otimes B});$$

$$\begin{aligned} t(1 \Rightarrow f) &= \lambda x^B (f(x^{A\otimes B})_{x^{A\otimes B}}[(x^{B\Rightarrow A\otimes B}, x^B)]) \\ &= \lambda x^B f((x^{B\Rightarrow A\otimes B}, x^B)); \end{aligned}$$

$$t(d_{AB}) = \lambda x^B \langle x^A, x^B \rangle;$$

$$\begin{aligned} t(\pi_{ABC}(f)) &= t((1 \Rightarrow f)d) \\ &= (\lambda x^B f((x^{B\Rightarrow A\otimes B}, x^B))_{x^{B\Rightarrow A\otimes B}}[\lambda x^B \langle x^A, x^B \rangle] \\ &= \lambda x^B f((\lambda y^B \langle x^A, y^B \rangle, x^B)) \end{aligned}$$

(note that in the last step we rename the bound variable x^B in the term $\lambda x^B \langle x^A, x^B \rangle$).

If $\Gamma = S_1, \ldots, S_n$, by $K(\Gamma)$ we shall denote the expression

$$(\ldots(\mathrm{I} \otimes S_1) \otimes S_2) \otimes \ldots \otimes S_n).$$

For every λ-term $t : \Gamma \to S$, let us define recursively a formal morphism $C_\Gamma(t^S) : K(\Gamma) \to S$; by ζ and ζ' we shall denote appropriate central isomorphisms (see Remark 1):

$C_S(x^S) = b_S c_{\mathrm{I}S} : \mathrm{I} \otimes S \to S;$

$C_\emptyset(\mathrm{I}) = 1_\mathrm{I} : \mathrm{I} \to \mathrm{I};$

$C_\Gamma(\lambda x^S t^T) = \pi(C_{\Gamma S}(t^T)) : K(\Gamma) \to S \Rightarrow T$, where $t : \Gamma S \to T$;

$C_{\Gamma\Sigma}((t, s)) = e_{ST}(C_\Gamma(t)) \otimes C_\Sigma(s))\zeta : K(\Gamma\Sigma) \to T$,
for $t : \Gamma \to S \Rightarrow T$, $s : \Sigma \to S$ and $\zeta : K(\Gamma\Sigma) \to K(\Gamma) \otimes K(\Sigma)$;

$C_{\Gamma\Sigma}(\langle t, s \rangle) = (C_\Gamma(t) \otimes C_\Sigma(s))\zeta : K(\Gamma\Sigma) \to T \otimes S$,
for $t : \Gamma \to T$, $s : \Sigma \to S$, and ζ as above;

$C_{\Gamma\Pi\Sigma}(t_{x,y}[ls, rs])$
$= C_{\Gamma ST\Sigma}(t)\zeta((1_{K(\Gamma)} \otimes C_\Pi(s)) \otimes 1_{K(\Sigma)})\zeta' : K(\Gamma\Pi\Sigma) \to T$,
for $t : \Gamma RS\Sigma \to T$, $s : \Pi \to R \otimes S$, $x \in R$, $y \in S$,
$\zeta' : K(\Gamma\Pi\Sigma) \to (K(\Gamma) \otimes K(\Pi)) \otimes K(\Sigma)$,
and $\zeta : (K(\Gamma) \otimes (S \otimes T)) \otimes K(\Sigma) \to K(\Gamma ST\Sigma)$;

$C_{\Gamma\Pi\Sigma}(\{t, s\}) = b(C_{\Gamma\Sigma}(t) \otimes C_\Pi(s))\zeta : K(\Gamma\Pi\Sigma) \to T$,
for $t : \Gamma\Sigma \to T$, $s : \Pi \to \mathrm{I}$ and $\zeta : K(\Gamma\Pi\Sigma) \to K(\Gamma\Sigma) \otimes K(\Pi)$;

$$C_\Gamma(f_A^B(t^A)) = fC_\Gamma(t^A),\text{ for } t:\Gamma\to A \text{ and } f:A\to B\in \mathrm{M(V)}.$$

Lemma 2. *If t is* V*-term, then $C(t)$ is a formal* V*-morphism and there exists an $f\in \mathrm{M(V)}$ such that $C(t)\sim f$.*

Proof. We show that $C(t)$ is a formal V-morphism by induction on the construction of t (note that the types of all variables in V-terms belong to O(V) and the construction uses only the rules (2.1), (2.3), (2.5) and (2.6)). If $C(t)$ is a formal V-morphism, then $\sharp C(t) = f \in \mathrm{Mor(V)}$ and $C(t)$ is equivalent to the corresponding $f\in \mathrm{M(V)}$.

Lemma 3. $C((\lambda x^T t, s^T)\sim C(t_{x^T}[s^T])$ ([2]).

5 Equivalence of λ-Terms

Let us define inductively an equivalence relation on quasi-terms such that its restriction to terms will correspond to the equivalence of formal morphisms in F(V):

(i) $t\sim \lambda x^S(t,x^S)$, for $t\in S\Rightarrow T$ and x a new variable;

(ii) $t\sim \langle lt, rt\rangle$, for $t\in S\otimes T$;

(iii) $(\lambda xt, s)\sim t_x[s]$, where $x\in R$, $s\in R'$ and $R\sim R'$;

(iv) $t_x[s]\sim \{t_x[\mathrm{I}], s\}$, for $x,s\in \mathrm{I}$;

(v) $t_{x,y}[l\langle u,v\rangle, r\langle u,v\rangle]\sim t_{x,y}[u,v]$, where $x,u\in R$, $y,v\in S$ and x and y are free variables in t;

(vi) $\{t,\mathrm{I}\}\sim t$, $\{\mathrm{I},t\}\sim t$;

(vii) if s and s' are V-terms with the same free variables and $C(s)\sim C(s')$, then $s_{x_1,\ldots,x_n}[u_1,\ldots,u_n]\sim s'_{x_1,\ldots,x_n}[u_1,\ldots,u_n]$;

(viii) $t\sim t'$ if t and t' are congruent (differ only by names of bound variables);

(ix) $\dfrac{s\sim s'}{t\sim t'}$, where t' is obtained from t by replacement of some occurrences of the quasi-term s by s';

(x) $\dfrac{t\sim t' \qquad t\sim t''}{t'\sim t''}$.

Remark 3. The equivalence relation generated by (i)–(vi) and (viii)–(x) corresponds to the equality of morphisms in the free SMC category in the language enriched by functional symbols. Let us denote this relation by $\sim_0$.

Lemma 4. *If $t \sim t'$ and t is a term (not a quasi-term), then t' is also a term.*

The following lemmas may be proved by a slight modification of proofs in [1,2].

Lemma 5. *For every term t and s, $t \sim s$ implies $C(t) \sim C(s)$.*

Lemma 6. *For every formal morphism $\varphi, \psi : S \to T$, $\varphi \sim \psi$ implies $t(\varphi) \sim t(\psi)$ if free variables in $t(\varphi)$ and $t(\psi)$ are chosen to be the same.*

Lemma 7. *For every formal morphism $\varphi : S \to T$, $\varphi bc \sim C(t(\varphi))$.*

Theorem 1. *For every formal morphism $\varphi, \psi : S \to T$, $\varphi \sim \psi$ iff $t(\varphi) \sim t(\psi)$.*

6 Equivalence of Terms and Decidability

Of course, the relation $\sim$ is not in general decidable. In [1,2] a decision procedure for equality of morphisms in the free SMC category was obtained. Modifying unessentially this procedure we may obtain a decision procedure for $\sim_0$ (this may be done easily because the definition of $\sim$ does not include equalities between functional symbols reflecting the fact that V is not free).

In this section, we describe the decision procedure for $\sim_0$, and give a necessary reformulation of results of [2]. The decision procedure described in [2] is a kind of normalization.

Let R be a binary relation between terms. An R-reduction is any (finite or infinite) sequence $t_0, t_1, \ldots$, where $t_i R t_{i+1}$ for all i. This sequence is called an R-reduction of t to t' if it is finite, its first member is t and the last is t'. If there is an R-reduction of t to t', we say that t is *R-reducible* to t', and write $t \models_R t'$ $(t \models t')$. By $t \to t'$ we understand tRt'. The term t is R-normal if $t \models t'$ implies $t = t'$. The results of [2] are based on the following well-known theorem (a form of the Church-Rosser theorem).

Theorem 2. *Let:*

(I) *E be an equivalence relation on terms such that if tEt' and t is R-normal, then t' is R-normal,*

(II) *R be well-founded (i.e., all R-reductions are finite),*

(III) *the following (Church-Rosser property) is satisfied: if tEt', $t \to s$, and $t' \to s'$, then u and u' exist, such that uEu' and $s \models u$ and $s' \models u'$.*

Then, for every term t, its R-normal form exists and it is unique up to the equivalence E; i.e., from tEt', $t \models \bar{\bar{t}}$, $t' \models \bar{\bar{t}}'$, where t and t' are R-normal, it follows that $\bar{\bar{t}}E\bar{\bar{t}}'$.

Two reducibility relations, R_1 and R_2, are introduced. The relation R_1 is defined by conversions $s \to s'$ corresponding to (i), (ii) and (iv) above; i.e., tR_1t' means that t' is obtained from t by replacement of its part s by s', where $s \to s'$ has one of the following forms:

(1) $s \to \lambda x^A(s, x^A)$, where $s \in A \Rightarrow B$, s is not of the form λys and the occurrences of the quasi-term s replaced by $\lambda x^A(s, x^A)$ are not parts of occurrences of the form:

$$(\{\ldots\{s, u_1\}, u_2\}, \ldots, u_n\}, v), \qquad n > 0, \;\; u \in \mathrm{I}, \;\; v \in A;$$

(2) $s \to \langle ls, rs\rangle$, where $s \in A \otimes B$, s is not of the form $\langle s_1, s_2\rangle$ and the occurrences of the quasi-term s replaced by $\langle ls, rs\rangle$ are not parts of occurrences of the form:

$$\delta\{\ldots\{s, u_1\}, u_2\}, \ldots, u_n\}, \qquad n > 0, \;\; \delta = l, r;$$

(3) $s_{x^{\mathrm{I}}}[u] \to \{s_{x^{\mathrm{I}}}[\mathrm{I}], u\}$, where u does not coincide with I and x is in the scope of a λ, ( ) or $\langle\ \rangle$.

A term is called *R-normal* if none of the conversions (1)–(4) is applicable.

The relation R_2 is defined by the following conversions:

(4) $\{s, \mathrm{I}\} \to s$, $\{\mathrm{I}, s\} \to s$;

(5) $s = (\lambda x^A u, v) \to u_{x^A}[v] = s'$;

(6) $s = u_{x^A, y^B}[l\langle v_1, v_2\rangle, r\langle v_1, v_2\rangle] \to u_{x^A, y^B}[v_1, v_2] = s'$.

We take E to be the smallest equivalence relation defined by the following conditions:

tEt' if t and t' are congruent (differ only by names of bound variables);

$\{\{u,v\},w\}E\{u,\{v,w\}\}$;

$\{u,v\}E\{v,u\}$;

$$\frac{tEs \quad tEs'}{sEs'} \text{ (transitivity);}$$

$\dfrac{sEs'}{tEt'}$, where t is obtained from t' by replacement of its part s by s'.

Theorem 3. *Every term t has an R_1-normal form. Every R_1-normal term t has an R_2-normal form, which is also R_1-normal. Let t' and t'' be R_2-normal forms of some R_1-normal forms of t. Then $t'Et''$.*

Let t be a term. We shall denote by $\overline{\overline{t}}$ the R_2-normal form of the R_1-normal form of t.

Theorem 4. $t \sim_0 t'$ *iff* $\overline{\overline{t}}E\overline{\overline{t}}'$.

7 The Category CL(Finord)

Let us consider the following category V. The objects of V are natural numbers, and morphisms $f : m \to n$ are the functions from the set $\{1,\ldots,m\}$ to $\{1,\ldots,n\}$. The distinguished object I is 0. We suppose that there is a unique morphism $0^m : 0 \to m$ (the empty function).

There is a well-known cartesian closed structure on V, but we shall consider a symmetric monoidal structure with sum as tensor: $m \otimes n = m+n$. The morphism $f\otimes g : m\otimes n \to m'\otimes n'$ is defined in the following way: if $1 < k < m$, then $(f \otimes g)(k) = f(k)$; if $m < k < m+n$, then $(f \otimes g)(k) = g(k-m)+m'$. In V, the morphisms a, a^{-1}, b and b^{-1} are identities and the values of $c_{mn} : m \otimes n \to n \otimes m$ on $1,\ldots,m$ are $n+1,\ldots,n+m$, and on $m+1,\ldots,m+n$ they are $1,\ldots,n$. The ordinary hom-functor does not provide a closed structure for such a tensor; so, the construction described above is especially interesting.

Remark 4. We could consider instead of V the category with the same objects, but order-preserving functions as morphisms. The commutativity c of tensor does not preserve order; so, that category would be monoidal but non-symmetric. Nevertheless, using the system of λ-terms for the non-symmetric case [7], the decidability for the equivalence of morphisms in CL(V) could be obtained.

8 A Specialization of Formal Morphisms and Corresponding λ-Terms

Note that a_{mnp} and b_m in Finord are identities. So, in every formal morphism φ they may be replaced by 1's or merely erased if they occur in a composition. If φ is a formal V-*morphism*, then the corresponding λ-term does not contain $\{,\}$. It is easily seen that if $s_1, s_2, \ldots, s_n$ are V-terms, then all terms obtained from them by iteration of $\langle , \rangle$ $(n-1$ times) are equivalent; this allows us to write the resulting term in the symmetric form $\langle s_1, \ldots, s_n \rangle$.

It is known that every $f : m \to n$ in Finord is equal to $(f_1 \otimes \ldots \otimes f_n)\sigma$ where $f_i : k_i \to 1$, and $\sigma : m \to m$ is a permutation of the set $\{1, \ldots, m\}$. So, we may suppose that all functional symbols represent a permutation or are of the form f_k^1 (if $k = 0$, they are the constant 0^1).

Let us define inductively the combinations δ_m^i of l and r corresponding formally to the projections of product on its i-th factor:

$$\delta_1^1 = \emptyset; \qquad \delta_m^m = r_{(m-l)1}; \qquad \delta_m^i = \delta_{m-l}^i l_{(m-l)1} \quad (i < m).$$

Lemma 8. *If $\sigma : m \to m$ is a permutation, then $t(\sigma)$ is $\langle \delta_n^{\sigma(l)} x^n, \ldots, \delta_n^{\sigma(n)} x^n \rangle$. If $f : m \to n$, then $t(f) \sim \langle s_1, \ldots, s_n \rangle$, where s is of the form 0^1 or $f_k^1(\langle u_1, \ldots, u_k \rangle)$, with $u_j = \delta_m^q x^m$ (all q's in $s_1, \ldots, s_n$ are different).*

Below we shall consider the decision problem for formal morphisms in CL(Finord). We shall show that it may be solved by a kind of normalization for the corresponding λ-terms. Obviously, it is sufficient to deal with such formal morphisms φ that do not contain a_{mnp} and b_m and in which every f is replaced in the way described above. In the corresponding λ-term every $f_m^n(s)$ may be replaced by $\langle s_1, \ldots, s_n \rangle_x[s]$, where $s = t(\sigma)$ is of the form described in Lemma 8 (so, all functional symbols have 1 as upper index). If $t(\varphi)$ contains a part of the form $s_{x_1,\ldots,x_n}[u_1, \ldots, u_n]$, where s is a V-term, then s does not contain $\{,\}$. As above, in s we may write all iterated $\langle , \rangle$ in a symmetric form. We shall call such λ-terms *restricted.*

9 Restricted Equivalence for Restricted λ-Terms

In this section we show that for restricted λ-terms the equivalence $\sim$ may be replaced by a weaker equivalence relation. Recall (vii) of the definition of the relation $\sim$:

(vii) if s and s' are V-terms with the same free variables and $C(s) \sim C(s')$, then $s_{x_1,\ldots,x_n}[u_1, \ldots, u_n] \sim s'_{x_1,\ldots,x_n}[u_1, \ldots, u_n]$.

Replace the condition that $C(s) \sim C(s')$ by its restricted form:

(a) s is $f^1_m(\langle \ldots, 0^1, \ldots \rangle)$ and s' is $f^1_{m-l}(\langle \ldots \rangle)$;

(b) s is $f^l_m(\langle \ldots, g^l_k(u), \ldots \rangle)$ and s' is $f_{m+k-l}(\langle \ldots, u, \ldots \rangle)$.

Note that in both cases $C(s) \sim C(s')$.

By (vii′) we denote (vii) where the condition $C(s) \sim C(s')$ has the special form (a) or (b).

Let $\sim_\wedge$ be the equivalence relation generated by (vii′) and (i)–(vi), (viii)–(xi) from Section 5.

Let R_3 be the reduction relation on restricted V-terms based on the conversions (Ra), (Rb) corresponding to (a) and (b) (the left-hand side is reduced to the right-hand side), and on the conversion (6) from Section 6:

$$t_{x,y}[l\langle u, v\rangle, r\langle u, v\rangle] \to t_{x,y}[u, v].$$

The following lemma is obvious:

Lemma 9. *If t is an R_3-normal restricted V-term, then t has the form $\langle s_1, \ldots, s_n\rangle$, where s_i is 0^1, or it has the form $f^l_m(\langle u_1, \ldots, u_m\rangle)$, with u of the form δx (δ is a 'projection' defined above; all necessary combinations of l and r are present).*

So, R_3-normal V-terms correspond directly to the morphisms $f : n \to n'$ of CL(Finord).

Let the relation E' be defined by the same conditions as E in Section 6 with the addition of the condition reflecting the fact that in CL(Finord) the order of $\langle , \rangle$ for V-terms and terms obtained from them by substitution is not essential.

It is easily checked that for restricted terms and relations R and E' the conditions of Theorem 2 are fulfilled (in particular we have the Church-Rosser property). So we have:

Lemma 10. *Every restricted term t has a unique R_3-normal form (up to E').*

Let us denote the R_3-normal form of t by $!t$.

Lemma 11. *Two V-terms t and t' are $\sim$-equivalent iff $!tE'!t'$. The same is true for $\sim_\wedge$ instead of $\sim$.*

Proof. Obviously, $!tE'!t'$ implies $t \sim t'$. Now, we easily check that if not $!tE'!t'$, then $C(!t)$ and $C(!t')$ are different V-morphisms (see Lemma 11), and hence not $!t \sim !t'$. Because $!s \sim s$ for all s, $t \sim t'$ does not hold. The proof for $\sim_\wedge$ is analogous.

Lemma 12. *For any restricted terms t and t', $t \sim t'$ iff $t \sim_\wedge t'$.*

Proof. By Lemma 11, for V-terms, $s \sim s'$ iff $s \sim_\wedge s'$. This implies that every application of (vii) to restricted terms may be replaced by a sequence of applications of (a), (b) and maybe (v) of the definition of $\sim$. The converse implication is obvious.

10 Decision Procedure for $\sim_\wedge$

Lemma 13. *Let us consider two conversions $t \xrightarrow{\mu} t'$ and $t \xrightarrow{\nu} t''$, where μ is one of* (1)–(6) *and ν is* (Ra) *or* (Rb). *Then there exists a sequence ν' of conversions of the same type as ν, such that the following square commutes:*

$$\begin{array}{ccc} t & \xrightarrow{\nu} & t'' \\ \mu \downarrow & & \downarrow \mu \\ t' & \xrightarrow{\nu'} & t''' \end{array}$$

Proof. Let us consider a quasi-term u subjected to ν. It has the form $f^1_m(\langle \ldots g^1_k(s) \ldots \rangle)$ or $f^l_m(\langle \ldots 0^l \ldots \rangle)$. The conversions (1)–(6) do not change this configuration because u is not of a type of the form $A \Rightarrow B$, $A \otimes B$ or I. All possible substitutions may change s but not the positions of f, g or 0. Hence, ν will be applicable after μ. Analogously, μ will be applicable after ν. Some conversions (e.g. λ-conversions) may change the number of occurrences of u, and then we need to apply ν several times to obtain the term obtained by μ from t''.

The following lemma is an immediate consequence:

Lemma 14. *If $t \xrightarrow{\nu} t'$, then $\overline{\overline{t}} \xrightarrow{\nu'} \overline{\overline{t}}'$ (ν and ν' are as in Lemma 13).*

Theorem 5. *If t and t' are restricted terms, then $t \sim_\wedge t'$ iff $!\overline{\overline{t}}E'!t'$.*

Proof. Obviously, $!\overline{\overline{t}}E'!\overline{\overline{t}}'$ implies $t \sim_\wedge t'$. Let us consider the chain of applications of axioms and rules defining $\sim_\wedge$ that proves $t \sim_\wedge t'$. We may use induction on the length of this chain. The basis is trivial. If the last application is not (Ra) or (Rb), then for its mates t and (say) t'', $\overline{\overline{t}}E\overline{\overline{t}}''$; hence, $!\overline{\overline{t}}E'!\overline{\overline{t}}''$. This allows us to apply the induction hypothesis. If it is (Ra) or (Rb), we use Lemma 1; hence, $\overline{\overline{t}} \xrightarrow{\nu'} \overline{\overline{t}}''$ (ν' is as in Lemma 13). But this implies $!\overline{\overline{t}}E'!\overline{\overline{t}}''$, and then we apply again the induction hypothesis.

Remark 5. Now we may obtain a similar result for the category CL(Perm), which has the same objects as CL(Finord); however, the morphisms of Perm are not all functions $f : m \to n$ but only the permutations $\sigma : n \to n$. The restricted terms for such a category will not contain functional symbols at all. This result was announced in [9].

Acknowledgement. The author is very grateful to Peter Schroeder-Heister and Kosta Došen as organizers of the Conference on Logics with Restricted Structural Rules in Tübingen. He is also very obliged to Kosta Došen for many improvements in the manuscript and to C. Barry Jay for useful discussions.

References

[1] Babaev, A. A., Equality of morphisms in closed categories, I (in Russian), *Izvestiya Akademii Nauk Azerbaĭdzhanskoĭ SSR, Seriya fiziko-tekhnicheskikh i matematicheskikh nauk*, 1981, 1, 3–9.

[2] ———, Equality of morphisms in closed categories, II (in Russian), *ibid.*, 1981, 2, 3–9.

[3] Bruce, K. B., Di Cosmo, R., Longo, G., Provable isomorphisms of types, *Proceedings of Symposium on Symbolic computation*, ETH, Zurich, 1990.

[4] Jay, C. B., The structure of free closed categories, *Journal of Pure and Applied Algebra 66* (1990), 271–285.

[5] Lambek, J., Multicategories revisited, in: J. W. Gray and A. Scedrov (eds), *Categories in Computer Science and Logic*, American Mathematical Society, Providence, 1989, 217–239.

[6] Mac Lane, S., *Categories for the Working Mathematician*, Springer, Berlin, 1971.

[7] Mints, G. E., Closed categories and proof theory (in Russian), *Zapiski Nauchnykh Seminarov Leningradskogo Otdeleniya Matematicheskogo Instituta 68* (1977), 83–114 (English translation in *Journal of Soviet Mathematics 15* (1981), 45–62).

[8] Soloviev, S. V., The category of finite sets and closed categories (in Russian), *Zapiski Nauchnykh Seminarov Leningradskogo Otdeleniya Matematicheskogo Instituta 105* (1981), 174–194 (English translation in *Journal of Soviet Mathematics 22* (1983), 1387–1400.)

[9] ———, Lambda terms with functional symbols and closed categories (abstract), Logic Colloquium '90, Helsinki, July 15–22, 1990.

Tutorial on Linear Logic

A. S. Troelstra

Faculty of Mathematics and Computer Science
University of Amsterdam

1 Introduction

In this tutorial[1] we present a very elementary introduction to linear logic, consisting of a description of the system, a sketch of cut-elimination, a discussion of the relationship with intuitionistic logic (embedding theorems), a sketch of completeness for algebraic semantics, and some remarks on the computational interpretation.

From a technical point of view linear logic appears as a refinement of ordinary logic; in a sequential formulation in Gentzen-style this is obtained by omitting the rules of contraction and weakening (thinning); afterwards weakening and contraction are reintroduced for specific formulas with help of special operators storage (!) and costorage (?).

Ordinary logic deals with truth, and if a formula A expresses a hypothesis assumed to be true, it does not matter whether we use it many times, use it only once, or not at all. That is to say, $A \Rightarrow B$ is simply read as: from the truth of A follows the truth of B, and $A, A \Rightarrow B$ means exactly the same. Moreover, if one has $A \Rightarrow C$, then one also has $A, B \Rightarrow C$.

If $A, B, C, \ldots$ stand for resources, things become different. Suppose you can buy for \$1 a Coke, you may express this by

$$\$1 \Rightarrow \text{Coke}$$

1 This paper was presented at the Joint International Conference and Symposium on Logic Programming, held November 9–14, 1992 in Washington, D.C. and does not assume much preliminary knowledge of subjects such as domain theory and category theory. Many interesting aspects are not touched upon, e.g. there is no discussion of proofnets. A more allround introduction may be obtained from [T]. We have included very few references. Many more references may be found in [T] and [Sc].

but after exchanging \$1 for the Coke, the dollar is not available any more. Let us further assume you can also buy for one dollar a packet of cigarettes:

$$\$1 \Rightarrow \text{Cig}$$

Buying both cigarettes and Coke is now expressible as

$$\$1 \text{ and } \$1 \Rightarrow \text{Coke and Cig}$$

In this statement 'and' is not idempotent: \$1 and \$1 is not equivalent to a single dollar. But we may consider another type of conjunction, say 'AND', such that we combine to

$$\$1 \Rightarrow \text{Coke AND Cig}$$

Formally, AND obeys the familiar rule for ordinary truth-functional conjunction: $A \Rightarrow B$ and $A \Rightarrow C$ yield $A \Rightarrow B \wedge C$. But in terms of resources, it corresponds to a choice we can make: with one dollar in our pocket, we can choose to buy either a Coke or a cigarettes, but not both. Both actions are possible, but we cannot execute both of them. (So AND has a bit of the flavour of a disjunction; but a real disjunction would correspond to two possible outcomes of an action of ours, without being able to determine which one.)

Exercise. If \$ $\Rightarrow$ Coke, may we then conclude that \$ and \$ $\Rightarrow$ Coke? No, but why not?

We shall now study this more formally in the context of a 'baby example', the calculus of conjunction.

2 Intuitionistic Conjunction Calculus

We describe the conjunction fragment of a Gentzen-style sequent calculus for intuitionistic logic. Deductions are trees, with sequents at the nodes; the sequents are expressions $\Gamma \Rightarrow \Delta$, with Γ, Δ finite sequences of formulas constructed from proposition letters with $\wedge$.

Axioms:

$$A \Rightarrow A$$

The structural rules of Weakening (W), Contraction (C), Exchange (E):

$$\frac{\Gamma \Rightarrow A}{\Gamma, B \Rightarrow A}\,\mathrm{W} \qquad \frac{\Gamma, B, B \Rightarrow A}{\Gamma, B \Rightarrow A}\,\mathrm{C} \qquad \frac{\Gamma, A, B, \Delta \Rightarrow C}{\Gamma, B, A, \Delta \Rightarrow C}\,\mathrm{E}$$

Cut rule:

$$\frac{\Gamma \Rightarrow A \qquad \Gamma', A \Rightarrow B}{\Gamma, \Gamma' \Rightarrow B}\,\mathrm{Cut}$$

We consider two variants of the usual conjunction rules:

Conjunction rules, variant 1:

$$\frac{\Gamma, A_i \Rightarrow C}{\Gamma, A_0 \wedge A_1 \Rightarrow C}\,\mathrm{L}\wedge \;\; (i = 0, 1) \qquad \frac{\Gamma \Rightarrow A_0 \qquad \Gamma \Rightarrow A_1}{\Gamma \Rightarrow A_0 \wedge A_1}\,\mathrm{R}\wedge$$

Conjunction rules, variant 2:

$$\frac{\Gamma, A_0, A_1 \Rightarrow C}{\Gamma, A_0 \wedge A_1 \Rightarrow C}\,\mathrm{L}\wedge' \qquad \frac{\Gamma_0 \Rightarrow A_0 \qquad \Gamma_1 \Rightarrow A_1}{\Gamma_0, \Gamma_1 \Rightarrow A_0 \wedge A_1}\,\mathrm{R}\wedge'$$

Variant 1 and Variant 2 yield equivalent calculi, in the sense that the same sequents are derivable; but in order to prove this equivalence, we essentially need the rules of contraction and weakening.

For example, assume the rules of variant 1. We derive $\mathrm{R}\wedge'$ by repeated weakening (indicated by double lines) and $\mathrm{R}\wedge$ as follows:

$$\frac{\dfrac{\Gamma_0 \Rightarrow A_0}{\overline{\Gamma_0, \Gamma_1 \Rightarrow A_0}} \qquad \dfrac{\Gamma_1 \Rightarrow A_1}{\overline{\Gamma_0, \Gamma_1 \Rightarrow A_1}}}{\Gamma_0, \Gamma_1 \Rightarrow A_0 \wedge A_1}$$

Exercise. Complete the proof of equivalence of the two variants.

Remark. By interpreting the Γ and Δ in sequents $\Gamma \Rightarrow \Delta$ as *finite sets* instead of sequences, one builds in the rules of contraction and exchange, so that they do not appear explicitly among the rules; and by generalizing the axioms to $\Gamma, A \Rightarrow A$ (intuitionistic case) or $\Gamma, A \Rightarrow \Delta, A$ (classical case) one also can omit weakening.

In the sequel *we shall assume neither weakening nor contraction*, on the other hand we shall always assume exchange; this may be built into the rules by reading the Γ, Δ in sequents as *multisets* (sets with multiplicities, sequences modulo permutations of the elements).

We may also consider the symmetric, classical version, where zero or more formulas may appear on the right hand side.

3 The Classical Conjunction Calculus

The classical conjunction calculus is very similar, but treats the left and right hand side of sequents (antecedent and succedent) in a completely symmetric way. We have left out the exchange rules since we now regard $\Gamma, \Delta, \ldots$ as multisets.

Axioms:

$$A \Rightarrow A$$

Structural rules:

$$\frac{\Gamma \Rightarrow \Delta}{\Gamma, A \Rightarrow \Delta}\,\mathrm{LW} \qquad \frac{\Gamma \Rightarrow \Delta}{\Gamma \Rightarrow A, \Delta}\,\mathrm{RW}$$

$$\frac{\Gamma, A, A \Rightarrow \Delta}{\Gamma, A \Rightarrow \Delta}\,\mathrm{LC} \qquad \frac{\Gamma \Rightarrow A, A, \Delta}{\Gamma \Rightarrow A, \Delta}\,\mathrm{RC}$$

Cut Rule:

$$\frac{\Gamma \Rightarrow A, \Delta \qquad \Gamma', A \Rightarrow \Delta'}{\Gamma, \Gamma' \Rightarrow \Delta, \Delta'}$$

Conjunction rules, variant 1:

$$\frac{\Gamma, A_i \Rightarrow \Delta}{\Gamma, A_0 \wedge A_1 \Rightarrow \Delta}\,\mathrm{L}\wedge\ (i = 0, 1) \qquad \frac{\Gamma \Rightarrow A_0, \Delta \qquad \Gamma \Rightarrow A_1, \Delta}{\Gamma \Rightarrow A_0 \wedge A_1, \Delta}\,\mathrm{R}\wedge$$

Conjunction rules, variant 2:

$$\frac{\Gamma, A_0, A_1 \Rightarrow \Delta}{\Gamma, A_0 \wedge A_1 \Rightarrow \Delta}\,\mathrm{L}\wedge' \qquad \frac{\Gamma_0 \Rightarrow A_0, \Delta_0 \quad \Gamma_1 \Rightarrow A_1, \Delta_1}{\Gamma_0, \Gamma_1 \Rightarrow A_0 \wedge A_1, \Delta_0, \Delta_1}\,\mathrm{R}\wedge'$$

4 The $\sqcap$- and the $\star$-Calculus

In the absence of weakening and contraction, the two variants of the conjunction rules are not equivalent, and characterize distinct operators; let us therefore introduce new symbols:

- $\sqcap$, 'and' for the context-sensitive variant 1;
- $\star$, 'tensor' for the context-free variant 2.

We shall use 'conjunction' for $\sqcap$, and 'tensor' for $\star$. 'Context-free': in variant 2, the side formulas $\Gamma_0, \Gamma_1, \Delta_0, \Delta_1$ are simply added together in the conclusion of R$\wedge'$. 'Context-sensitive': in variant 1, the side-formulas in both premises of R$\wedge$ must coincide for the rule to be applicable.

The $\sqcap$-calculus: Cut plus

$$\frac{\Gamma \Rightarrow A_0, \Delta \quad \Gamma \Rightarrow A_1, \Delta}{\Gamma \Rightarrow A_0 \sqcap A_1, \Delta}\,\mathrm{R}\sqcap \qquad \frac{\Gamma, A_i \Rightarrow \Delta}{\Gamma, A_0 \sqcap A_1 \Rightarrow \Delta}\,\mathrm{L}\sqcap\ (i = 0, 1)$$

The $\star$-calculus: Cut plus

$$\frac{\Gamma_0 \Rightarrow A_0, \Delta_0 \quad \Gamma_1 \Rightarrow A_1, \Delta_1}{\Gamma_0, \Gamma_1 \Rightarrow A_0 \star A_1, \Delta_0, \Delta_1}\,\mathrm{R}\star \qquad \frac{\Gamma, A_0, A_1 \Rightarrow \Delta}{\Gamma, A_0 \star A_1 \Rightarrow \Delta}\,\mathrm{L}\star$$

These calculi may also be combined into a single system. The result is the $\{\star, \sqcap\}$-fragment of linear logic. It is not hard to show that the cut rule is in fact redundant. What would happen if we combined the left rule of variant 1 with the right rule of variant 2, for example? This would lead to an unbalanced system, in which either weakening or contraction become derivable.

It is now time to introduce the full system.

5 Full Linear Logic

Tensor, and, par, disjunction

First of all linear logic contains the rules for $\star, \sqcap$ already listed, but also the left-right duals of $\star, \sqcap$: $+$ (par) and $\sqcup$ (disjunction); the rules for these new constants look the same as for their duals, but with the roles of left and right in the rules interchanged.

Negation

The rules for negation shunts formulas from left to right and vice versa, prefixing them with $\sim$. Thus $\Gamma, A \Rightarrow \Delta$ yields $\Gamma \Rightarrow {\sim}A, \Delta$ etc.

Linear implication

For 'linear implication' $\multimap$ there is no additive counterpart listed, the rules are

$$\frac{\Gamma, A \Rightarrow B}{\Gamma \Rightarrow A \multimap B} \qquad \frac{\Gamma \Rightarrow A, \Delta \quad \Gamma', B \Rightarrow \Delta'}{\Gamma, \Gamma', A \multimap B \Rightarrow \Delta, \Delta'}$$

It is possible to formulate an additive counterpart; the rules ought to be

$$\frac{\Gamma \Rightarrow B, \Delta}{\Gamma \Rightarrow A \leadsto B, \Delta} \qquad \frac{\Gamma, A \Rightarrow \Delta}{\Gamma \Rightarrow A \leadsto B, \Delta} \qquad \frac{\Gamma \Rightarrow A, \Delta \quad \Gamma, B \Rightarrow \Delta}{\Gamma, A \leadsto B \Rightarrow \Delta}$$

Constants

There are two 'multiplicative' constants, **1** and **0**, with rules and axioms

$$\mathbf{0} \Rightarrow \qquad \qquad \Rightarrow \mathbf{1}$$

$$\frac{\Gamma \Rightarrow \Delta}{\Gamma \Rightarrow \Delta, \mathbf{0}} \qquad \frac{\Gamma \Rightarrow \Delta}{\Gamma, \mathbf{1} \Rightarrow \Delta}$$

If we read $A_1, A_2, \ldots A_n \Rightarrow B_1, B \ldots B_m$ as $A_1 \star A_2 \star \cdots \star A_n \Rightarrow B_1 + B_2 + \cdots + B_m$, where the empty tensor product of A's is taken to be **1**, and the empty sum is taken to be **0**, we see that **0** is a 'unit' for the monoidal operator $+$, and **1** of the tensor $\star$, since the rules and axioms ensure equivalence of $\Gamma \Rightarrow \Delta$ with $\Gamma, \mathbf{1} \Rightarrow \Delta$ and with $\Gamma \Rightarrow \Delta, \mathbf{0}$:

$$\frac{\Gamma \Rightarrow \Delta, \mathbf{0} \qquad \mathbf{0} \Rightarrow}{\Gamma \Rightarrow \Delta} \text{Cut}$$

In particular,

$$A \Leftrightarrow A + \mathbf{0} \Leftrightarrow A \star \mathbf{1},$$

since

$$\dfrac{\dfrac{A \Rightarrow A}{A \Rightarrow A, \mathbf{0}}}{A \Rightarrow A + \mathbf{0}} \qquad \dfrac{A \Rightarrow A \qquad \mathbf{0} \Rightarrow}{A + \mathbf{0} \Rightarrow A}$$

etc. There are also two additive constants, $\perp$ and $\top$, with axioms

$$\Gamma, \perp \Rightarrow \Delta \qquad\qquad \Gamma \Rightarrow \Delta, \top$$

(falsehood implies everything and truth is implied by everything). $\perp$ and $\top$ behave as unit for $\sqcup$ and $\sqcap$ respectively.

Exercise. Show $A \Leftrightarrow A \sqcup \perp \Leftrightarrow A \sqcap \top$.

Storage and costorage

The most interesting rules are those for ! and ? With a formula $!A$ we can weaken and contract on the left:

$$\dfrac{\Gamma \Rightarrow \Delta}{\Gamma, !A \Rightarrow \Delta} \qquad \dfrac{\Gamma, !A, !A \Rightarrow \Delta}{\Gamma, !A \Rightarrow \Delta}$$

and also, if a single use of A is needed, it can be obtained from an arbitrary supply $!A$:

$$\dfrac{\Gamma, A \Rightarrow \Delta}{\Gamma, !A \Rightarrow \Delta}$$

Furthermore, if we can obtain A from a suitable number of instances of elements of Γ, we can obtain $!A$ from $!\Gamma$.

$$\dfrac{!\Gamma \Rightarrow A}{!\Gamma \Rightarrow !A}$$

$?A$ behaves as the left-right dual of A. But if we want to have actually provable $?A := \sim !\sim A$ we need a generalization of the last rule:

$$\dfrac{!\Gamma \Rightarrow A, ?\Delta}{!\Gamma \Rightarrow !A, ?\Delta}$$

(The definition of $?A$ together with the original rule yields the more general form.)

Example. As an example of a deduction we prove $!(A \sqcap B) \Rightarrow !A \star !B)$

$$\dfrac{\dfrac{\dfrac{\dfrac{\dfrac{A \Rightarrow A}{A \sqcap B \Rightarrow A}}{!(A \sqcap B) \Rightarrow A}}{!(A \sqcap B) \Rightarrow !A} \qquad \dfrac{\dfrac{\dfrac{B \Rightarrow B}{A \sqcap B \Rightarrow B}}{!(A \sqcap B) \Rightarrow B}}{!(A \sqcap B) \Rightarrow !B}}{!(A \sqcap B), !(A \sqcap B) \Rightarrow !A \star !B}}{!(A \sqcap B) \Rightarrow !A \star !B}$$

Exercise. Prove $!A \star !B \Rightarrow !(A \sqcap B)$.

Here is the table of the full system:

The system **CLL**

Logical axiom and Cut rule:

$$\text{Ax} \quad A \Rightarrow A \qquad\qquad \text{Cut}\ \dfrac{\Gamma \Rightarrow A, \Delta \qquad \Gamma', A \Rightarrow \Delta'}{\Gamma, \Gamma' \Rightarrow \Delta, \Delta'}$$

Rules for the propositional constants:

$$\text{L}{\sim}\ \dfrac{\Gamma \Rightarrow A, \Delta}{\Gamma, {\sim}A \Rightarrow \Delta} \qquad\qquad \text{R}{\sim}\ \dfrac{\Gamma, A \Rightarrow \Delta}{\Gamma \Rightarrow {\sim}A, \Delta}$$

$$\text{L}\sqcap\ \dfrac{\Gamma, A \Rightarrow \Delta}{\Gamma, A \sqcap B \Rightarrow \Delta} \quad \dfrac{\Gamma, B \Rightarrow \Delta}{\Gamma, A \sqcap B \Rightarrow \Delta} \qquad\qquad \text{R}\sqcap\ \dfrac{\Gamma \Rightarrow A, \Delta \qquad \Gamma \Rightarrow B, \Delta}{\Gamma \Rightarrow A \sqcap B, \Delta}$$

$$\text{L}\star\ \dfrac{\Gamma, A, B \Rightarrow \Delta}{\Gamma, A \star B \Rightarrow \Delta} \qquad\qquad \text{R}\star\ \dfrac{\Gamma \Rightarrow A, \Delta \qquad \Gamma' \Rightarrow B, \Delta'}{\Gamma, \Gamma' \Rightarrow A \star B, \Delta, \Delta'}$$

$$\text{L}\sqcup\ \dfrac{\Gamma, A \Rightarrow \Delta \qquad \Gamma, B \Rightarrow \Delta}{\Gamma, A \sqcup B \Rightarrow \Delta} \qquad\qquad \text{R}\sqcup\ \dfrac{\Gamma \Rightarrow A, \Delta}{\Gamma \Rightarrow A \sqcup B, \Delta} \quad \dfrac{\Gamma \Rightarrow B, \Delta}{\Gamma \Rightarrow A \sqcup B, \Delta}$$

$$\text{L}+\ \dfrac{\Gamma, A \Rightarrow \Delta \qquad \Gamma', B \Rightarrow \Delta'}{\Gamma, \Gamma', A + B \Rightarrow \Delta, \Delta'} \qquad\qquad \text{R}+\ \dfrac{\Gamma \Rightarrow A, B, \Delta}{\Gamma \Rightarrow A + B, \Delta}$$

$$\text{L}{\multimap}\ \dfrac{\Gamma \Rightarrow A, \Delta \qquad \Gamma', B \Rightarrow \Delta'}{\Gamma, \Gamma', A \multimap B \Rightarrow \Delta, \Delta'} \qquad\qquad \text{R}{\multimap}\ \dfrac{\Gamma, A \Rightarrow B, \Delta}{\Gamma \Rightarrow A \multimap B, \Delta}$$

$$\text{L}\mathbf{1}\ \dfrac{\Gamma \Rightarrow \Delta}{\Gamma, \mathbf{1} \Rightarrow \Delta} \qquad\qquad \text{R}\mathbf{1} \quad \Rightarrow \mathbf{1}$$

$$(\text{no L}\top) \qquad\qquad \text{R}\top \quad \Gamma \Rightarrow \top, \Delta$$

$$\text{L}\mathbf{0} \quad \mathbf{0} \Rightarrow \qquad\qquad \text{R}\mathbf{0}\ \dfrac{\Gamma \Rightarrow \Delta}{\Gamma \Rightarrow \Delta, \mathbf{0}}$$

$$\text{L}\bot \quad \Gamma, \bot \Rightarrow \Delta \qquad\qquad (\text{no R}\bot)$$

Rules for the quantifiers (y not free in Γ, Δ):

$$\mathrm{L}\forall\ \frac{\Gamma, A[x/t] \Rightarrow \Delta}{\Gamma, \forall x\, A \Rightarrow \Delta} \qquad \mathrm{R}\forall\ \frac{\Gamma \Rightarrow A[x/y], \Delta}{\Gamma \Rightarrow \forall x\, A, \Delta}$$

$$\mathrm{L}\exists\ \frac{\Gamma, A[x/y] \Rightarrow \Delta}{\Gamma, \exists x\, A \Rightarrow \Delta} \qquad \mathrm{R}\exists\ \frac{\Gamma \Rightarrow A[x/t], \Delta}{\Gamma \Rightarrow \exists x\, A, \Delta}$$

Rules for the exponentials:

$$\mathrm{W!}\ \frac{\Gamma \Rightarrow \Delta}{\Gamma, !A \Rightarrow \Delta} \qquad \mathrm{L!}\ \frac{\Gamma, A \Rightarrow \Delta}{\Gamma, !A \Rightarrow \Delta} \qquad \mathrm{R!}\ \frac{!\Gamma \Rightarrow A, ?\Delta}{!\Gamma \Rightarrow !A, ?\Delta} \qquad \mathrm{C!}\ \frac{\Gamma, !A, !A \Rightarrow \Delta}{\Gamma, !A \Rightarrow \Delta}$$

$$\mathrm{W?}\ \frac{\Gamma \Rightarrow \Delta}{\Gamma \Rightarrow ?A, \Delta} \qquad \mathrm{L?}\ \frac{!\Gamma, A \Rightarrow ?\Delta}{!\Gamma, ?A \Rightarrow ?\Delta} \qquad \mathrm{R?}\ \frac{\Gamma \Rightarrow A, \Delta}{\Gamma \Rightarrow ?A, \Delta} \qquad \mathrm{C?}\ \frac{\Gamma \Rightarrow ?A, ?A, \Delta}{\Gamma \Rightarrow ?A, \Delta}$$

Definition. Intuitionistic linear logic **ILL** is obtained by dropping all references to $+, ?, \mathbf{0}$ and restricting the consequent of all sequents to multisets with at most one element.

ILZ (**ILL** with zero) is like **ILL**, except that one adds $\mathbf{0}$ with axiom $\mathbf{0} \Rightarrow$ and the rule

$$\frac{\Gamma \Rightarrow}{\Gamma \Rightarrow \mathbf{0}}$$

$\mathbf{ILL}_{\mathrm{q}}$ and $\mathbf{CLL}_{\mathrm{q}}$ are used for the systems with quantifiers, but without exponentials.

Remarks. *Notation.* There is no uniformity in the notation for the operators of linear logic; Girard's original notation is widely used, but modifications abound. Here I have deviated from Girard's original notation, so as to make 'similar' operations also look similar. In particular, the notation should make duality between operators and constants visible. This is shown in a little table.

The first group constitutes the so-called 'multiplicative' operators, the second group the 'additives', the third the quantifiers, the fourth the exponentials. The terminology 'multiplicative', 'additive' derives from the behaviour of these operators in the type-theoretic model of the so-called Girard domains (cf. [T]).

This tutorial	Girard's notation	names in common use
$\star$	$\otimes$	tensor
$+$	$\wp$	par
$\multimap$	$\multimap$	(linear) implication, lollipop
$\sim$	$(\cdot)^\perp$	negation
$\mathbf{1}$	$\mathbf{1}$	unit
$\mathbf{0}$	$\perp$	zero, false (Girard)
$\sqcap$	$\&$	and, conjunction
$\sqcup$	$\oplus$	or, disjunction
$\top$	$\top$	true
$\perp$	$\mathbf{0}$	false, zero (Girard)
$\forall$	$\forall$	all-quantifier, universal quantifier
$\exists$	$\exists$	exists-quantifier, existential quantifier
!	!	storage, bang, ofcourse (Girard)
?	?	costorage, reuse, whynot (Girard)

Grouping of the operators. In the following two lines the operators appearing in the same column correspond to each other under the Morgan-duality:

$$\begin{array}{ccccccc} \star & \sim & \mathbf{1} & \sqcap & \top & \forall & ! \\ + & \sim & \mathbf{0} & \sqcup & \perp & \exists & ? \end{array}$$

$\multimap$ has no the Morgan-dual. A dual might be defined as $A \multimap^{\text{dual}} B \equiv A \star {\sim}B$. In the following table we have indicated the classical analogues to the linear operators.

linear logic operators	classical analogue
$\star, \sqcap$	$\wedge$
$+, \sqcup$	$\vee$
$\multimap$	$\rightarrow$
$\sim$	$\neg$
$\mathbf{1}, \top$	$\top$
$\mathbf{0}, \perp$	$\perp$
$\forall$	$\forall$
$\exists$	$\exists$
$!, ?$	identity

6 The One-Sided Calculus

The strong symmetries in **CLL** permit us to use *one-sided sequents*, that is we consider the one-side sequent $\Rightarrow {\sim}\Gamma, \Delta$ instead of $\Gamma \Rightarrow \Delta$; in this case the initial '$\Rightarrow$' may be dropped. We shall assume the prime formulas come in pairs $P, \sim P$. Negation on prime formulas is then $\sim(P) := \sim P$, $\sim(\sim P) := P$. Negation $\sim$ for compound formulas is defined by duality, e.g. $\sim(A \sqcap B) := (\sim A \sqcup \sim B)$, $\sim(A \star B) := (\sim A + \sim B)$. Implication is regarded as defined: $A \multimap B := \sim A + B$. The rules of the one-sided calculus are:

$$\text{Axiom: } A, \sim A$$

$$\text{Cut: } \frac{\Gamma, A \qquad \Delta, \sim A}{\Gamma, \Delta}$$

Rules for propositional operators:

$$\frac{\Gamma, A \qquad \Delta, B}{\Gamma, \Delta, A \star B} \qquad \frac{\Gamma, A \qquad \Gamma, B}{\Gamma, A \sqcap B} \qquad \frac{\Gamma, A}{\Gamma, A \sqcup B} \qquad \frac{\Gamma, B}{\Gamma, A \sqcup B}$$

$$\frac{\Gamma, A, B}{\Gamma, A + B} \qquad \frac{\Gamma}{\Gamma, 0} \qquad \Gamma, \top \qquad \mathbf{1} \qquad \text{no rules for } \bot$$

Quantifier rules (y not free in Γ):

$$\frac{\Gamma, A[x/y]}{\Gamma, \forall x\, A} \qquad \frac{\Gamma, A[x/t]}{\Gamma, \exists x\, A}$$

Rules for exponentials:

$$\frac{\Gamma, ?A, ?A}{\Gamma, ?A} \qquad \frac{\Gamma}{\Gamma, ?A} \qquad \frac{\Gamma, A}{\Gamma, ?A} \qquad \frac{?\Gamma, A}{?\Gamma, !A}$$

---

7 Cut Elimination

The cut rule in its simplest form:

$$\frac{A \Rightarrow B \qquad B \Rightarrow C}{A \Rightarrow C}$$

intuitively expresses that succesive actions may be combined. If we look at the collection of rules, we note that the cut rule is the only rule where a formula may *disappear* without leaving a trace in the conclusion; in all other cases the formulas of the premises reappear in the conclusion, either as themselves or as subformulas. For this reason one might also describe proofs involving cuts as 'indirect proofs': the proofs contain formulas not appearing as subformulas in the conclusion. The cut elimination theorem states that the calculus with Cut is *conservative* over the calculus without it, i.e. with Cut we cannot prove more sequents than without it. Better still, there is an algorithmic procedure for eliminating cuts from proofs; this algorithm also presents us with a possible way of giving computational meaning to linear logic.

The crucial idea for this process is the same as in Gentzen's original proof or even more closely, Kleene's version of cut elimination in his 'Introduction to Metamathematics'. Let m be the maximal complexity of cut formulas in a given proof; among the cuts of maximal complexity we look for an uppermost one. The level of this cut is the sum of the levels of the cut formula in the premisses; the level of a formula occurrence is most easily defined recursively: if the formula occurrence in the conclusion of deduction $\mathcal{D}$ has been introduced by the last rule application, the height of the occurrence is 1; otherwise the height is the height of the corresponding occurrence in the premise of the last rule application + 1. We look only at proofs which use only axioms for atomic formulas. If the height of a cut formula is two, the cut formula is principal in both premises, and we can reduce the cut to cuts of lower degree, so that the total number of cut formulas of maximal degree in the deduction diminishes. Example: one replaces

$$\dfrac{\dfrac{\Gamma \Rightarrow A, \Delta \qquad \Gamma', B \Rightarrow \Delta'}{\Gamma, \Gamma', A \multimap B \Rightarrow B, \Delta, \Delta'}\,\text{L}\multimap \qquad \dfrac{\Gamma'', A \Rightarrow B, \Delta''}{\Gamma'' \Rightarrow A \multimap B, \Delta''}\,\text{R}\multimap}{\Gamma, \Gamma', \Gamma'' \Rightarrow \Delta, \Delta', \Delta''}\,\text{Cut}$$

by

$$\dfrac{\dfrac{\Gamma \Rightarrow A, \Delta \qquad \Gamma'', A \Rightarrow B, \Delta''}{\Gamma, \Gamma'' \Rightarrow B, \Delta, \Delta''} \qquad \Gamma', B \Rightarrow \Delta'}{\Gamma, \Gamma', \Gamma'' \Rightarrow \Delta, \Delta', \Delta''}\,\text{Cut}$$

A complication arises in the following case: consider a deduction of the following type, with $!A$ active in both premises of the cut.

$$\dfrac{\dfrac{\begin{matrix}\mathcal{D}\\ \Gamma, !A, !A \Rightarrow \Delta\end{matrix}}{\Gamma, !A \Rightarrow \Delta} \qquad \dfrac{\begin{matrix}\mathcal{D}'\\ !\Gamma' \Rightarrow A, ?\Delta'\end{matrix}}{!\Gamma' \Rightarrow !A, ?\Delta'}}{\Gamma, !\Gamma' \Rightarrow \Delta, ?\Delta'}\,\text{Cut}$$

One is tempted to replace this by the following deduction:

$$
\dfrac{\dfrac{\mathcal{D}}{\Gamma, !A, !A \Rightarrow \Delta} \quad \dfrac{\dfrac{\mathcal{D}'}{!\Gamma' \Rightarrow A, ?\Delta'}}{!\Gamma' \Rightarrow !A, ?\Delta'}}{!A, \Gamma, !\Gamma' \Rightarrow \Delta, ?\Delta'}\,\text{Cut} \quad \dfrac{\dfrac{\mathcal{D}'}{!\Gamma' \Rightarrow A, ?\Delta'}}{!\Gamma' \Rightarrow !A, ?\Delta'}
$$

$$
\dfrac{\dfrac{\cdots}{\Gamma, !\Gamma', !\Gamma' \Rightarrow \Delta, ?\Delta', ?\Delta'}\,\text{Cut}}{\Gamma, !\Gamma' \Rightarrow \Delta, ?\Delta'}\,\text{C!,C?}
$$

where the double line stands for a number of applications of the rules C! and C? This does not work because the lower cut will have the same height as the original one. The solution is to permit a derivable generalization of the Cut rule:

$$
\dfrac{\Gamma, (!A)^n \Rightarrow \Delta \qquad \Gamma' \Rightarrow !A, \Delta}{\Gamma, \Gamma' \Rightarrow \Delta, \Delta'}\,\text{Cut}^*
$$

$$
\dfrac{\Gamma, ?A \Rightarrow \Delta \qquad \Gamma' \Rightarrow (?A)^n, \Delta}{\Gamma, \Gamma' \Rightarrow \Delta, \Delta'}\,\text{Cut}^*
$$

Now we can tranform the deduction above into

$$
\dfrac{\dfrac{\mathcal{D}}{\Gamma, !A, !A \Rightarrow \Delta} \quad \dfrac{\dfrac{\mathcal{D}'}{!\Gamma' \Rightarrow A, ?\Delta'}}{!\Gamma' \Rightarrow ?A, ?\Delta'}}{\Gamma, \Gamma' \Rightarrow \Delta, \Delta'}\,\text{Cut}^*
$$

Closure under Cut is a property we absolutely want, since it intuitively expresses that 'actions are closed under sequential composition'; on the other hand in many arguments it is essential to know that sequents, if provable at all, have cut-free proofs.

A striking example is provided by the proof of the undecidability of **ILL** and **CLL** (without quantifiers). The proof is based on the following ideas.

(i) On the logical side, we need a slight generalization of the cut-elimination argument sketched above, showing that we can eliminate all cuts except cuts involving certain simple axioms.

(ii) The acceptance problem for two-counter machines is undecidable. A two-counter machine has two registers A, B say, containing numbers a, b, and a finite number of states, some of which are terminal. Basic instructions are: adding or subtracting 1 from a register and changing the state, and the testing for content zero of a register and changing the state. A programme for the machine is a finite set of basic instructions.

(iii) The aim is to encode a two-counter machine into a set of axioms for the logic, but since the direct encoding of zero-test instructions is difficult, one first replaces the two-counter machines by a variant, the ACM ('and-branching counter machine').

(iv) Instantaneous descriptions or configurations S of ACM's are encoded by sets of sequents $\theta(S)$ of the logic; one then shows that the sequent $\theta(S)$ is provable from the axioms encoding the machine $\Leftrightarrow$ the machine accepts the configuration S. For the direction from left to right one heavily relies on the cut-elimination theorem.

8 Permutability of Rules

Definition. Rule R_1 is said to be *permutable* below R_2, if for all proofs ending with R_1, introducing A say, immediately above the last rule R_2 introducing B, A not used by R_2 as subformula of B, there is a deduction with the same conclusion with R_2 introducing B immediately above the last rule R_1 introducing A.

8.1. Example. $\sqcap$ permutes below $\star$:

$$\dfrac{\dfrac{C,\Gamma,A \qquad C,\Gamma,B}{C,\Gamma,A\sqcap B} \qquad D,\Gamma'}{C\star D,\Gamma,\Gamma',A\sqcap B} \quad \text{goes to} \quad \dfrac{\dfrac{C,\Gamma,A \qquad D,\Gamma'}{C\star D,\Gamma,\Gamma',A} \qquad \dfrac{C,\Gamma,B \qquad D,\Gamma'}{C\star D,\Gamma,\Gamma',B}}{C\star B,\Gamma,\Gamma',A\sqcap B}$$

the converse is false. In fact, the following is a full list (circulated by P. Lincoln on the Linear Logic News Net) of impermutabilities, with counterexamples:

- $\star$ does not permute below $+$, counterexample: $\vdash A+B, {\sim}A\star{\sim}B$.
- $\star$ does not permute below $\sqcap$, counterexample: $\vdash A\sqcap B, \mathbf{0}\star\top, {\sim}A, {\sim}B$.
- $\star$ does not permute below ?C: $\vdash ?A, {\sim}A\star{\sim}A$.
- $\sqcup$ does not permute below $\sqcap$: $\vdash A\sqcap B, {\sim}A\sqcup{\sim}B$.
- ?W does not permute below $\sqcap$: $\vdash \mathbf{1}\sqcap A, ?{\sim}A$.
- ?D does not permute below !S: $\vdash !({\sim}A\sqcup B), ?A$.
- $\exists$ does not permute below $\sqcap$: $\vdash {\sim}At\sqcap{\sim}Au, \exists x Ax$.
- $\exists$ does not permute below $\forall$: $\vdash \forall y(Ay\sqcup B), {\sim}\exists x Ax$.

- !S does not permute below $\star, +, \sqcap, \sqcup, \mathbf{0}, \forall, \exists$.

  The reason is obvious, !S can only be applied if all side formulas start with ?

By way of illustration, here is the proof of the derivability of the second counterexample:

$$\dfrac{\dfrac{\dfrac{\sim A, A}{\sim A, A, \mathbf{0}} \quad \sim B, \top}{\mathbf{0} \star \top, \sim A, \sim B, A} \qquad \dfrac{\dfrac{\sim B, B}{\sim B, B, \mathbf{0}} \quad \sim A, \top}{\mathbf{0} \star \top, \sim A, \sim B, B}}{A \sqcap B, \mathbf{0}\star, \top, \sim A, \sim B}$$

To see that we cannot derive this sequent by means of a proof with $\star$ introduced at the last rule application, note that a final rule application introducing $\mathbf{0} \star \top$ requires premises $\mathbf{0}, \Gamma_1$ and $\top, \Gamma_2$, where $\Gamma_1, \Gamma_2 = A \sqcap B, \sim A, \sim B$. Suppose e.g. that $\Gamma_2 = \emptyset$. Derivability of $\mathbf{0}, \Gamma_1$ is equivalent to derivability of $\Gamma_1 = A \sqcap B, \sim A, \sim B$; this requires derivability of both $A, \sim A, \sim B$ and $B, \sim A, \sim B$; but both are in fact underivable. Similarly with all other possibilities for splitting $A \sqcap B, \sim A, \sim B$ into Γ_1 and Γ_2.

Does it make sense to ask about permuting Cuts? Not globally, since we can always eliminate cuts; but locally the question makes sense, and locally, Cut does not permute below the rules $+, \sqcap, C, !$.

Exercise. Give detailed arguments for the other counterexamples.

9 Embeddings

Do we lose anything of the expressive power if we drop the structural rules but introduce the storage operator and its dual instead? Intuitively one would expect not. But we can be more precise, we shall describe some embeddings.

First of all we mention an embedding due to V.N. Grishin, originally formulated for the calculus with weakening added ('affine logic') but slightly modified by H. Ono to fit linear logic.

It is actually an infinite sequence of translations; the n-version translates $\Gamma \Rightarrow \Delta$ as $\Gamma^{-n} \Rightarrow \Delta^{+n}$ (so antecedent and succedent translate differently) such that

$$\mathbf{CL} \vdash \Gamma \Rightarrow \Delta \text{ iff } \mathbf{CLL} \vdash \Gamma^{-n} \Rightarrow \Delta^{+n} \text{ for some } n.$$

Note: this is an embedding into the calculus *without* exponentials! Moreover, inspection of the proof of the soundness theorem, which proceeds by induction on the length of proofs in classical logic, shows that only applications of L$\forall$ or $R\exists$ rule of classical logic increase the n; the height of the n needed is a measure for the maximal depth of nestings of L$\forall$ and R$\exists$ occurring in the original classical proof. Here is the definition of the embedding:

(a) *The modified Grishin embedding.*

$$
\begin{array}{llll}
\text{(a1)}\ P^{+n} & := \mathbf{0} \sqcup P, & P^{-n} & := \mathbf{1} \sqcap P \quad \text{for } P \text{ prime},\\
\text{(a2)}\ (\neg A)^{+n} & := {\sim}(A^{-n}), & (\neg A)^{-n} & := {\sim}(A^{+n}),\\
\text{(a3)}\ (A \to B)^{+n} & := A^{-n} \multimap B^{+n}, & (A \to B)^{-n} & := {\sim}A^{+n} \sqcup B^{-n},\\
\text{(a4)}\ (A \wedge B)^{+n} & := A^{+n} \sqcap B^{+n}, & (A \wedge B)^{-n} & := A^{-n} \star B^{-n},\\
\text{(a5)}\ (A \vee B)^{+n} & := A^{+n} + B^{+n}, & (A \vee B)^{-n} & := A^{-n} \sqcup B^{-n},\\
\text{(a6)}\ (\forall x\, A)^{+n} & := \forall x(A^{+n}), & (\forall x\, A)^{-n} & := (\forall x(A^{-n}))^{\mathbf{n}},\\
\text{(a7)}\ (\exists x\, A)^{+n} & := \mathbf{n}(\exists x(A^{+n})), & (\exists x\, A)^{-n} & := (\exists x(A^{-n})),
\end{array}
$$

where for all B

$$B^{\mathbf{1}} := B,\ \ \mathbf{1}B := B,\ \ B^{\mathbf{n+1}} := B^{\mathbf{n}} \star B,\ \ (\mathbf{n+1})B := \mathbf{n}B + B.$$

(b) *Grishin's embedding with exponentials.* Clause (b1)–(b5) are as (a1)–(a5), with $^-$ replacing $^{-n}$, $^+$ replacing $^{+n}$. (a6) and (a7) are replaced by

$$
\begin{array}{llll}
\text{(b6)}\ (\forall x\, A)^{+} & := \ \forall x(A^{+}), & (\forall x\, A)^{-} & := \ !\forall x(A^{-}),\\
\text{(b7)}\ (\exists x\, A)^{+} & := \ ?\exists x(A^{+}) & (\exists x\, A)^{-} & := \ \exists x(A^{-}),
\end{array}
$$

The proof of the faithfulness of the translation is trivial, by replacing each linear operator in the translated sequent by it classical analogue, we get the original classical sequent back, modulo equivalence. The correctness of the Grishin embedding can be proved smoothly using a special calculus for classical logic, in which contraction and weakening have been built into the logical rules.

It is to be noted that the increase in the 'n' of the translation is exclusively caused by the right $\exists$-rule and the left $\forall$-rule. The maximal n needed for the correctness proof relative to any given classical deduction is a measure of the nesting of these two rules in the given proof.

The following system for classical logic is suitable for the proof of correctness of Grishin's embedding:

Axioms: $\Gamma, P \Rightarrow P, \Delta$ with all formulas atomic.

Rules

$$\frac{\Gamma \Rightarrow A, \Delta \qquad \Gamma, B \Rightarrow \Delta}{\Gamma, A \to B \Rightarrow \Delta} \qquad \frac{\Gamma, A \Rightarrow B, \Delta}{\Gamma \Rightarrow A \to B, \Delta}$$

$$\frac{\Gamma, A, B \Rightarrow \Delta}{\Gamma, A \wedge B \Rightarrow \Delta} \qquad \frac{\Gamma \Rightarrow A, \Delta \quad \Gamma \Rightarrow B, \Delta}{\Gamma \Rightarrow A \wedge B, \Delta}$$

$$\frac{\Gamma, A \Rightarrow \Delta \qquad \Gamma, B \Rightarrow \Delta}{\Gamma, A \vee B \Rightarrow \Delta} \qquad \frac{\Gamma \Rightarrow A, B, \Delta}{\Gamma \Rightarrow A \vee B, \Delta}$$

$$\frac{\Gamma \Rightarrow A, \Delta}{\Gamma, \neg A \Rightarrow \Delta} \qquad \frac{\Gamma, A \Rightarrow \Delta}{\Gamma \Rightarrow \neg A, \Delta}$$

$$\frac{\Gamma, \forall x\, A, A[x/t] \Rightarrow \Delta}{\Gamma, \forall x\, A \Rightarrow \Delta} \qquad \frac{\Gamma \Rightarrow A[x/y], \Delta}{\Gamma \Rightarrow \forall x\, A, \Delta}$$

$$\frac{\Gamma, A[x/y] \Rightarrow \Delta}{\Gamma, \exists x\, A \Rightarrow \Delta} \qquad \frac{\Gamma \Rightarrow A[x/t], \exists y\, A, \Delta}{\Gamma \Rightarrow \exists y\, A, \Delta}$$

(y not free in Γ, Δ)

One of the original motivations for the formulation of linear logic was Girard's discovery that in the special version of domain (in the sense of denotational semantics) considered by him, the analogue of (intuitionistic) implication $\to$ permitted a decomposition into two other operators $!$, $\multimap$ on domains: $A \to B := !A \multimap B$. This suggested the following embedding map:

Embedding of **IL** *into* **ILL** *and* **CLL**

P°	$:= P$ for P atomic,
$\bot^\circ$	$:= \bot$,
$(\neg A)^\circ$	$:= !A^\circ \multimap \bot$,
$(A \wedge B)^\circ$	$:= A^\circ \sqcap B^\circ$,
$(A \vee B)^\circ$	$:= !A^\circ \sqcup !B^\circ$,
$(A \to B)^\circ$	$:= !A^\circ \multimap B^\circ$,
$(\forall x\, A)^\circ$	$:= \forall x\, A^\circ$,
$(\exists x\, A)^\circ$	$:= \exists x !A^\circ$

It is quite easy to see that this is a faithful embedding of **IL** into **ILL**.

Theorem. **IL** $\vdash \Gamma \Rightarrow A$ *iff* **ILL** $\vdash !\Gamma^\circ \Rightarrow A^\circ$.

N.B. The translations of instances of the Cut rule are valid!

With much more hard work we can show that the embedding is also faithful as an embedding in **CLL**.

More crude versions of such embeddings are obtained by simply prefixing every subformula in sight with !, a very wasteful procedure.

Finally, let us mention an analogue of Kolmogorov's and Gödel's well-known 'negative translation' of **CL** into **IL**. Here we can embed **CLL** into **ILZ** in a similar way. **ILZ** is **ILL** enriched with **0** and axiom/rule

$$\mathbf{0} \Rightarrow \qquad\qquad \frac{\Gamma \Rightarrow}{\Gamma \Rightarrow \mathbf{0}}$$

The Kolmogorov-style translation

Definition. We define A^k by induction on the complexity of A. We let $\sim A := A \multimap \mathbf{0}$.

(1) $P^k := \sim\sim P$ for P prime, not $\mathbf{0}$;

(2) $\mathbf{0}^k := \mathbf{0}$;

(3) $(A \circ B)^k := \sim\sim(A^k \circ B^k)$ for $\circ \in \{\star, \sqcap, \multimap\}$;

(4) $(A + B)^k := \sim(\sim A^k \star \sim B^k)$;

(5) $(A \sqcup B)^k := \sim(\sim A^k \sqcap \sim B^k)$;

(6) $(!A)^k := \sim\sim !A^k$;

(7) $(?A)^k := \sim !\sim A^k$;

(8) $(\forall x\, A)^k := \sim\sim\forall x A^k$;

(9) $(\exists x\, A)^k := \sim\forall x \sim A^k$.

Theorem. $\mathbf{CLL} \vdash \Gamma \Rightarrow A$ *iff* $\mathbf{ILZ} \vdash \Gamma^k \Rightarrow A^k$.

ILZ is just what you need to achieve that **ILZ**+ axiom schema = **CLL**; for the axiom schema we can take $\sim\sim A \multimap A$.

10 Do-it-Yourself Semantics

We outline a semantics for linear logic which is completely analogous to the usual truth-value semantics for classical and intuitionistic logic (valuations in boolean algebras and Heyting algebras respectively). In the literature it appears under various names; Girard used the term 'phase semantics'. Many authors have given completeness proofs for this semantics; as far as I could ascertain, all proofs are in a sense equivalent; we choose here a presentation (in outline) which closely follows the pattern for intuitionistic logic.

We concentrate on **ILL** with quantifiers $\mathbf{ILL}_{\mathrm{q}}$, but leave out the modalities.

Definition. $\mathcal{X} \equiv (X, \sqcap, \sqcup, \bot, \multimap, \star, \mathbf{1})$ is an *IL-algebra* (*intuitionistic linear algebra*) if

(i) $(X, \sqcap, \sqcup, \bot)$ is a lattice with bottom $\bot$;

(ii) $(X, \star, \mathbf{1})$ is a commutative monoid with unit $\mathbf{1}$;

(iii) if $x \leq x'$, $y \leq y'$, then $x \star y \leq x' \star y'$ and $x' \multimap y \leq x \multimap y'$;

(iv) $x \star y \leq z$ iff $x \leq y \multimap z$.

An IL-algebra with an additional constant $\mathbf{0}$ (zero) is called an *IL-algebra with zero* or *ILZ-algebra.* We write

$$\sim x := x \multimap \mathbf{0}$$

An IL-algebra with zero is a *CL-algebra* (*classical linear algebra*) iff

(v) $x = \sim\sim x$ for all x.

A *complete* IL-algebra (ILZ-algebra, CL-algebra) is an IL-algebra (ILZ-algebra, CL-algebra) which is complete as a lattice. □

Remark. Viewing posets as a special case of categories, (iii) expresses that $\star : \mathcal{X} \times \mathcal{X} \longrightarrow \mathcal{X}$ and $\multimap$: $\mathcal{X}^{\mathrm{op}} \times \mathcal{X} \longrightarrow \mathcal{X}$ are functors. (iv) expresses that the functor $- \star y$ is left adjoint to $y \multimap -$; so $- \star y$ preserves all colimits.

10.2. Lemma. *In any IL-algebra* $\mathcal{X} \equiv (X, \sqcap, \sqcup, \bot, \multimap, \star, \mathbf{1})$, *for all* x, y, z *in* X:

(i) $z \star (x \sqcup y) = (z \star x) \sqcup (z \star y)$ *and moreover, if the join* $\bigsqcup_{i \in I} y_i$ *exists, then* $x \star \bigsqcup_{i \in I} y_i = \bigsqcup_{i \in I} (x \star y_i)$.

(ii) $x \multimap (y \multimap z) = x \star y \multimap z$.

(iii) $\bot \multimap \bot$ *is top of* X,

and in any CL-algebra

(iv) $x \sqcup y = \sim(\sim x \sqcap \sim y)$, $x \multimap y = \sim(x \star \sim y)$.

Remark. (i) of the lemma holds because, in categorical terms, joins are colimits, and the functor $- \star z$ having a right adjoint preserves colimits. (ii) An alternative definition of IL-algebra is obtained replacing definition 8.2(iii) by $z \star (x \sqcup y) = (z \star x) \sqcup (z \star y)$.

Complete IL-algebras may be introduced in another way, namely as quantales.

Definition. $\mathcal{X} \equiv (X, \bigsqcup, \star, \mathbf{1})$ is a *commutative quantale with unit* if $(X, \bigsqcup)$ is a complete lattice with infinitary join operator $\bigsqcup$, and $(X, \star, \mathbf{1})$ is a monoid, such that for all $x \in X$, $\{y_i : i \in I\} \subset X$

$$x \star \bigsqcup_{i \in I} y_i = \bigsqcup_{i \in I}(x \star y_i).$$

We shall use simply *quantale* for commutative quantale with unit. □

Proposition. *A quantale becomes a complete IL-algebra if we define*

$$\bigsqcap X := \bigsqcup\{y : \forall x \in X(y \leq x)\}, \quad x \multimap y := \bigsqcup\{z : x \star z \leq y\}.$$

So quantales are the same objects as IL-algebras; nevertheless, from an algebraic or category-theoretic point of view there is a difference, since the obvious notion of (homo)morphism for a quantale must preserve $\bigcup, \star, \mathbf{1}$, but a (homo)morphism for IL-algebras should preserve $\cap, \cup, \perp, \multimap, \star, \mathbf{1}$.

Definition. (IL-model) An *IL-structure* consists of a domain D and an IL-algebra. An *IL-model* is an IL-structure with a valuation $[\![\;]\!]$ assigning a value $[\![P]\!]$ to each atomic sentence P in the language extended with constants for elements of the domain D, and such that for the extension $[\![\;]\!]$ to arbitrary formulas defined below all relevant infinite meets and joins are defined. The definition of the extension to arbitrary formulas is

$$\begin{array}{lll}
[\![\mathbf{1}]\!] & := & \mathbf{1} \\
[\![\perp]\!] & := & \perp \\
[\![A \diamond B]\!] & := & [\![A]\!] \diamond [\![B]\!] \text{ for } \diamond \in \{\star, \sqcup, \sqcap, \multimap\} \\
[\![\forall x\, A(x)]\!] & := & \bigsqcap_d [\![A(d)]\!] \\
[\![\exists x\, A(x)]\!] & := & \bigsqcup_d [\![A(d)]\!].
\end{array}$$

$[\![\;]\!]$ is extended to multisets by

$$\begin{array}{lll}
[\![\Lambda]\!] & := & \mathbf{1} \text{ (Λ is the empty multiset)} \\
[\![\Gamma, \Delta]\!] & := & [\![\Gamma]\!] \star [\![\Delta]\!]
\end{array}$$

A sequent $\Gamma \Rightarrow A$ is *valid* iff $[\![\Gamma]\!] \leq [\![A]\!]$.

An *ILZ-structure* (*CL-structure*) is a domain with a ILZ-algebra (CL-algebra). An *ILZ-model* (*CL-model*) is defined as an IL-model, but now we require in addition for $[\![\;]\!]$

$$[\![\mathbf{0}]\!] := \mathbf{0},$$

and we stipulate that $\Gamma \Rightarrow \Lambda$ is valid iff $\Gamma \Rightarrow \mathbf{0}$ is valid iff $[\![\Gamma]\!] \leq [\![\mathbf{0}]\!]$. □

By a routine induction on the length of derivations one proves

Theorem. (Soundness) *If* $\vdash \Gamma \Rightarrow A$ *in* $\mathbf{ILL}_\mathrm{q}$ *then* $\Gamma \Rightarrow A$ *is valid in every IL-model. Similarly for* $\mathbf{ILZ}_q(\mathbf{CLL}_\mathrm{q})$ *and ILZ-models (CL-models).*

We also have the converse:

10.4. Theorem. (Completeness) *If* $\Gamma \Rightarrow A$ *is valid in every IL-model, then* $\Gamma \Rightarrow A$ *is provable in* $\mathbf{ILL}_\mathrm{q}$. *Similarly for ILZ-models (CL-models) and* $\mathbf{ILZ}_1$ $(\mathbf{CLL}_\mathrm{q})$.

Now we can take for $\mathcal{X}$ the Lindenbaum algebra consisting of equivalence classes of formulas modulo $\Leftrightarrow$, i.e. classes

$$[A] := \{B : \vdash A \Rightarrow B \text{ and } \vdash B \Rightarrow A\}.$$

Then we construct a special valuation assigning to prime sentences P the value $[P]$, so

$$[\![A]\!] := [A]$$

and we have completeness in the obvious sense:

$$\mathbf{1} \leq [\![A]\!] \text{ iff } \mathbf{1} \leq [A] \text{ iff } \vdash \mathbf{1} \Rightarrow A \text{ iff } \vdash A.$$

The resulting model is not as nice as we should like however; in particular it is ugly that not all joins and meets exist, but only those that are needed for interpreting the quantifiers.

However, there is a standard method for embedding an IL-algebra into a complete IL-algebra, preserving joins and meets. For this, we consider the collection of 'complete ideals' of $\mathcal{X}$.

Definition. $Y \subset \mathcal{X}$ is a *complete ideal* (*c-ideal*) iff

$$Y := \{z : \forall x(\forall y \in Y(y \leq x) \to z \leq x)\}.$$

Or in words: any element below all upper bounds of Y is in Y.

Alternatively, a complete ideal is a fixed point of the following operation on subsets Y of X:

$$\mathrm{C}(Y) := \{z : \forall x(\forall y \in Y(y \leq x) \to z \leq x)\}. \quad \square$$

We note that

$$\mathrm{CC}(Y) = Y, \quad X \subset Y \Rightarrow \mathrm{C}(X) \subset \mathrm{C}(Y).$$

so the set of c-ideals of $\mathcal{X}$ is $\{\mathrm{C}(Y) : Y \subset X\}$. Let us define some operations on elements of $\mathcal{P}(\mathcal{X})$:

$$\begin{array}{lcl} X \sqcup Y & := & \mathrm{C}(X \cup Y), \\ X \sqcap Y & := & X \cap Y, \\ X \star Y & := & \mathrm{C}(\{x \star y : x \in X, y \in Y\}), \\ X \multimap Y & := & \{z : \forall x \in X(z \star x \in Y)\}. \end{array}$$

The following properties are not hard to check:

1. c-ideals are closed under $\bigcap$ and $\multimap$;
2. c-ideals form a complete lattice under inclusion as partial ordering.

Proposition. $h(a) := \{x : x \leq a\}$ *is an order-preserving embedding of* X *into the set* $\mathrm{C}[\mathcal{P}(\mathcal{X})]$ *of c-ideals preserving all operations.*

Proof. We consider some cases only and leave the rest to the reader.
Case 1. Preservation of $\multimap$.

$$\begin{aligned} h(a) \multimap h(b) &= \{z \leq a\} \multimap \{z' \leq b\} \\ &= \{x : \forall z \leq a(x \star z \leq b\} \\ &= \{x : x \star a \leq b\} \text{ (monotonicity of } \star \text{ in both arguments)} \\ &= \{x : x \leq a \multimap b\} = h(a \multimap b) \text{ (adjointness of } \multimap \text{ and } \star) \end{aligned}$$

Case 2. Preservation of $\bigsqcup$. Let $A \subset X$, $\bigsqcup A = a^*$. Then

$$\begin{aligned} &\bigsqcup h(A) = \bigsqcup\{h(a) : a \in A\} = \mathrm{C}(\bigcup h[A]) \\ &= \{z : \forall x(\forall y \in \bigcup h[A](y \leq x) \rightarrow z \leq x)\} \\ &= \{z : \forall x(\forall a \in A \forall y \leq a(y \leq x) \rightarrow z \leq x)\} \end{aligned}$$

hence taking a^* for x we see that $h(a^*) \subset \bigsqcup h[A]$. Conversely, since $h(a) \subset h(a^*)$ for all $a \in A$, $\bigcup h[A] \subset h(a^*)$, hence $\mathrm{C}(\bigcup h[A]) \subset Ch(a^*) \equiv h(a^*)$.

The construction extends to **ILZ** with quantifiers without problems. The algebra of c-ideals of a CL-algebra is again a CL-algebra, it suffices to check that

$$(Y \multimap \mathrm{C}\{\mathbf{0}\}) \multimap \mathrm{C}\{\mathbf{0}\} = \mathrm{C}Y$$

Proof. Let $v \in ((Y \multimap \mathrm{C}\{\mathbf{0}\}) \multimap \mathrm{C}\{\mathbf{0}\})$, i.e.

$$(1)\ \forall u(\forall y \in Y(y \star u \leq \mathbf{0}) \to v \star u \leq \mathbf{0}).$$

We must show

$$(2)\ \forall u(\forall y \in Y(y \leq u) \to v \leq u).$$

Use (1) for $u = {\sim}w$, then

$$\begin{array}{l} \forall y \in Y((y \star {\sim}w \leq 0) \to v \star {\sim}w \leq 0), \text{ so} \\ \forall y \in Y((y \leq {\sim}w \multimap 0) \to v \leq {\sim}w \multimap 0), \text{ so} \\ \forall y \in Y((y \leq w) \to v \leq w) \end{array}$$

using ${\sim}{\sim}w = w$. □

For the extra work needed to include the operator ! see 'Lectures on linear Logic'.

Can we do something useful with these models? Avron gave a nice example of the use of soundness for some particular concrete models of this kind.

Let us consider the fragments of **CLL** consisting of $\{\star, \sqcup, \sqcap, +, {\sim}, \mathbf{1}, \mathbf{0}\}$ and $\{\star, \sqcup, {\sim}, \mathbf{1}, \mathbf{0}\}$, say $\mathbf{CLL}_a$ (with additives) and $\mathbf{CLL}_m$ (multiplicatives only). Because we are in the classical situation with full symmetry, we can use the Morgan-duality to define some operations in terms of others. For $\mathbf{CLL}_a$ it suffices to take $\sqcup, +, {\sim}, \mathbf{1}$, for $\mathbf{CLL}_m$ $\sqcup, {\sim}, \mathbf{1}$ suffice.

Definition. The integer structures $\mathcal{Z}_c$ for $c \in \mathbb{Z} \setminus \{0\}$ are structures

$$\mathcal{Z}_c := (\mathbb{Z}, +, c, \max)$$

where the operations have their usual meaning. A model in $\mathcal{Z}_c$ is a valuation of the sentences of **CLL** such that

$$\begin{array}{lcl} [\![{\sim}A]\!] & = & c - [\![A]\!], \\ [\![A + B]\!] & = & [\![A]\!] + [\![B]\!], \\ [\![A \sqcup B]\!] & = & \max([\![A]\!], [\![B]\!]), \\ [\![\mathbf{1}]\!] & = & c. \end{array}$$

By definition we get

$$\begin{array}{lcl} [\![A \star B]\!] & = & [\![A]\!] + [\![B]\!] - c, \\ [\![A \multimap B]\!] & = & c - [\![A]\!] + [\![B]\!], \\ [\![A \sqcap B]\!] & = & \min([\![A]\!], [\![B]\!]), \\ [\![\mathbf{0}]\!] & = & 0. \end{array}$$

A sequent $\Gamma \Rightarrow \Delta$ is valid [strongly valid] iff ${\sim}(\star\Gamma) \multimap (+\Delta)$ is valid [strongly valid]. A is valid [strongly valid] iff $c \leq [\![A]\!]$ ($c = [\![A]\!]$). □

Note that 'valid' is just a specialization of our general notion above; 'strongly valid' is a sharpening which is useful in the present context.

Theorem. *If* $\vdash_a A$ *then* $\models A$. *If* $\vdash_m A$ *then* $\models_s A$.

Easy corollaries of the soundness theorem are

Theorem. 1. *For no* A *in the multiplicative fragment* $\vdash_m A + \cdots + A$ *for more than one summand.*

2. *If* $\vdash_m \Gamma_i \Rightarrow \Delta_i$ *for* $1 \leq i \leq n$, *then* $\vdash_m \Gamma_1, \ldots, \Gamma_n \Rightarrow \Delta_1, \ldots, \Delta_n$ *iff* $n = 1$.

3. *In the additive fragment, it follows that no sequent of the form* $\vdash A_1, \ldots, A_n \Rightarrow A_1, \ldots, A_n$ *is provable.*

Proof. Consider any $\mathbb{Z}_c$ model. Set $A_i = \star\Gamma_i$, $B_i = +\Delta_i$, then $\vdash_a \Gamma_i \Rightarrow \Delta_i$ means $c \leq [\![\sim A_i + B_i]\!]$, i.e. $[\![A_i]\!] \leq [\![B_i]\!]$. Then $[\![A_i, \ldots, A_n \Rightarrow B_1, \ldots B_n]\!]$ $= nc - ([\![A_1]\!] + \cdots + [\![A_n]\!]) + [\![B_1]\!] + \cdots + [\![B_n]\!] \geq nc$. If we are in the multiplicative fragment, we have in fact strong validity so the result is nc; for negative c and $n > 1$ this is always less than c.

11 Computation Content

We cannot really treat this topic thoroughly here, a few remarks must suffice.

There are two obvious methods to extract computational content from linear logic: the first is based on the algorithm of cut elimination, the second is suggested by bottom-up proof search exemplified in logic programming. The second approach will be dealt with elsewhere in this conference; here we make a few remarks about the first one.

The algorithm of cut elimination is an indication of the computational content of the various sequent calculi, including linear logic. However, as a rule different strategies for eliminating cuts may lead to different cutfree proofs. Moreover, different cutfree proofs of the same sequent are often not *essentially* different, since they differ at most in the order of application of certain rules (cf. our examples of permutabilities of rules).

So if we wish to view sequent proofs as expressions to be evaluated to cutfree form, we are led to make more identifications of proofs than are suggested by the steps of the cut-elimination procedure alone.

Instead of the sequent calculus, we may try to find another calculus, reducing the multiple presentation of what ought to be considered as essentially identical proofs.

To some extent this is achieved by natural deduction (ND) presentations, at least for intuitionistic and intuitionistic linear logic, and may be suitably illustrated by the calculi for intuitionistic implicational logic and the calculus of intuitionistic linear implication.

The prooftrees in ND calculi are trees with nodes labelled by formulas, and have assumptions at the top nodes, which may be open (i.e. still in force as assumptions at the conclusion) or closed (= discharged) by some rule application in the tree.

Such prooftrees may also be written as terms of a typed lambda calculus, with proposition letters as type variables, and implications as functional types. (If more logical operators than just implication are present, the calculus has to be extended with corresponding term-constructing operations.) Intuitionistic implication logic now corresponds to the general case; for the linear case the subterms of any term contain any free variable at most once, and in a subterm $\lambda x.t$ the variable x occurs exactly once in t.

Cut elimination in the sequent calculus corresponds to a process of normalization in ND calculi; in the case of implicational logic, this is just normalizing w.r.t. beta-conversion; the result of normalization is unique, regardless of the strategy (Church-Rosser theorem).

The term notation is suggestive: it suggests that we should think of a formula (proposition) as the set of its proofs (represented by terms); this is the formulas -as -types (propositions -as -types) paradigma.

Instead of considering ND calculi with the associated terms, we may also look at term-calculi of a combinatorial nature (analogous to combinatory logic), without variables for assumptions, and which are suggested by certain special types of categories (such as bicartesian closed categories for intuitionistic propositional logic, or symmetric monoidal closed categories for the tensor fragment of linear logic). Now the formulas correspond to objects of the category, and proofs of 1-sequents $A \Rightarrow B$ to arrows from A to B in the category. Propositional operators correspond to operations on objects producing new objects; as a rule, such constructs on the category side are assumed to have functorial properties (the details are irrelevant here).

Example. $\mathrm{id}_A : A \longrightarrow A$ is the identity arrow corresponding to the axiom $A \Rightarrow A$. If $f : A \longrightarrow B$, $B \longrightarrow C$, then $g \circ f : A \longrightarrow C$ (closure of morphisms under composition) corresponds to the rule

$$\frac{A \Rightarrow B \qquad B \Rightarrow C}{A \Rightarrow C}$$

The wish to interpret proofs as arrows imposes identifications between proofs, e.g. $h \circ (g \circ f) = (h \circ g) \circ f$.

In the case of intuitionistic and linear logic, it is natural to let conjunction correspond to categorical product, and disjunction to co-product; but $\star$ has only the properties of a tensor in the sense of symmetric monoidal categories.

These classes of categories provide us with a notion of model for the calculi. But what corresponds to logical validity? We may say that $A \Rightarrow B$ is *valid* in the category $\mathcal{C}$ if there is an arrow $f : A \longrightarrow B$ in $\mathcal{C}$. (In many cases completeness becomes then trivial by looking at the free category of the appropriate type constructed from the calculus itself.) However, there are many interesting and highly non-trivial categories in the classes considered in which there are arrows between every pair of objects; therefore these models are logically 'inconsistent', even if non-trivial as models of the term-calculus associated with the logic.

Example. The category of relations Rel is a very simple instance of such a model. A rough sketch must suffice.

The objects are sets, the arrows from A to B are binary relations between A and B. id_A is the diagonal relation $a\mathrm{id}_A a' := (a = a')$. Composition of arrows is the relational product:

$$a(S \circ R)b := \exists x(aRx \wedge xRb).$$

$A \sqcap B$ is $A \dot{\cup} B := (\{0\} \times A) \cup (\{1\} \times B)$ (disjoint union), $A \star B$ is $A \times B$, $\top$ and $\bot$ are both the empty set, $\mathbf{0} = \mathbf{1}$ is $\{*\}$ (the singleton set); the axioms $\mathbf{p}_0 : A \sqcap B \Rightarrow A$ and $\mathbf{p}_1 : A \sqcap B \Rightarrow B$ are projections, namely $\mathbf{p}_0 := \{((0, a), a) : a \in A\}$, $\mathbf{p}_1 := \{((1, b), b) : b \in B\}$ etc. etc. Note that between any two sets there is always the empty relation as arrow!

Although natural deduction is a bit more satisfactory than the sequent calculus for expressing the algorithmic idea in cut elimination, it works only well in the intuitionistic case. For classical linear logic Girard introduced a graphic representation of proofs, *proofnets*, which take the symmetries of classical linear logic into account. But this would really carry us to far here, and we must refer to [T].

12 Corrections to [T]

22^{22} K. Došen in his paper 'The First Axiomatization of Relevant Logic' (*Journal of Philosophical Logic 21* (1992), 339–356) discusses an early paper from 1928 on relevance logic, by I. E. Orlov.

24 In the rules of the Lambek calculus the antecedents of sequents must always be non-empty.

29 Replace in the left displayed deduction tree the line ‘ $!(A \sqcap B), !(A \sqcap B) \Rightarrow !(A \sqcap B)$ ’ by ‘ $!(A \sqcap B), !(A \sqcap B) \Rightarrow !(A \star B)$ ’.

35_3 In the displayed prooftree at the bottom, the sequent $!A, \Gamma, !\Gamma' \Rightarrow \Delta, ?\Delta$ should read $!A, \Gamma, !\Gamma' \Rightarrow \Delta, ?\Delta'$.

49^3 Read ‘$\mathbf{CLL}_{\mathrm{q}}$’ for ‘$\mathbf{CLL}_0$’.

57 In the system of table 7, the rule !I ought to be replaced by the following version:

$$\frac{\Delta_1 \vdash !A_1 \quad \dots \quad \Delta_n \vdash !A_n \qquad !A_1, \dots, !A_n \vdash B}{\Delta_1, \dots, \Delta_n \vdash !B}$$

The rule as originally stated generates the same set of derivable sequents, but the resulting system is not closed under substitution of deductions for assumptions (and hence does not permit normalization of deductions).

63 Corresponding to the correction indicated for page 56, the rule for !I should here become

$$\frac{\Delta_1 \vdash t_1 : !A_1 \quad \dots \quad \Delta_n \vdash t_n : !A_n \qquad y_1 : !A_1, \dots, y_n : !A_n \vdash s : B}{!_{y_1,\dots,y_n}(t_1, \dots, t_n; s) : !B}$$

where $y_1, \dots, y_n$ in s are bound in the term for the conclusion, and the variables for $\Delta_1, \dots, \Delta_n$ and the $y_1, \dots, y_n$ are all distinct.

65 Add after **7.2**
Remark. In the presence of axiom C, the axiom B is equivalent to

B′ $\qquad (A \multimap B) \multimap ((B \multimap C) \multimap (A \multimap C))$;

we shall quote this as ‘schema B′ ’.
Formula (1) should read

$$(B \multimap A \star B) \multimap D \ \ \text{(schema B}'\text{)}$$

and in the last line of the page read ‘again by schema B′ ’.

67 In the top line of table 9, read ‘$\mathbf{CLL}_{\mathrm{q}}$’ and ‘$\mathbf{ILL}_{\mathrm{q}}$’ for ‘**CLL**’ and ‘**ILL**’ respectively.
Axiom B should read ‘$(B \multimap C) \multimap ((A \multimap C) \multimap (A \multimap C))$’ (so as to correspond to the combinator B under the formulas-as-types isomorphism.

68^{13} Read ‘schema B′ ’ for ‘axiom B’.

68, section **7.7.** The proof is more perspicuously formulated as follows. 'Proof by induction on the number of formulas in $!\Gamma$. Let $!\Gamma \equiv !\Delta, !C$, and suppose $!\Delta, !C \vdash !A$. Then $!\Delta \vdash !C \multimap A$ (deduction theorem); by the IH we have $!\Delta \vdash !(!C \multimap A)$; use (19) and modus ponens to obtain $!\Delta \vdash !!C \multimap !A$; by (21) and transitivity (2) we obtain $!\Delta \vdash !C \multimap !A$; again using modus ponens and $!C \vdash !C$ we find $!\Delta, !C \vdash !A$.

70_1 read '$\mathbf{T} \vdash \mathrm{S}$' for '$\mathrm{T} \vdash \mathrm{S}$', and '$B$' for 'B'.

71_4 Read 'the constants $\bot, \mathbf{1}$'.

78_1 Read '$!x \leq !!x$' for '$!x \leq !!y$'.

108 In the big rectangle the dotted lines should be placed between $(AB)C$ and $C(AB)$, and between $(BA)C$ and $C(BA)$ respectively.

120 Drop '1' from the two lists of canonical forms.

121 2 lines above section **13.2**, read '$\kappa_i c$' for '$\kappa_i s$'.

121^3 Read '$t'[x/c]$' for '$t[x/s]$'in the third displayed formula.

121^{10} Read 'It' for '(it'.

121^{11} Read a stop for the closing parenthesis.

123^5 Read 'sets $\mathbf{1}, \top, \bot$' for 'sets $!, \top, \bot$'.

123^{13} Read '$t \rhd \kappa_1 c$' for '$t \lhd \kappa_1 c$'.

121–123 Because of the greater complexity of the revised rule !I, the proof in section **13.3** cannot be straightforwardly adapted. Therefore it is best to drop all references to terms $!t$ and the corresponding evaluation rules (last group on page 122).

124_{18} Last lines of *Case 2*, read '$\lhd c'$' for '$\lhd c$' (twice).

$141^{3,4}$ Delete 'of products'.

146 In the picture at the bottom of the page, the axiom links should connect $\sim A$ with A, and $\sim B$ with B.

151_8 Read 'As to (iii),'.

173^3 Read 'q_j' for 'q_i' (twice).

174^{12} Interchange 'δ_2' and 'δ_3'.

References

[A] A. Avron, Some properties of linear logic proved by semantic methods. Preprint, 1992.

[LMSS] P. D. Lincoln, J. Mitchell, A. Scedrov, N. Shankar, Decision problems for propositional linear logic, *Annals of Pure and Applied Logic 56* (1992), 239–311.

[Gi] J.-Y. Girard, Linear Logic, *Theoretical Computer Science 50* (1987), 1–102.

[Sc] A. Scedrov, A Brief Guide to Linear Logic, to appear in a collection of articles edited by G. Rozenberg and published by World Scientific. Preprint from 1992.

[T] A. S. Troelstra, *Lectures on Linear Logic*, CSLI Lecture Notes 29, CSLI, Stanford 1992 (distributed by the University of Chicago Press).

The Landscape of Deduction

Johan van Benthem
Institute of Logic, Language and Computation
University of Amsterdam

1 The Landscape of Deduction

Recent logical research is showing an increasing sensitivity to the existence of what may be called different 'styles of deduction', organized in logical proof calculi that may vary even in the basic *structural rules* governing their general handling of premises and conclusions. Standard structural rules, such as Monotonicity, Contraction, Cut or Permutation, may be either lost completely in the process, or be modified to subtler variants (such as 'leftward' or 'cautious' forms of Monotonicity). Any system of structural rules may be seen as characteristic of a certain deductive style: 'classical', 'relevant', 'dynamic', or 'categorial'. In general, these styles will be more sensitive to the syntactic presentation of premises, whose ordering and multiplicity become crucial: most inference is 'occurrence-oriented'. Starting from different motivations, students of categorial logic, relevance logic and linear logic have been exploring this space of possibilities, using various formats, including Hilbert-style axiomatic systems and Gentzen-style calculi of sequents (cf. [Lambek 1958], [Dunn 1985], [Girard 1987], [van Benthem 1991]). In particular, the latter format seems to be gaining the upper hand these days as a natural form of organization, bringing out what are perceived to be the major options in choosing one inferential style or another. There are also other research lines, however, which tend in the same direction, such as the current work on varieties of inference in artificial intelligence and their logical properties (cf. [Makinson 1991]), which has brought in 'preferential' or 'default' styles of heuristic reasoning, again with a spectrum of different structural rules.

Substructural systems have usually been motivated from a proof-theoretic point of view. Indeed, one can understand their genesis retrospectively as follows. Upon reflection, standard proof theory for predicate logic is not an inseparable package deal. It turns out to consist of a 'logical core' of rules for connectives surrounded by a more negotiable apparatus of structural rules, embodying various decisions as to handling premises in inference. Some crucial meta-properties of predicate logic even come out more clearly without the latter apparatus. E.g., in the absence of Contraction, Cut Elimination becomes a simple linear operation on derivations without any exponential blow-up in size. Other standard meta-properties turn out to be accidental effects of adding certain structural rules to the logical core. E.g., Undecidability is mainly due to the interplay of the quantifier rules with Contraction. (The latter remark is already to be found in [Curry 1963], whose formalist insistence on 'occurrence logic' was an obvious precursor of current trends.)

Nevertheless, there are also independent model-theoretic perspectives feeding the substructural movement, representing more concrete motivations for dropping structural rules from the point of view of some described reality or cognitive practice. For instance, as we shall see below, classical structural rules will fail with natural notions of inference in more 'resource-oriented' or 'dynamic' modelings. And in artificial intelligence, classical structural rules like Monotonicity turn out to fail in a universe of so-called 'minimal models' along some preference ordering, which is appropriate for heuristic reasoning. Finally, in science, probabilistic reasoning involves similar failures of classical structural rules, while 'quantum logic' has been an occurrence-based style of inference from the start.

So far, proof-theoretic and model-theoretic perspectives upon general reasoning have remained somewhat different 'cultures', with the former often written in 'fundamentalist' literary style (how one can 'discover the genuine' logical constants, or why one should 'always strive for' cut elimination) and the latter often in a more applied spirit (asking which logical operators express significant patterns in the relevant models, and which structural properties can be expected for reasonable inferential procedures over the latter). In this discursive paper, we shall contrast the two approaches, relying heavily on [van Benthem 1991] for technical background and more concrete examples. In the process, we also hope to highlight some of the main logical issues concerning inference in general, in particular, its 'logical architecture' given the plurality of the substructural landscape.

2 Proof Theory versus Model Theory

In general logical analysis, there are two main directions of thought. At a more combinatorial, syntactic level, one may organize intuitions of reasoning in formal theories, stating proof rules and axioms, and then search only afterwards for models of the resulting principles. This direction is exemplified by the early history of modal logic, the type-free lambda calculus, or more recently, linear logic. Here, one often designs models only for auxiliary purposes, 'loading the dice' so as to obtain a match with the initial formalism. In the opposite direction, one defines a field of research via suitable independently motivated classes of model structures, and the question is rather to design appropriate formalisms and deductive systems describing these models. This has been the main trend, e.g., in temporal logic or provability logic of arithmetic, where axiomatizations are sought for flows of time or specific arithmetical theories, and where the initial formal language has undergone several changes in the process. Eventually, mature areas of research will witness an *interplay* between the two directions. For instance, the discovery of models for some proof-theoretic calculus may in its turn suggest modifications and extensions of the latter. There are many illustrations of this process in the development of logic. Presumably, the same will happen in the above areas, so that we shall witness the emergence of, e.g., completeness theorems axiomatizing the complete categorial logic or linear logic of special models of independent conceptual interest. In the process, the connections between the relevant semantic intuitions and proof-theoretic ones may turn out less than straightforward: and by itself, there is no compelling reason for preferring one perspective over the other. To take a concrete example from our later discussion: the fact that 'relation algebras', a natural model for dynamic procedures, are only partial models for linear logic carries no moral connotation whatsoever. If relation algebras are not very good at describing a natural proof-theoretic calculus like linear logic, then ipso facto, linear logic is not very good at describing what is an important mathematical structure with a good deal of independent computational motivation.

We shall start here from two specific kinds of model, one more 'linguistic', one more 'procedural'. These seem particularly relevant, as the newer systems of logic in the above landscape all seem involved, in one way or another, with information encoding and information processing. The two modelings suggest various general logical questions which may be compared with those arising in Gentzen-style proof-theoretic calculi, while often allowing for more sensitive discussion of the relevant issues. It is easier to be dogmatic about 'the correct calculus of logic' or 'the proper logical constants' as long as one remains at a global level, with no particular semantic situation in mind, but such a fundamentalist attitude becomes harder to maintain in the light of complex realities to be described. In the end, one should probably aim for 'logicality' as a general phenomenon, rather than any single canonical manifestation thereof.

3 Language Models

Perhaps the most intuitive kind of model in categorial logic consists of families of languages, i.e., sets of expressions over some alphabet, together with certain natural mathematical operations over these. Taking a cue from mathematical linguistics, the latter may be divided roughly into general Boolean operations and special 'ordering' operations reflecting the action of the basic syntactic operation of concatenation. Thus, we are thinking of *language models* $\{L_a \mid a \in A\}$ with the following structural repertoire:

Boolean operations:	$-, \cap, \cup$	$\bot, \top$	
Order operations:	$\bullet, \backslash, /$	1	$*$

As for the latter, we have (with juxtaposition indicating concatenation):

$L_a \bullet L_b$	$= \{xy \mid x \in L_a, y \in L_b\}$	product
$L_a \backslash L_b$	$= \{x \mid \forall y \in L_a : yx \in L_b\}$	left inverse
L_b / L_a	$= \{x \mid \forall y \in L_a : xy \in L_b\}$	right inverse
L_1	$= \{\langle\ \rangle\}$	empty sequence

and the Kleene star $*$ denotes finite *iteration* as usual. These operations are natural, e.g., in the sense that the family of all *regular* languages is closed under them. Moreover, they have emerged time and again in computer science too.

What would be an appropriate description language for these models? Perhaps the first proposal that comes to mind is the first-order language of concatenation. But this formalism turns out both too rich and too poor. It is too rich, because it can be shown that all of True Arithmetic lies effectively embedded in it, so that we would be burdening our 'logic' with a lot of specialized mathematical knowledge, leading to non-axiomatizability. On the other hand, the formalism would also be too poor, in that it does not refer to languages as such. For the latter purpose, at least, unary predicates $A, B, \ldots$ are needed, so that we can state connections between different descriptions of languages in the model. Thus, the art becomes to design a suitably expressive formalism expressing interesting relationships between languages, while remaining weak enough to leave the full arithmetic of syntax aside. And the 'logical constants' of that formalism will reflect the natural ways of constructing new languages out of old ones. Now, the standard categorial languages may be viewed as one excellent compromise for precisely this purpose. Let primitive types stand for arbitrary languages, with complex types being computed through type-forming operators matching the above operations. To be sure, this computation involves concatenation as above, but only to a modest degree. Then, significant relations between languages may be expressed while still maintaining a hope of complete axiomatizability.

To carry out this program, we need a suitable notion of valid inference over these models. Again, a subtlety arises then. 'Valid inference' in such an applied setting is not a matter of imposing some uniform canon, but rather, of locating some intuition that should be appropriate to working in language models. The following would seem to be a plausible candidate, reflecting successive concatenation of incoming premises:

> A sequent of types $X \Rightarrow a$ is *valid* in language models if, for every interpretation $[\![\ ]\!]$ sending primitive types to arbitrary languages and complex types to the matching compounds, $[\![\bullet X]\!] \subseteq [\![a]\!]$ holds.

Here, '$\bullet X$' is the concatenation product of X (with the stipulation that $\bullet\emptyset$ be 1). Valid principles on this account include all Boolean laws, as well as typical categorial laws such as $a \bullet (a \setminus b) \Rightarrow b$ or $a \Rightarrow (b/a) \setminus b$. In fact, this is no coincidence, witness the following observation about what is probably the key system of categorial grammar:

Fact. The Lambek Calculus is sound for interpretation in language models.

But the present interpretation also extends to other logical constants. For instance, one may compare the behaviour of our two 'conjunctions': with categorial product $\bullet$ satisfying the Gentzen laws of the Lambek Calculus, and Boolean intersection $\wedge$ rather those of the additive conjunction of linear logic. In fact, under the obvious interpretation of its non-exponential operators, one even has that, in its directed version with two implications:

Fact. Intuitionistic Linear Logic is sound for interpretation in language models.

This still leaves the Kleene iteration * unaccounted for. The latter behaves somewhat like a modality, though the analogy is not perfect. Even so, much of its behaviour is known from existing descriptions of the logic of 'regular expressions', involving $\cup$, $\bullet$ and *.

Now, various general logical questions arise. The first concerns deductive completeness. Can one find complete effective proof-theoretic descriptions for valid inference in language models? The situation here is less than clear, and there are only some partial results. It is shown in [Buszkowski 1986] that the Lambek Calculus with the two directed slashes $\setminus$ and $/$ is complete, but it is not even known if this result extends to the language with product. (This result has been claimed recently by Dov Gabbay.) In [Roorda 1991] one finds an extension to the case of Boolean disjunction of types. Finally, [Pratt 1991] proposes an 'action logic' for the regular operations enriched with the two slashes that might be complete here. Of course, there is an interplay between the special needs of our models and the business of general logic. Probably, the categorial logic of language models falls into a more general part of broader significance plus various 'specifics of syntactic concatenation'. This may be judged, for instance, by seeing which valid principles on language models transfer to other natural model structures, such as the 'relational models' of the next Section.

Even the specifics of a special application may offer interesting new types of logical question, that may have a broader logical significance eventually. Here are two 'linguistic' examples in the present setting. First, categorial derivation of one type from a sequence of others amounts to a well-known grammatical process, namely, categorial parsing, showing how the component categories of some compound expression admit of grammatical derivation to a well-formed sentence. Thus, grammatical analysis may be reduced to proof search in categorial logics ([Moortgat 1988]). And what becomes vital then is the development of various perspicuous and efficient representations for logical search spaces. Here is where special proof-theoretic formats of implementation, such as the 'proof nets' of linear logic, may become useful (cf. [Roorda 1991], [Moortgat & Morril 1991]). Also, there may be various special mathematical invariants of derivations that may be exploited here, when trying to avoid barren parts of the tree. Examples are the numerical 'counts' for the Lambek Calculus, or the use of 'small model checks' in rejecting branches in theorem provers for relevant logics. In other words, linguistic parsing and logical deduction may mutually influence each other.

Another basic linguistic concern is that of 'recognizing power' for grammatical paradigms. What is the grammatical complexity of the languages that can be recognized via categorial assignment to their primitive symbols in the type structures corresponding to language models? There are some long-standing open mathematical problems here, such as 'Chomsky's Conjecture' stating that the Lambek Calculus recognizes precisely the context-free languages. The next obvious question is how addition of structural rules, or of new logical constants, changes this picture. As to the former, we know that strengthening structural rules may affect recognizing power rather drastically. In particular, adding Contraction to the Lambek Calculus effects a collapse into the realm of regular languages. Moreover, [Kanazawa 1991] shows that adding intersective conjunction will recognize all intersections of context-free languages (some of which are not context-free themselves).

The point of these two topics is merely to illustrate a more general matter. Logical calculi may serve not just as engines for 'deduction', but also as paradigms for certain styles of grammatical analysis. That is, they may perform cognitive tasks for which they were not originally intended. This observation also opens the way to unorthodox uses of proof theory, beyond its original motivation in the foundations of mathematics. Linguistic issues of recognizing power now turn out to require logical proof analysis, via well-known techniques of cut elimination or general normalization: and much of 'mathematical linguistics' really becomes a branch of applied proof theory.

Finally, let us return from proof theory to model theory, as linguistics involves not only syntax, but also semantics. Then, a more refined view of the parsing process is needed: since different derivations of the same expression may correspond to different 'readings', giving it distinct meanings. For instance, the scope order of operators in a flat string may still be undetermined, and only become fixed through the choice of one particular derivation. 'The unit of meaning is not the sentence but the derivation'. This view is familiar from category-theoretic interpretations of proof calculi: the 'arrows' are what counts. From a linguistic point of view, this raises at least two issues. First, one needs an efficient semantically sound calculus for establishing equivalence of proofs, allowing one to determine numbers of different readings for one derivable sequent. A standard way of doing this employs terms in a suitable typed lambda calculus matching derivations in the particular categorial logic involved. Also relevant here is the 'non-ambiguous' version of categorial logic proposed in [Hendriks 1992], where different proofs always have different readings. And next, there is an interest in the general question of determining whether given derivations describe denotations having certain desired semantical properties, such as 'conservativity', 'monotonicity' or being a 'Boolean homomorphism' in the relevant semantic domain. There seem to be few relevant results so far: even in category theory, results are scarce on just when the shape of a given arrow allows us to predict some desirable mathematical property for its denotations.

4 Procedural Models

Categorial derivation is one prime form of natural language processing, and with it, of more general procedural mechanisms of information flow. But then, one can also focus directly on what are arguably the most general models of procedures across a wide range of applications, namely, families of 'transition relations' for procedures, provided with an appropriate mathematical superstructure reflecting composition and control of subprocedures. More precisely, *relational models* will be structures of the form $(S, \{R_p \mid p \in P\})$, being families of transition relations on some carrier set S of states, closed under suitable operations. These states can still be quite diverse: cognitive or physical, or mixtures of both — and likewise, the actions in P may range from value assignment in registers to digging for gold in one's garden. For the moment, we are after their common procedural 'superstructure'.

Here, the existing work on *Relational Algebra* provides a better grasp of the relevant logical operations. In the theory of binary relations, one generalizes the Boolean Algebra of unary propositions by creating a suitable richer similarity type, namely:

Boolean operations	$-, \cap, \cup$	$\bot, \top$
Order operations	*composition* $\bullet$ as well as *converse* $^{\cup}$	
Special relation	*identity* id.	

Moreover, there are also analogues of our earlier slashes

$$R \setminus S = \{(x,y) \mid \forall z : (z,x) \in R \Rightarrow (z,y) \in S\}$$
$$S/R = \{(x,y) \mid \forall z : (y,z) \in R \Rightarrow (x,z) \in S\}.$$

Such further operations are often available by definition. For instance:

$$R \setminus S = -(R^{\cup} \bullet -S)$$
$$R/S = -(R \bullet (-S)^{\cup}).$$

Finally, there are also natural *infinitary* operators on binary relations. These correspond to 'unlimited structures' in action or programming, such as endless repetition. The most prominent example is the ordering operation of *transitive closure*:

$$R^* = \{(x,y) \mid \text{some finite sequence of successive } R \text{ transitions links } x \text{ to } y\}.$$

Again, what would be a natural formalism for reasoning about these structures? Traditionally, in Relational Algebra, an equational format is chosen, and one chooses basic principles including all Boolean identities, together with such identities as

$$(R \cup S)^{\cup} = R^{\cup} \cup S^{\cup}$$
$$(R \cup S) \bullet T = R \bullet T \cup S \bullet T$$
$$(R \bullet S)^{\cup} = S^{\cup} \bullet R^{\cup}$$
$$R^{\cup} \bullet -(R \bullet S) \leq -S.$$

And iteration will satisfy various algebraic principles too, witness an identity like

$$(R \cup S)^* = (R^* \bullet S^*)^*.$$

Also, Pratt's 'action logic' has typical 'Induction Axioms'

$$(R \setminus R)^* = (R \setminus R)$$
$$(R/R)^* = (R/R).$$

These principles fit rather well with our categorial or linear calculi. This time, the natural notion of valid inference for sequents is composition of instructions for the successive premises, seeing if the resulting transitions are admissible for the conclusion:

$X \Rightarrow R$ is *valid* if, for all relational interpretations $[\![\ \ ]\!], [\![\bullet X]\!] \subseteq [\![R]\!]$.

In the absence of premises, this is meant to reduce to the inclusion $id \subseteq [\![R]\!]$ (the empty sequence of premises is interpreted as the identity relation). In particular, we have again that, in its directed version:

Fact. Intuitionistic linear logic is sound for interpretation in relational models.

A completeness result has just been found too, for the Lambek Calculus with \, /, •, in [Mikulás 1992].

Again, let us discuss some features of this modeling that may be of more general logical interest. First, Relational Algebra raises some interesting issues of axiomatization. Its valid principles on set-based models are non-finitely axiomatizable (in particular, the above finite set is incomplete). In a sense, the formalism encodes too much specific mathematical information, this time concerning the underlying set-theoretic structure of ordered pairs. But then, the analogy with language models suggests another perspective. One might take 'transitions' themselves as primitive dynamic objects, structured by some basic operations of partial 'composition' and 'conversion'. What the above algebraic principles then describe might be viewed as a mixture. We should isolate a basic underlying *Arrow Logic*, together with appropriate abstract models for it, on top of which further principles will then reflect the mathematics of specific transition models (cf. [van Benthem 1991,1992], [Venema 1991]). On this view, categorial or linear logics might provide perspicuous, decidable *fragments* of arrow logic, which are perhaps even complete with respect to relational models.

Let us now turn to some general issues again that are raised by this special model structure. For instance, relational models have natural operations that do not seem to admit of any obvious Gentzen-style analysis in terms of introduction rules, the most striking example being 'converse'. So, assuming that the relational structure is natural in its own right — and conversion of procedures is a ubiquitous and natural phenomenon — there is really a substantive issue as to whether all fundamental procedural phenomena will be visible through Gentzen-style lenses. (This issue of the adequacy of 'gentzenization' for various logics, viewed as a topic of research rather than a presupposition, has a long history in the development of relevance logic: cf. [Brady 1990].) And a similar question may be raised with procedural iteration. In other words, some logical constants will show up naturally in proof-theoretic formats while others will not. What we need then is a non-partisan comparison of various different approaches.

The relational modeling also suggests an interesting interaction with proof-theoretic viewpoints. Arrow Logic is a deductively simple 'base calculus', admitting of semantic completeness results via simple representation theorems, while optional axioms on top of this will reflect special assumptions about specific relational models. This perspective is quite familiar from Modal Logic, whose base logic is complete for arbitrary possible worlds models, with special axioms expressing more specific frame conditions. Likewise, in categorial or linear logic, the deductive minimum is the common core which makes the semantic representation methods of [Girard 1987], [Ono 1990], [Abrusci 1988] or [Sambin 1990] work, with further principles expressing special assumptions on specific mathematical models. One intriguing question then, whose discussion we must forego here, is as follows. Is this deductive core really the same as that suggested by the Gentzen-style paradigm in the field?

5 Logical Constants

One of the striking phenomena in the landscape of deduction is that changing structural rules has repercussions for the repertoire of logical constants. In particular, relaxing the standard structural rules makes room for a variety of new logical operators that had all been 'equalized' under the ancien régime. This phenomenon comes in two varieties. In some cases, old classical constants split up into natural alternatives, such as 'sequential' and 'parallel' conjunction. But also, genuinely new notions emerge, such as 'exponentials' or 'conversion'. Now, one natural question is this. At least in classical propositional logic, there is functional completeness for the standard Boolean connectives. And even in intuitionistic logic, there is a reasonable consensus on the basic repertoire. But what is the natural class of logical constants in this broader setting? In particular, is there still some finite bound on the natural candidates?

This issue forces one to think more clearly about what is meant by the notion of 'logicality' in a broader setting. Here, the literature shows a great diversity of options, some of them proof-theoretic, others model-theoretic. From a proof-theoretic point of view, it seems natural to continue the style of analysis proposed by Dag Prawitz, Peter Schroeder-Heister, Kosta Došen and others, searching for a natural definition of a 'Gentzen-style introduction rule', and then determining logical constants by sets of such rules. Thus, their purpose is that of 'strategic interpunction' in proof patterns. Such an approach to functional completeness for linear logic has been attempted, amongst others, in [Wansing 1989], resulting in the conclusion that the system is essentially 'complete' as it stands. Moreover, the proof theorist will often insist on the possibility of Cut Elimination as a check on the adequacy of the logical operators introduced.

Even so, the proof-theoretic combinatorial viewpoint has its limitations. These are demonstrated by Arthur Prior's 'Tonk' example of a binary logical operator T having the introduction rules of disjunction but the elimination rules of conjunction. This would lead to a collapse into triviality in three steps:

$$A \qquad \text{introduction} \qquad A\,\mathrm{T}\,B \qquad \text{elimination} \qquad B.$$

Presumably, one needs some independent intuition to rule out such combinatorial monstra. Something similar happens with new logical constants in substructural logics, where we often lack a guiding semantic intuition to choose between various possible proof patterns. Consider the case of conjunction. Product conjunction gets the following introduction rules

$$\frac{X, A, B, Y \Rightarrow C}{X, A \bullet B, Y \Rightarrow C} \qquad \frac{X \Rightarrow A \quad Y \Rightarrow B}{X, Y \Rightarrow A \bullet B},$$

while intersective conjunction has

$$\frac{X, A, Y \Rightarrow C}{X, A \cap B, Y \Rightarrow C} \qquad \frac{X, A, Y \Rightarrow C}{X, B \cap A, Y \Rightarrow C} \qquad \frac{X \Rightarrow A \quad X \Rightarrow B}{X \Rightarrow A \cap B}.$$

But why this particular arrangement? What about a new conjunction $\sharp$ having the left introduction rules for $\cap$ and the right introduction rule for $\bullet$? In this case, we would be able to derive a new, undesired structural rule, viz. Monotonicity:

$$\frac{\dfrac{A \Rightarrow A \qquad B \Rightarrow B}{A, B \Rightarrow A\sharp B} \qquad \dfrac{X, A, Y \Rightarrow C}{X, A\sharp B, Y \Rightarrow C}}{X, A, B, Y \Rightarrow C}$$

Thus, adding certain logical constants is not 'conservative' over the underlying structural rules. But this cannot be the only criterion of adequacy. For instance, it may be shown that adding another conjunction with the opposite mixture of rules for $\bullet$ and $\cap$ is indeed conservative for structural rules, and yet this notion would not be a natural candidate either. But then, what would be a non-ad-hoc criterion of logicality here?

A model-theoretic perspective on logicality brings quite different intuitions into play. One recurrent idea in the literature has been that logical constants are those expressions whose denotations show a certain mathematical robustness, in that they are 'invariant under permutations' of the models in which they are interpreted. Thus, only the formal structure counts: a logical constant is completely neutral to individual peculiarities of its arguments. There are various ways of making this intuition precise, depending on what structures are employed. For instance, in relational models, one may call a propositional operator O 'invariant' if it commutes with arbitrary permutations π of the underlying state set S, in the sense that $\pi[O(X)] = O(\pi[X])$, for all subsets X of S. And extensions to polyadic propositional operators are obvious. In particular, all basic operators from relational algebra are invariant. But in fact, there is an *infinite* space of invariant possibilities for ever more complex constructions of procedures. Thus, logicality is a quite general semantic phenomenon, with logical constants forming an infinite family of 'pattern descriptions', coming in a *hierarchy* with layers of greater or lesser complexity. This fine-structure may be brought out as follows. Over our intended models, most proposed operators so far had their action described by schemata in a simple first-order language. (For instance, parallel conjunction amounts to an intersection $\lambda xy \bullet (Rxy \wedge Sxy)$ while sequential conjunction corresponds to the composition schema $\lambda xy \bullet \exists z(Rxz \wedge Szy)$.) Thus, the complexity of proposed logical operations may be measured by the number of variables essentially involved, i.e., the number of states employed in tracing their action. For intersection, this number is 2, for composition, it is 3, and so on. In fact, all of the usual relational algebra lies at complexity level 3. This finite-variable hierarchy of definitional complexity for logical operators again extends infinitely, and no finite base choice will exhaust it. Thus, there is an infinite supply of logical constants in information processing, out of which most proof-theoretic systems select only a few for special treatment.

What remains to be understood is the connection between these semantic analyses of logicality and standard proof-theoretic ones. Both seem natural — but are, say, criteria like 'Permutation Invariance' and 'Cut Elimination' really getting at the same phenomena? Clearly, the two approaches do not coincide extensionally. The relational operation of conversion is permutation-invariant, but it does not admit of any obvious Gentzen-style characterization. And conversely, predicate logic with restricted quantifiers 'some A', 'all A' has natural Gentzen rules and Cut Elimination, even though these operators are not permutation-invariant, due to their restricting non-logical predicates. Still, this example is not yet conclusive, in that it shows a parallel between what may be called 'relativized' forms of invariance and 'modified' Gentzen rules. Indeed, like the semantic perspective, proof theory suggests a natural layering into less or more complex rule formats, involving checks on premises of different kinds. For instance, intuitively, the right introduction rule for product conjunction is less demanding than that for intersective conjunction, which requires an identity check. (Another example is the spectrum of variable conditions on quantifier rules with vastly different complexity in [van Lambalgen 1991].) So, there is still hope for some deeper connection between proof-theoretic and semantic views of logicality.

6 Styles of Inference

What our special models also suggest is the existence of natural 'styles of inference'. These styles may differ in their treatment of premises and conclusions. For instance, with language models, one naturally writes down expressions exemplifying consecutive premises viewed as languages, and requires that 'the product language of the premises must be contained in that of the conclusion'. And in procedural models, one naturally carries out successive instructions realizing successive premises, now viewed as procedures, obtaining the earlier 'dynamic style': 'the composition of the premise relations is contained in that of the conclusion'. Thus, one crucial decision here may be described as one of 'text logic': how is one to interpret the *commas* as well as the *turnstile* in a sequent? (This view of inference has been put forward in great generality within relevance logic by Mike Dunn and Bob Meyer.) Other reasonable styles have been proposed too, such as 'successive processing of all premises, followed by a test for the conclusion in the resulting state'. The emergence of such options makes sense: ordinary reasoning seems to come in different modes, depending on the intended application. Thus, the task of logic becomes to chart different natural styles of inference and their mutual relations.

Here is precisely where the structural rules of the proof-theoretic landscape come in. For, these may be viewed as stating some basic defining traits of a 'reasoning practice'. This perspective on structural rules is reinforced by similar ideas in the foundations of artificial intelligence, prompted by the necessities of intellectual life after the loss of standard 'monotonicity'. Ordinary logical consequence is concerned with all models of the premises, or all states where they hold, checking whether the conclusion follows. This feature is typically reflected in the Cut Rule, which sanctions unlimited 'transmission of validity' through intermediate steps in reasoning. But in heuristic reasoning, when immediate conclusions are required from perhaps limited data, one may be content with less, asking just more limited truth of the conclusion in some special class of models or states: say, the most likely ones, or more generally, the most 'preferred' ones along some preference order. (One might 'prefer' situations which are most similar to ours, or which exemplify most of our prejudices, and so on.) Thus, *preferential consequence* means that the conclusion is true in all most preferred models of the premises. 'Preferential reasoning' is certainly a natural style of inference, but again, it shows great divergence from standard logical consequence in its structural rules.

Two kinds of question arise now, relating model-theoretic and proof-theoretic perspectives here. The first direction is to find complete proof-theoretic descriptions of all structural properties for a semantically given inferential style. Here are some results in this direction:

> The complete structural theory of classical inference is axiomatized by {Cut, Reflexivity, Monotonicity, Contraction}.
>
> The complete structural theory of dynamic inference is axiomatized by {Cut, Reflexivity}.

These outcomes may still be misleading. The semantic approach to styles of inference does more than just choose between structural rules offered by standard proof theory. It may also suggest entirely new (versions of) structural rules: new religions need not be definable by merely listing the old dogmas which they retain or reject. One example is the third style mentioned above ('processing followed by test').

> The complete structural theory of the processing/test style is axiomatized by

Left Monotonicity $\dfrac{X \Rightarrow C}{P, X \Rightarrow C}$

Left Cut $\dfrac{X \Rightarrow D \qquad Y, X, D, Z \Rightarrow C}{Y, X, Z \Rightarrow C}$.

As another example of this greater subtlety, here is the complete set of structural properties for preferential reasoning, which trades standard monotonicity and cut for more sensitive versions allowing replacement by certain types of premise only:

> The complete structural theory of preferential inference is axiomatized by {Permutation, Contraction} plus

Strong Reflexivity $X, A, Y \Rightarrow A$

Cautious Cut $\dfrac{X \Rightarrow B \qquad X, B \Rightarrow C}{X \Rightarrow C}$

Cautious Monotonicity $\dfrac{X \Rightarrow B \qquad X \Rightarrow C}{X, B \Rightarrow C}$.

Going in the opposite direction, one may analyze a proof-theoretic calculus in the landscape of deduction, and try to determine some concrete reasoning practice for it in model-theoretic terms. By way of illustration, here is such a concrete model for commutative linear logic:

> {Cut, Reflexivity, Permutation} axiomatize all structural properties of the process of adding successive elements from a sequence of bags, and then checking whether the resulting vector belongs to the conclusion.

The proof of such a result is essentially a mathematical representation theorem for abstract relations satisfying the listed structural properties ([van Benthem 1991]). Finding 'complete sets' of inferential properties may turn out to be much more complicated for proof-theoretic calculi, even when these have been defined in terms of a certain cluster of explicit structural rules. For instance, what is the set of all implicit *derived* structural rules of the Lambek Calculus, or of Linear Logic? No answer to this question is known yet. For instance, [Rybakov 1985] has shown that the set of all admissible rules for intuitionistic propositional logic is decidable. Would a similar result obtain here?

Thus, there is more to structural rules than might have appeared from the standard Gentzen approach. For instance, we may now ask which proof-theoretic systems admit of some form of 'modified cut elimination' so as to obtain at least Cautious Cut. More generally, the landscape of substructural deduction becomes much richer, since acceptance or rejection of classical structural rules need no longer be an all-or-nothing matter.

7 Trade Routes in the Landscape

Different styles of inference do not live in isolation: they co-exist in human reasoning, and therefore, it is natural to ask how they are related from a logical point of view. In particular, when can one style be 'simulated' on the basis of another? As has been pointed out by many authors, one powerful method to this end is the introduction of suitable logical operators. Thus, our two main themes so far are closely related. The above dynamic styles of inference provide an illustration. With both intersection and composition of relations explicitly available in the formalism, 'sequential' and 'parallel' styles of processing premises become mutually reducible. Or, in a language with an explicit 'fixed point' operator sending relations to their diagonal, the classical style may be reduced to the dynamic one as follows:

$$X_1, \ldots, X_n \models_{classical} C \quad \textit{iff} \quad \text{fix}(X_1), \ldots, \text{fix}(X_n) \models_{dynamic} \text{fix}(C).$$

Special structural properties of one style of inference will then follow from the logic of the logical constants involved in their definition (on the basis of the structural rules for the reducing notion, of course). No converse reduction is possible, however, because then, the dynamic style would inherit the Monotonicity of the classical one.

Nevertheless, there may be subtleties in the issue of 'reduction'. For instance, in linear logic, the exponentials are supposed to 'absorb the classical structural rules into the logical rules' (again, on the basis of the structural rules of linear logic itself). But it is not at all clear whether they are the only means of doing so, and whether their logical rules are uniquely determined by this purpose. Indeed, the fact that the exponentials as they stand, whose main purpose is to 'simulate Contraction', make linear logic undecidable, whereas the standard propositional logics with Contraction are not, suggests the possibility of a mismatch. Probably, the thesis of transfer from structural rules to logical rules needs precization, before its plausibility can be sensibly discussed.

Another subtlety is the possibility of having various types of reduction. The above examples were 'local', translating on a formula-to-formula basis. But there may also be more 'global' reductions from one style to another, even when the former kind is unavailable. An example is the earlier-mentioned preferential style of inference in Artificial Intelligence. What we have, e.g., is

$$X_1, \ldots, X_n \models_{preferential} C \quad \textit{iff} \quad \mu(X_1, \ldots, X_n) \models_{classical} C.$$

Here, μ is a global operator taking the intersection of (the model classes for) its arguments and then selecting the most preferred out of these. No local reduction would work here, for the same reason as earlier. (Incidentally, so far, current research into non-monotonic logics has rested content with the classical logical constants, without paying attention to the issue which new logical operators would be most appropriate to a preferential model theory.)

Thus, we need a deeper understanding of the different kinds of reduction connecting inferential styles and their logical properties. Particularly useful would be a delineation of this procedure which is sharp enough to admit of *counter-examples* to reducibility. Here is one, admittedly rather contrived, example. Classical inference has the form '*all models for the premises* are models for the conclusion', involving a restricted quantification over the 'premise universe'. Now, suppose that one wants to concentrate on 'absolute validity', stating that some conclusion holds in all *models* tout court. Then a well-known simulation is available using material implication:

$$X_1, \ldots, X_n \models_{classical} C \quad \textit{iff} \quad \models_{absolute} (X_1 \to (\ldots \to (X_n \to C) \ldots)).$$

But a probabilistic style of inference like '*most models for the premises* are models for the conclusion' could not be simulated within the absolute style, since it is known that 'most' is an irreducibly binary relation between sets, resisting definition in terms of a standard first-order logic with an added unary quantifier 'most things in the universe'. (Stated in more proof-theoretic terms, the 'Deduction Theorem' fails here.)

Finally, reductions in the above style may also be viewed purely proof-theoretically, as syntactic transformations from one proof calculus to another preserving derivability. Nevertheless, here again, a model-theoretic perspective has something to add of its own. Semantic methods for establishing reducibility often provide something extra. They show how, given one model and type of discourse, one can pass on to another related model with translated discourse. This kind of 'reduction' is likely to be more realistic. Intuitively, when switching, say, from classical to intuitionistic reasoning, one does not apply a miracle trick to some finished piece of classical reasoning in order to make it intuitionistic. One rather thinks of switching from discussing some semantic situation in classical terms to intuitionistic terms in a systematic fashion.

8 Logical Pluralism

The salient conclusions of the preceding discussion are as follows. In understanding the landscape of substructural logics, model-theoretic viewpoints are just as important as proof-theoretic ones, and the connection between the two deserves continuous attention. This double approach does not diminish the importance of proof theory, for the interaction with model theory even suggests new proof-theoretic questions. Indeed, one would like to see the emergence in the field of an 'Abstract Proof Theory', similar in aims and scope to Abstract Model Theory. Its brief would be to investigate connections between important meta-properties of proof calculi (Cut Elimination, Interpolation, Decidability and the like) and to tell us what clusters of (structural) properties may be expected for systems of deduction (there might even be Lindström theorems characterizing certain logics uniquely).

To conclude, we want to emphasize another general view behind our discussion. Some people regard the wider landscape of deduction as it is unfolding right now as one more step in 'foundational' research, looking for 'the correct base logic' governing human reasoning. Thus, people might convert from being a classical logician to a disciple of intuitionism, and then as the century comes to a close, to being a believer in linear logic. On our view, however, conversion is not the proper moral of what is going on. What we are seeing is rather a growing plurality of logical systems, replacing the traditional idea of uniquely anointed calculi like 'Frege's logic' or 'Heyting's logic' by the view of a family of systems, with possibly different merits for different purposes. Applying some common sense, it cannot be plausibly maintained that people are 'really' doing occurrence-based logic in their daily reasoning. In many contexts, what they are doing is still much more adequately described by classical or intuitionistic systems. But, for other purposes, linear logics or relevance logics may be much closer to the mark.

Indeed, to be quite honest, the overwhelming 'convergence' in text books toward the 'standard logics', that come out on top in many different analyses (proof-theoretic, model-theoretic, category-theoretic, ...) is to a large extent a matter of deliberate presentation. For instance, what 'constructive reasoning' really is remains a matter of debate, with various plausible options. Even the celebrated Brouwer-Heyting-Kolmogorov proof interpretation still leaves many natural choice points, leading either to standard intuitionistic systems, or to minimal logic, or relevant logics, or even linear logic. Likewise, what 'internal logic' comes out of mathematical categories is determined to a large extent by the definitions chosen for the logical constants, without enquiring into the possible effects on the induced logic of variation of the latter parameter. And similar comments apply to game-theoretical analyses of standard systems, where often the dice have to be loaded rather blatantly in rules for debate in order to make classical or intuitionistic logic the winning strategy.

This pluralist view is much closer to what has always been standard practice in modal logic, where no single system has ever commanded general allegiance. It is also closer to the spirit of the earlier-mentioned linguistic applications. Linguistic competence is usually measured through a variety of grammatical mechanisms, which may be different for different syntactic tasks. And the same might be the appropriate view in logic, viewed as a description of human cognitive competence. For instance, to mention just one obvious fact, human reasoning seems to have both short-term manifestations, reacting directly to syntactic cues and order of presentation, and more long-term ones, digesting information obtained whose representation may be long forgotten by now. It seems quite reasonable to expect that these two levels may have different styles of inference, governed by different structural rules. Then, co-existence of various logical calculi may be just what is needed.

In other words then, the secret of reasoning need not be hidden in any particular treasure chest, hidden somewhere in the above landscape, waiting for us to come with daggers and spades, in order to fight about preferences and priorities: the true marvel might well be that landscape itself.

References

V. M. Abrusci, 1988, Sequent Calculus for Intuitionistic Linear Logic, Department of Philosophy of Science, University of Bari.

J. van Benthem, 1985, *Modal Logic and Classical Logic*, Bibliopolis, Naples/ The Humanities Press, Atlantic Heights.

______, 1986, *Essays in Logical Semantics*, Reidel, Dordrecht (Studies in Linguistics and Philosophy 29).

______, 1991, *Language in Action. Categories, Lambdas and Dynamic Logic*, North-Holland, Amsterdam (Studies in Logic 130).

______, 1992, Dynamic Arrow Logic, in: Van Eyck & Visser, eds., to appear.

R. T. Brady, 1990, The Gentzenization and Decidability of RW, *Journal of Philosophical Logic 19*, 35–73.

W. Buszkowski, Completeness Results for Lambek Syntactic Calculus, *Zeitschrift für Mathematische Logik und Grundlagen der Mathematik 32*, 13–28.

W. Buszkowski, W. Marciszewski & J. van Benthem, eds., 1988, *Categorial Grammar*, John Benjamins, Amsterdam and Philadelphia.

H. B. Curry, 1963, *Foundations of Mathematical Logic*, McGraw Hill, New York.

J. van der Does & J. van Eyck, eds., 1992, *Generalized Quantifiers: Theory and Applications*, CSLI Publications, Stanford University (distributed by University of Chicago Press).

M. Dunn, 1985, Relevance Logic and Entailment, in: Gabbay & Guenthner, eds., 117–224.

J. van Eyck & A. Visser, eds., to appear, *Logic and Information Flow*, Kluwer, Dordrecht (Studies in Logic, Language and Information).

D. Gabbay & F. Guenthner, eds., 1985, *Handbook of Philosophical Logic, Vol. 3: Alternatives to Classical Logic*, Reidel, Dordrecht.

J.-Y. Girard, 1987, Linear Logic, *Theoretical Computer Science 50*, 1–102.

H. Hendriks, 1992, Normalization, Spurious Ambiguity and Atomic Categories in the Lambek Calculus, Institute for Logic, Language and Computation, University of Amsterdam.

M. Kanazawa, 1991, The Lambek Calculus Enriched with Additional Connectives, Institute for Language, Logic and Information, University of Amsterdam, Report LP–91–04.

M. van Lambalgen, 1991, Natural Deduction for Generalized Quantifiers, Institute for Logic, Language and Computation, University of Amsterdam. (To appear in: Van der Does & Van Eyck, eds., 1992.)

J. Lambek, 1958, The Mathematics of Sentence Structure, *American Mathematical Monthly 65*, 154–170.

D. Makinson, 1991, General Non-Monotonic Logic, to appear in D. Gabbay et al., eds., *Handbook of Logic in Artificial Intelligence and Logic Programming*, Oxford University Press.

S. Mikulás, 1992, The Completeness of the Lambek Calculus with Respect to Relational Semantics, Institute for Logic, Language and Computation, University of Amsterdam, Report LP–92–03.

M. Moortgat, 1988, *Categorial Investigations: Logical and Linguistic Aspects of the Lambek Calculus*, Foris, Dordrecht.

M. Moortgat & G. Morril, 1991, Heads and Phrases. Type Calculus for Dependency and Constituent Structure, Onderzoeksinstituut voor Taal en Spraak, Rijksuniversiteit, Utrecht.

H. Ono, 1990, Semantics for Substructural Logics, (in this book).

V. Pratt, 1991, Action Logic and Pure Induction, in J. van Eyck, ed., *Logics in AI. Proceedings European Workshop JELIA '90, Amsterdam*, Lecture Notes in Computer Science 478, Springer Verlag, Berlin, 97–120.

D. Roorda, 1991, *Resource Logics: Proof-Theoretical Investigations*, Dissertation, Institute for Logic, Language and Computation, University of Amsterdam.

V. V. Rybakov, 1985, A Criterion for Admissibility of Rules of Inference in Modal and Intuitionistic Logic, *Soviet Mathematical Doklady 32:2*, 452–455.

G. Sambin, 1990, Intuitionistic Formal Spaces and Linear Logic, Institute of Mathematics, University of Padova.

Y. Venema, 1991, Many-Dimensional Modal Logic, Dissertation, Institute for Logic, Language and Computation, University of Amsterdam.

H. Wansing, 1989, The Adequacy Problem for Sequential Propositional Logic, Institute for Language, Logic and Information, University of Amsterdam.

Index